American Institute of Chemical Engineers

Education

Presentations at the
2007 AIChE Annual Meeting

November 4-9, 2007
Salt Lake City, Utah, USA

Printed from e-media with permission by:

Curran Associates, Inc.
57 Morehouse Lane
Red Hook, NY 12571
www.proceedings.com

ISBN: 978-1-60560-002-4

Some format issues inherent in the e-media version may also appear in this print version.

ISBN: 978-1-60560-002-4

American Institute of Chemical Engineers
Proceedings
Three Park Avenue
New York, NY 10016-5991
Phone: 212-591-8100

www.aiche.org

American Institute of Chemical Engineers

Education
2007

TABLE OF CONTENTS

Nano-Particle Toughening of Epoxy Composites Using Polyethyleneimine Dendrimer Shell Materials

Aaron Saks, Chemical Engineering, University of Washington, 4338 N.E. 58th St., Seattle, WA 98105

A novel technique has been developed for toughening epoxy polymers used in composite materials. This technique uses specifically engineered core-shell nano-particles to both increase the ultimate strength and the stiffness of the polymer. This paper investigates the effects of the polyethyleneimine dendrimer shell material on the physical properties of the epoxy polymer.

Synthesis of Starch-G-Polymethylmethacrylate through Emulsion Photopolymerization

Dan Weber, Chemical and Biochemical Engineering, The University of Iowa, 4133 Seamans Center, Iowa City, IA 52242

Many researchers believe that grafting polymer chains to starch could be the best way to make non-toxic and biodegradable polymers. Emulsion photopolymerization was used to synthesize starch-g-PMMA copolymers. The variables investigated in this research were the amount of starch in the reaction, length of illumination, type of photoinitiator, amount of photoinitiator, and synthesis without photoinitiator. Data was analyzed gravimetrically to determine the reaction yield and grafting efficiency. Gel permeation chromatography was used to analyze the relative molecular weights of homopolymer and copolymer. The highest grafting efficiency occurred at lower irradiation times without photoinitiator and at higher irradiation times for samples when uvinul,,¥ 3048 is present. The highest relative molecular weights were seen at the lowest irradiation times in samples with and without photoinitiator. Because the lowest molecular weight and lowest grafting efficiencies occurred at high illumination times, this indicates that grafted and nongrafted chains are degrading in the presence of the UV light.

Transport in Epdm Elastomer: Molecular Simulation and Experimental Study

David T. Limmer, Chemical Engineering, New Mexico Institute of Mining and Technology, 801 Leroy Pl, Socorro, NM 87801

Author to enter abstract

Cellular Uptake of Modified Red Clover Necrotic Mosaic Virus and Small Molecule Release from the Virion

K. Wesley Overton[1], Joseph Ryan[2], LiNa Loo[2], Molly Speight[3], Richard Guenther[4], Steven Lommel[4], Stefan Franzen[2]

1. *Department of Chemical and Biomolecular Engineering, North Carolina State University*
2. *Department of Chemistry, North Carolina State University*
3. *Department of Biology, North Carolina State University*
4. *Department of Plant Pathology, North Carolina State University*

Abstract

The ability to efficiently and specifically target cancerous cells with therapeutic agents would be a monumental advance in the area of cancer treatment. This research seeks to investigate the possibility of using a novel synthetic drug delivery vector to target human cervical cancer (HeLa) cells. This delivery vector is constructed by infusing small, positively-charged molecules into a plant virion—*Red clover necrotic mosaic virus* (RCNMV)—and then labeling the infused virion with multiple copies of a targeting peptide or agent. The infused molecules can be dyes for diagnostic applications or anti-cancer for chemotherapeutic applications. The targeting peptides can be of viral origin such as the nuclear localization sequence from the adenoviral fiber peptide. This peptide has the ability to target the CD46 receptor. The peptide can also be fluorescently labeled to monitor uptake in cells. The viral construct is delivered to HeLa cells, and cellular uptake of the conjugate and intracellular rhodamine release is simultaneously monitored. Once the virus is internalized, the decreased divalent calcium ion concentration and salt content of the cellular environment cause pores between coat proteins in the virion to open, and small, positively-charged molecules such as rhodamine can pass through these pores. Both the cellular uptake of the conjugate and rhodamine release are monitored using fluorescence activated cell sorting (FACS) flow cytometry. It has been found that degree of cellular uptake of the modified RCNMV constructs is a function of incubation time with the maximum uptake after a nine-hour incubation time. Furthermore, it has been determined that the amount of rhodamine released from the virion inside the cell increases with incubation time, but this release time lags behind the uptake time. Again, the maximum amount of rhodamine release is seen after a nine-hour incubation time. Once this system is fully characterized, studies of *in vivo* virus localization as well as encapsulation of probe molecules for detection of genes or proteins inside cells can be completed for diagnostic and research applications.

Introduction

Red clover necrotic mosaic virus (RCNMV) is a plant virion comprised of 180 identical coat proteins that assemble into an icosohedron when the two strands of the viral RNA genome interact at the origin of assembly. Each of the 180 coat proteins contains an average of two reactive lysine residues that can be used for chemical coupling to organic molecules. RCNMV has a triangulation number of 3, and at each 3-protein interaction point there exists a small pore. The native virus is found in the soil where the divalent calcium ion concentration is high and the pH is typically below 6.0. At this pH value, the pores in the viral capsid remain closed. However, once the virus enters the cytoplasm of the root cells in a plant, the divalent calcium ion concentration drops in comparison to the soil, and the pH of the cell is higher relative to the soil. This change in divalent calcium ion concentration and pH causes the coat proteins to shift slightly, thus widening the pore between the three proteins at each interaction point. This slight widening of the pore allows the RNA genome inside the virion to diffuse out through the pores and take over the plant's translational machinery and replicate. It is this dynamic characteristic of the pores in the virion that makes it an exceptional drug delivery vector. The calcium ion concentration and pH of the fluid surrounding the virion can be adjusted *in vitro*, and the virus can be easily infused with small molecules, the pores can be closed, and the cargo-carrying virion can be delivered to cells where it will release its cargo in exactly the same manner as it does natively in plant cells. Since RCNMV is a plant virus it is completely harmless to humans so there is no possibility of infection from the virion.

Each of the coat proteins carries a slightly negative charge, so positively charged molecules diffuse into the virion through the pores more easily than negatively charged or neutral molecules. This research seeks to investigate the practicality of targeting human carcinoma cells by attaching receptor-specific peptide sequences to the infused RCNMV particle and the efficiency of cargo release—in this case a rhodamine fluorophore—from the virion inside a human carcinoma cell. HeLa cells are used in this study since they are well characterized and widely available. HeLa cells express the CD46 cell surface receptor which can be targeted with the adenoviral nuclear localization signal peptide CD46, which has the sequence CGGFSTSLRARKA* where the asterisk denotes the location of a fluorescein fluorophore used for tracking the uptake of the viral construct.

Methods

The first step in the process of delivering small molecules to cells using RCNMV is to infuse the small molecules into the virion. Briefly, native RCNMV is treated with a solution of EDTA in the absence of calcium ions to open the pores in the virion. After EDTA treatment, the small molecule—rhodamine in this study—is introduced into the buffer and allowed to incubate for several hours to allow the molecule to diffuse into the virion. Next, the virus is dialyzed in buffer containing a relatively high concentration of calcium ion to close the pores. After this incubation, excess rhodamine is removed from the bulk solution by dialysis. Once RCNMV has been infused with rhodamine dye, receptor-specific targeting peptides can be attached to each of the approximately 360 reactive lysine residues found on the surface of the virion. N-succinimidyl-4-(maleimidomethyl)cyclohexanecarboxylate, an NHS ester, is attacked at the carbonyl carbon by the amine group on the reactive lysine residue. The targeting peptide has a cysteine residue at the N-terminus, and the thiol of this cysteine attacks the organic linker and is chemically coupled to the virion. A schematic of the infusion and labeling process can be seen in Figure 1.

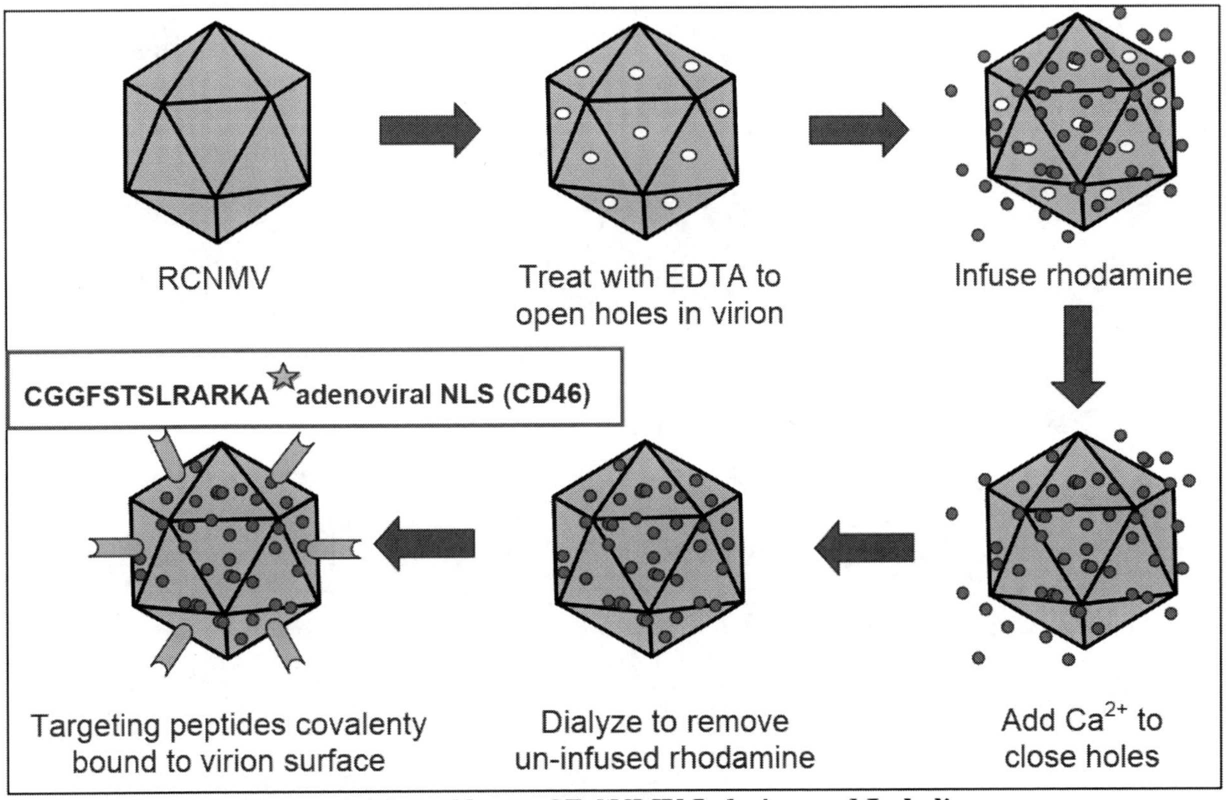

Figure 1. Flow Chart of RCNMV Infusion and Labeling

The proposed mechanism of uptake for the RCNMV construct is receptor-mediated endocytosis. It is expected that the virion is enclosed in an endosome once it gains entry to the cell, and this endosome travels through the cell until it approaches the nucleus. At this point, it is proposed that the endosome breaks open and releases the virion into the cytoplasm in the vicinity of the nucleus. When the endosome breaks open and the virus encounters the calcium ion-depleted cytoplasm with relatively high pH, the pores between the coat proteins will open and the cargo will be released. This uptake and shuttling process is illustrated in Figure 2.

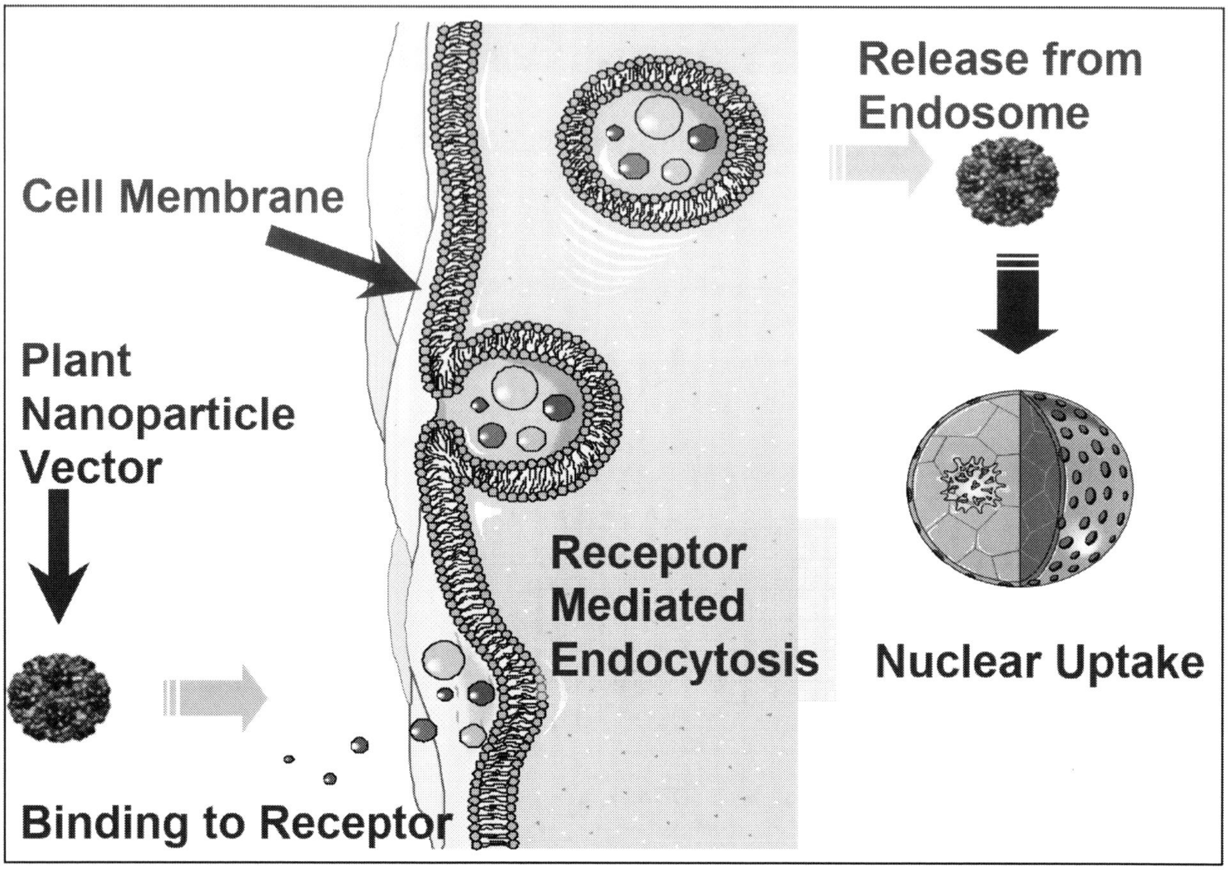

Figure 2. Cellular Targeting, Uptake, and Shuttling of RCNMV

To test the hypothesis that HeLa cells will internalize the RCNMV construct and that the rhodamine fluorophore encapsidated in the virion will be released in the cell, infused and labeled RCNMV constructs were incubated with HeLa cells for varying periods of time between 0.5 and 9 hours in a 37°C, 5% CO_2 incubator. Once the incubation was complete, the green and red fluorescence in the cells, which would indicate viral uptake and rhodamine release, respectively, was simultaneously monitored using fluorescence activated cell sorting (FACS). To begin FACS analysis, each cell is passed one by one through a detection laser that first determines the viability of the cell by measuring its size and granularity. A plot of the side scatter versus the forward scatter is constructed, and a gated region signifying healthy, viable HeLa cells is drawn.

Once this region is denoted, only the cells that are determined to be viable are analyzed further. The viable cells are next measured for fluorescence, and a plot of red fluorescence versus green fluorescence is generated. To control for autofluorescence of the cells, a sample of HeLa cells incubated without RCNMV construct is passed through the FACS system, and bars are drawn on the plot to denote the background fluorescence region. Any cells that have been incubated with RCNMV constructs that appear on the graph above the background region would indicate viral uptake and/or rhodamine release.

Results

The FACS plots of the HeLa cells incubated with infused, labeled RCNMV can be seen in Figures 3-7.

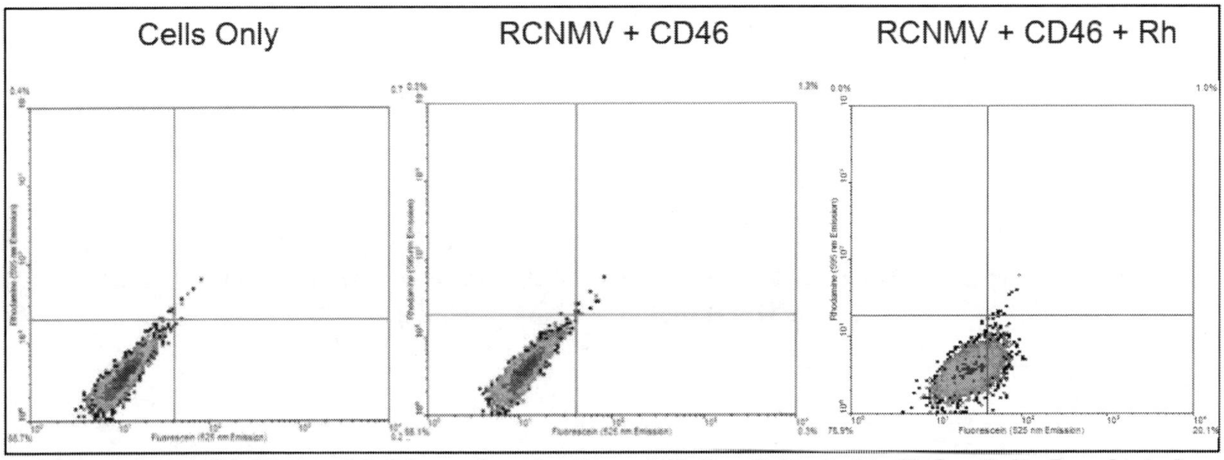

Figure 3. FACS Plots of 0.5 hour incubation of HeLa Cells (left), HeLa Cells Incubated with CD46-labeled RCNMV (middle), and HeLa Cells Incubated with Rhodamine-Infused, CD46-labeled RCNMV (right).

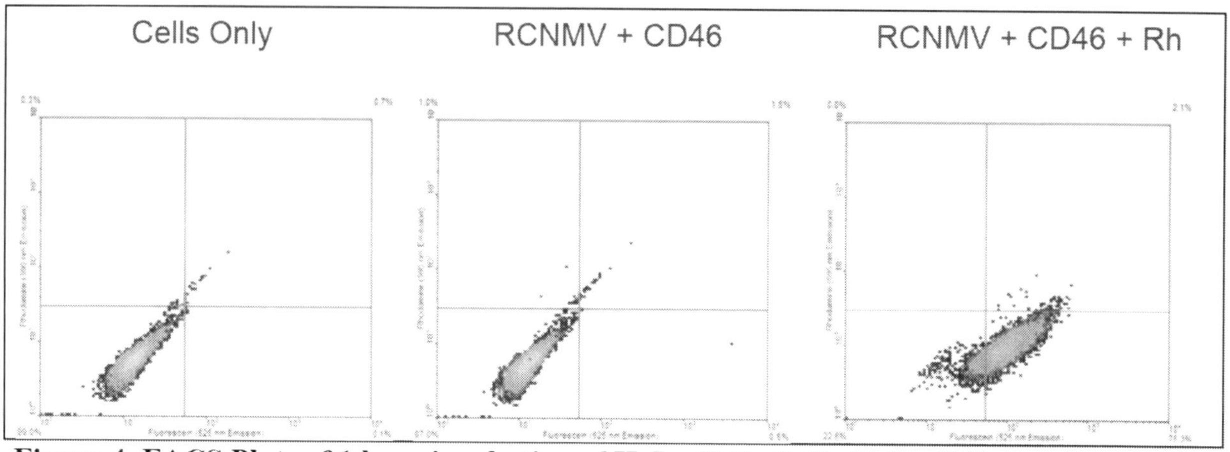

Figure 4. FACS Plots of 1 hour incubation of HeLa Cells (left), HeLa Cells Incubated with CD46-labeled RCNMV (middle), and HeLa Cells Incubated with Rhodamine-Infused, CD46-labeled RCNMV (right).

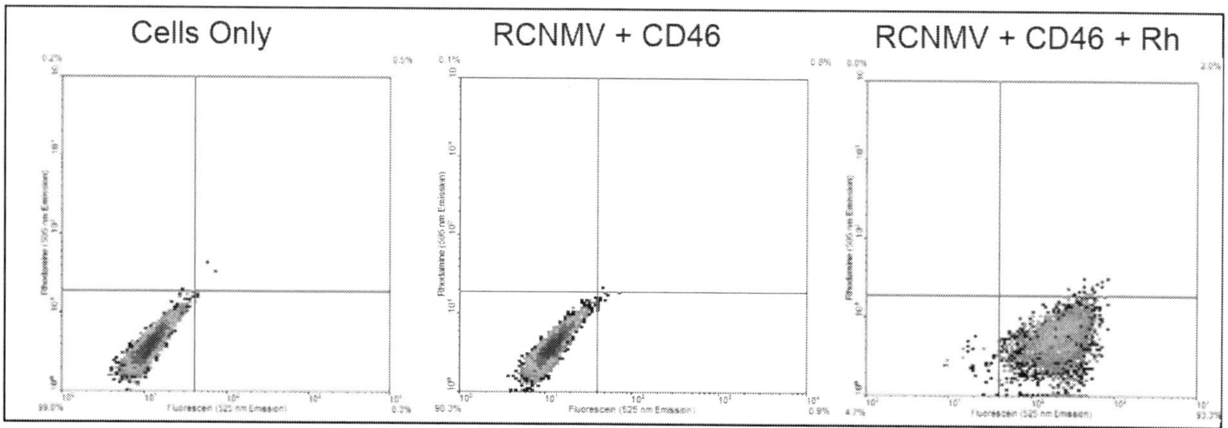

Figure 5. FACS Plots of 3 hour incubation of HeLa Cells (left), HeLa Cells Incubated with CD46-labeled RCNMV (middle), and HeLa Cells Incubated with Rhodamine-Infused, CD46-labeled RCNMV (right).

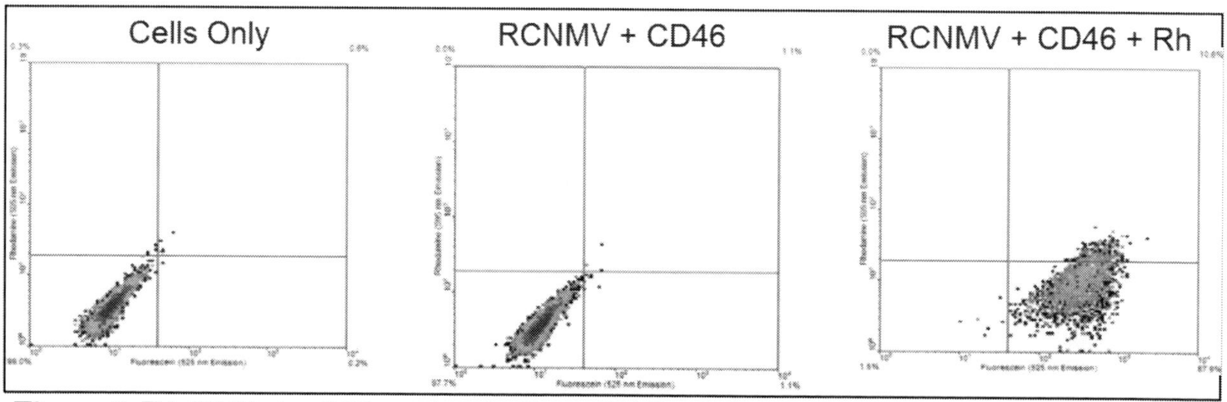

Figure 6. FACS Plots of 6 hour incubation of HeLa Cells (left), HeLa Cells Incubated with CD46-labeled RCNMV (middle), and HeLa Cells Incubated with Rhodamine-Infused, CD46-labeled RCNMV (right).

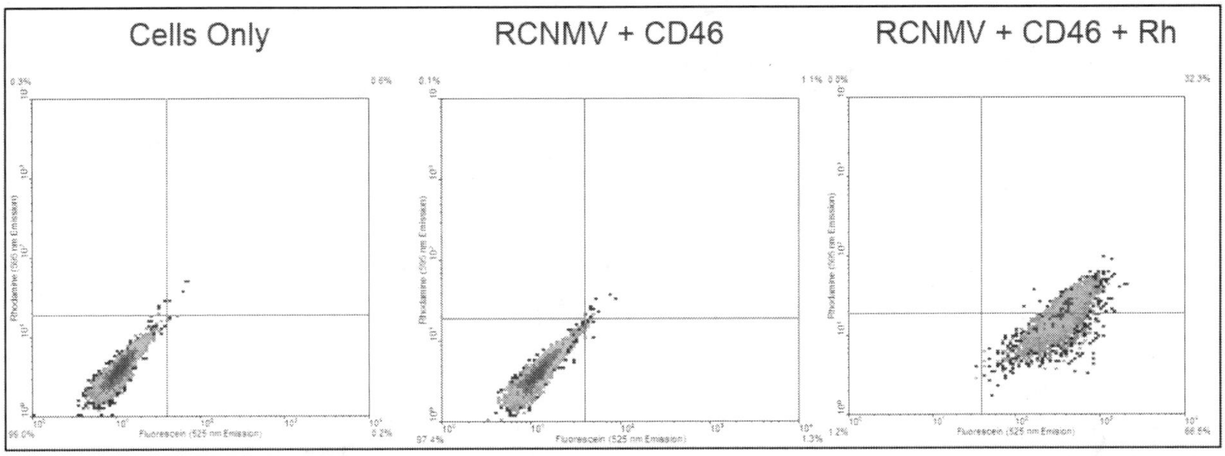

Figure 7. FACS Plots of 9 hour incubation of HeLa Cells (left), HeLa Cells Incubated with CD46-labeled RCNMV (middle), and HeLa Cells Incubated with Rhodamine-Infused, CD46-labeled RCNMV (right).

It can be seen in Figure 3 that after 0.5 hours, approximately 28% of HeLa cells incubated with rhodamine-infused, CD46-labeled RCNMV had taken up this construct while virtually no uptake is seen with the uninfused, CD46-labeled constructs. Figure 4 shows that approximately 75% of cells had taken up the rhodmaine-infused, CD46-labeled constructs after one hour of incubation while virtually no cells had taken up the uninfused, CD46-labeled RCNMV after this time. Figure 5 shows that after a 3-hour incubation about 95% of cells had taken up the rhodamine-infused, CD46-labeled RCNMV construct, but no rhodamine release is seen at this point. No uptake of the uninfused, CD46-labeled RCNMV constructs is seen after 3 hours. Figure 6 shows that after 6 hours of incubation with rhodamine-infused, CD46-labeled RCNMV constructs, approximately 100% of HeLa cells had taken up this construct but virtually no rhodamine has been released at this point. No uptake of uninfused, CD-46 labeled constructs is seen after 6 hours. Figure 7 shows that after 9 hours of incubation, approximately 100% of HeLa cells had taken up the rhodamine-infused, CD46-labeled RCNMV construct, and 33% of cells have rhodamine release inside them. No uptake of the uninfused, CD46-labeled constructs is seen after 9 hours. The results in Figure 7 show that after 9 hours, rhodamine is being released from the virion and is seen inside the cell. A plot of percent of cells that have taken up the rhodamine-infused, CD46-labeled RCNMV construct versus time and percent of cells that have rhodamine release from the virion inside the cell versus time can be seen in Figure 8.

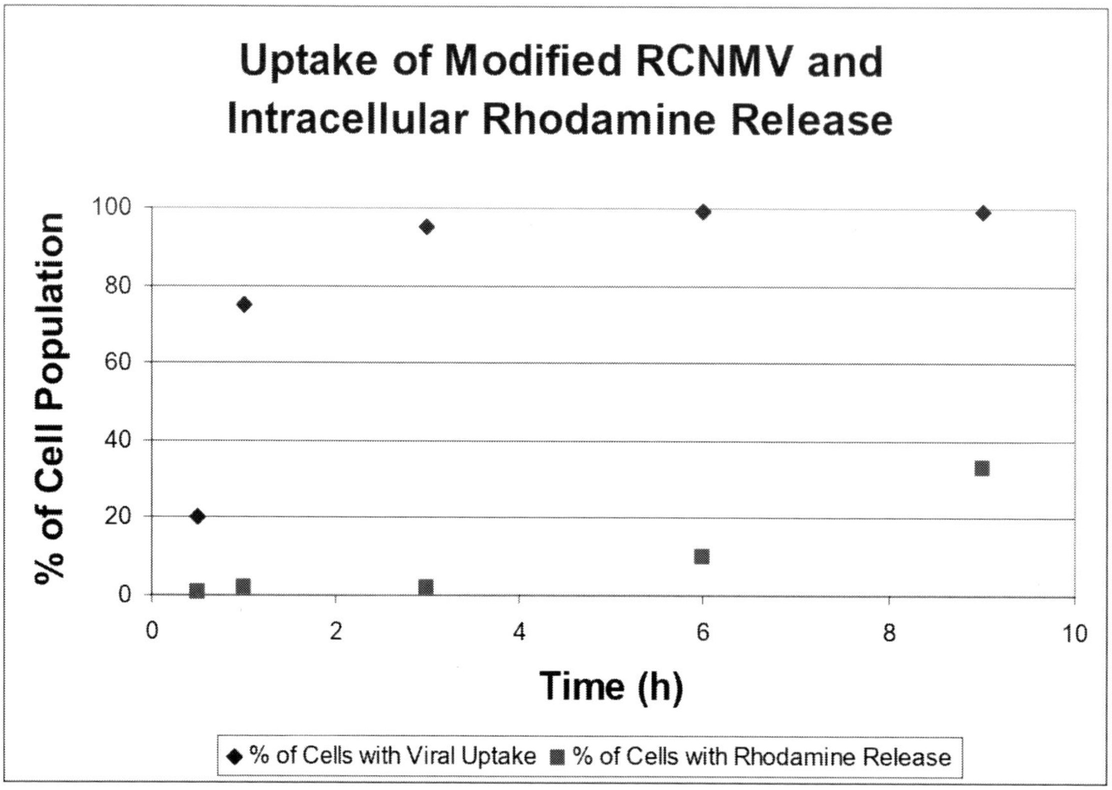

Figure 8. Percent of Cells with Rhodamine-Infused CD46-Labeled Viral Uptake (blue diamonds) versus Time of Incubation and Percent of Cells with Rhodamine Release from the Virion (red squares) versus Incubation Time.

It can be seen in Figure 8 that after three hours virtually 100% of cells have taken up the infused, labeled RCNMV construct. A time lag is seen between the viral uptake and rhodamine release of the virion. It is not until 9 hours that significant rhodamine release is seen. Experiments with longer time points should be performed to determine the time at which the maximum release is seen.

Visual confirmation of viral uptake and confirmation that the construct is inside the cell rather than attached to the surface of the cell can be seen using confocal fluorescence microscopy, as seen in Figure 9.

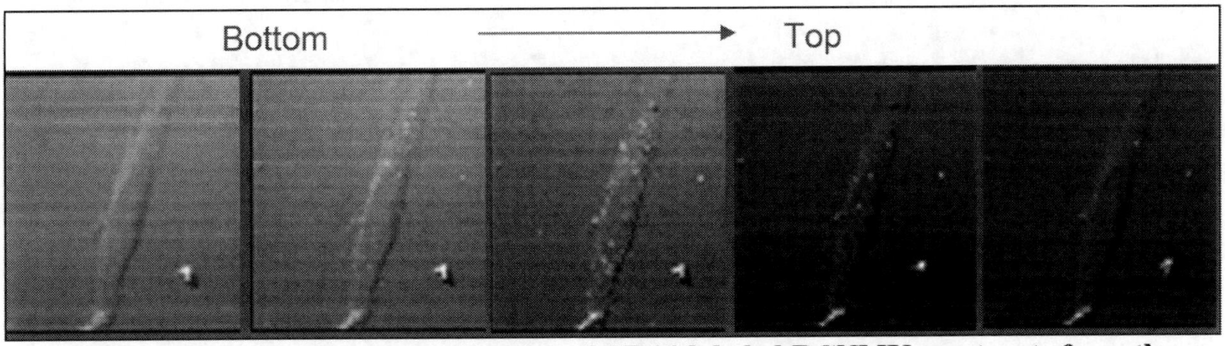

Figure 9. Confocal Microscopy Images of CD46-labeled RCNMV constructs from the Bottom (left) to the Top (right) of the cell. Images are overlays of DIC and fluorescence channels.

Figure 9 is an overlay of a differential interference contrast (DIC) image and a fluorescence image. No green fluorescence, which would indicate the presence of CD46-labeled RCNMV constructs, is seen on the bottom surface of the cell (left), but increasing amounts of green fluorescence is seen toward the center of the cell (middle). There is also no green fluorescence at the top of the cell (right). These data confirm that the viral construct has been internalized by the cell and is not simply associating with the cell membrane.

Conclusions

The data from this research indicate that RCNMV is capable of efficiently delivering small molecules to a cancer cell. Covalent coupling of a peptide that targets specific cell surface receptors has been characterized, and internalization of these viral constructs with peptides attached has been demonstrated. In addition, it has been demonstrated that small molecules are released from the viral vector once inside a cancer cell. The lack of uptake seen with the uninfused, CD46-labeled RCNMV construct is unexpected but could have easily been caused by any number of factors including variability of the cells or variability of the RCNMV constructs incubated with the cells. Experiments performed since the time of this research have confirmed that uptake of uninfused, CD46-labeled RCNMV occurs. In addition, the uninfused, CD46-

labeled RCNMV constructs used in the confocal microscopy experiment were internalized by the HeLa cells.

Cell targeting has a number of fundamental applications. RCNMV constructs can carry probe molecules into cells for the detection of intracellular conditions. Selectivity according to cell type can be achieved using appropriate targeting peptides. Tissue specific targeting is a desirable attribute in many medical applications including diagnostic imaging and drug delivery. If this type of specificity can be demonstrated, then using RCNMV as a delivery vector that targets only cancerous cells and not healthy cells could be a highly viable future cancer treatment option.

On-Board Hydrogen Storage and Production: An Application of Ammonia Electrolysis

Amy M. Weber*, Bryan K. Boggs, and Gerardine G. Botte[1]
Department of Chemical and Biomolecular Engineering
Stocker Center 183, Ohio University, Athens, OH 45701, USA
Phone[1]: 740-593-9670, Fax[1]: 740-593-0873, e-mail[1]: botte@ohio.edu

1. Introduction

1.1. Availability of Ammonia

Nearly 60 million metric tons of gas-phase ammonia is emitted each year worldwide [1]. As a result, nitrate levels in ground and drinking water, mainly attributed to ammonia emissions, are becoming a serious problem recently recognized by the U.S. Environmental Protection Agency (EPA) [2]. As a solution, 97% of this ammonia in ground and waste water can be captured using a simple stripping and absorption process [3]. Electrolyzing this captured ammonia, which generates pure hydrogen and nitrogen, is key to moving towards a hydrogen economy.

1.2. Ammonia and electrolysis

Liquid ammonia is a non-carbon containing hydrogen-dense (17.6 wt.%) fuel that can be stored at ambient temperature and pressure [4]. Ammonia electro-oxidation occurs at the anode (Eqn. 1) and has a potential of -0.77 V versus standard hydrogen electrode (SHE). Alkaline reduction of water occurs on the cathode (Eqn. 2) and requires -0.83 V versus SHE. The overall reaction (Eqn. 3) only requires 0.06 V versus SHE. This makes ammonia electrolysis attractive for producing hydrogen when comparing the required 1.23 V for water electrolysis, according to the thermodynamics at standard conditions [5].

$$2NH_3(aq)+6OH^- \rightarrow N_2(g)+6H_2O+6e^- \quad (1)$$

$$6H_2O+6e^- \rightarrow 3H_2(g)+6OH^- \quad (2)$$

$$2NH_3(aq) \rightarrow N_2(g)+3H_2(g) \quad (3)$$

Table 1 shows that hydrogen produced from ammonia electrolysis costs much less than the Department of Energy's (DOE) target. This is a direct result of the small energy consumption of ammonia electrolysis.

Table 1. Energy consumption and cost of hydrogen comparison between water and ammonia electrolysis.

	Water Electrolysis	Ammonia Electrolysis
Energy Consumption (Wh/g)	33.00	1.55
Cost of Hydrogen ($/kg)	7.10	0.89

1.3. On-board hydrogen production

Hydrogen fuel cell vehicles (HFCV) can obtain 60 miles per gallon compared to internal combustion engines (ICE), which can only travel 24 miles per gallon of gasoline. Also, HFCVs only emit pure water unlike ICE's greenhouse gas emissions. A hydrogen future is not only driven by the transportation sector but as a movement for reducing oil imports in general and eliminating the dependence on petroleum for daily energy production. HFCVs as personal transportation vehicles are believed to be the best alternative. However, storing enough hydrogen that allows a fuel-cell vehicle to travel the same range as ICEs between fueling is proving difficult and delaying the commercialization and market approval of HFCVs. Compared to gasoline, hydrogen has 2.7 times more energy content based on weight However, due to its tremendously low density, hydrogen has 25% less energy content than gasoline when based on volume [6]. Because of this, there is no storage technology currently available that allows a vehicle to travel the average 300-mile range that today's internal combustion engines obtain [7].

As a result, research on producing hydrogen on board has accelerated. Generating hydrogen on board in a manner that does not produce air pollutants and requires the small amount of energy available from renewable sources is the definite long-term solution [8]. This is where *in situ* ammonia electrolysis enters the picture. Within this context, there are two objectives:

1. Design and construct a static alkaline ammonia electrolytic cell. An AEC, that separates hydrogen from the cathode from the nitrogen generated at the anode, was constructed.
2. Determine the feasibility of using ammonia for on-board vehicular hydrogen storage. Synergistic analysis was performed on the AEC and proton exchange membrane fuel cell (PEMFC). This was carried out using polarization techniques allowing for energy consumption and generation data to be obtained.

2. Experimental Methodology

2.1. Ammonia electrolytic cell design and construction

Due to pending publication, a lot of detail about the design and construction of the AEC cannot be divulged. Materials were chosen based on their resistance to the alkaline media present and all materials were purchased from McMaster-Carr. Figure 1 shows a schematic representation of the sandwich-configured AEC. This design allows for pure hydrogen to be collected from the cathode and pure nitrogen from the anode. This hydrogen is to be used in a PEMFC for synergistic analysis of the AEC-PEMFC process to determine the feasibility of using ammonia for on-board vehicular hydrogen storage.

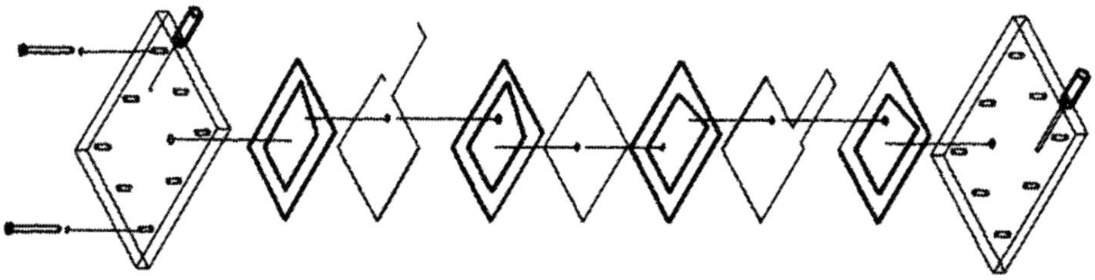

Figure 1. Schematic representation of the ammonia electrolytic cell (AEC).

2.2. AEC and PEMFC integration study

In order to determine the feasibility of *in situ* ammonia electrolysis as an on-board hydrogen generating technology, a synergistic analysis of the AEC and PEMFC was required. Figure 2 shows a schematic representation of the integration experiment. The study was performed using a multi-channel Arbin cycler BT2000 potentiostat. An electrolyte consisting of 5 M KOH was added to the cathode side of the AEC while a solution of 1 M NH_4OH and 5 M KOH was added to the anode. The gas storage columns displaced water and created a small hydrostatic pressure that help the PEMFC receive a continuous flow of hydrogen rather than intermediate bubbles created in the AEC. Currents were applied to the AEC in 100 mA steps. The voltage response was recorded by the potentiostat allowing for energy consumption data to be obtained.

To characterize the PEMFC, a constant current of 300 mA was applied to the AEC. The hydrogen generated in the AEC was sent to the PEMFC, and loads were applied in 20 mA increments. The maximum energy generated in the PEMFC is shown in Figure 4. This allowed energy consumption and generation data to be obtained.

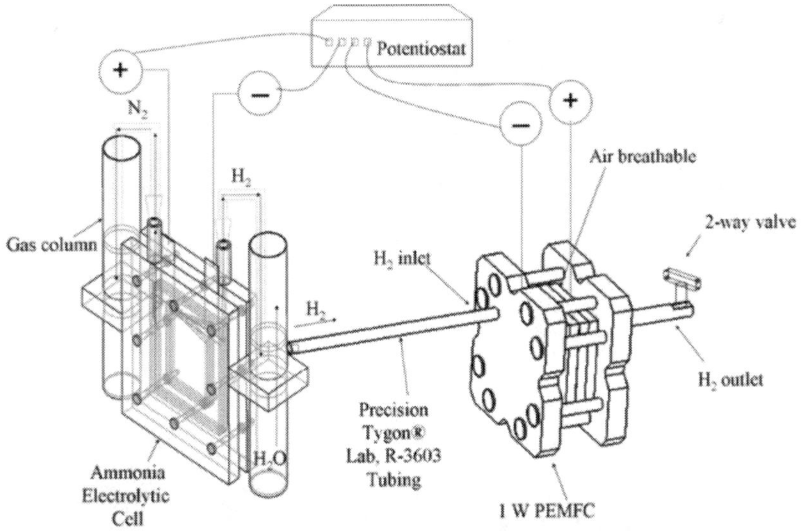

Figure 2. Schematic diagram of the AEC-PEMFC integration set-up. All synergistic experiments were performed with this configuration.

3. Results and Discussion

Figure 3 shows the data obtained from the characterization of the AEC. By choosing to run the AEC at 300 mA, it can be seen in Figure 3 that the AEC required 159 mW (13.10 Wh/g H_2) to operate. The PEMFC data, collected by applying the 300 mA current to the AEC, is shown in Figure 4. Current was drawn in 20 mA increments. The maximum power from the fuel cell was determined from the last step before reverse polarization occurred, which is shown by the data in Figure 4. Under these conditions, the PEMFC generated 227 mW (18.70 Wh/g H_2). This means that a net energy of 5.60 Wh/g H_2 was created from the coupled AEC-PEMFC experiment.

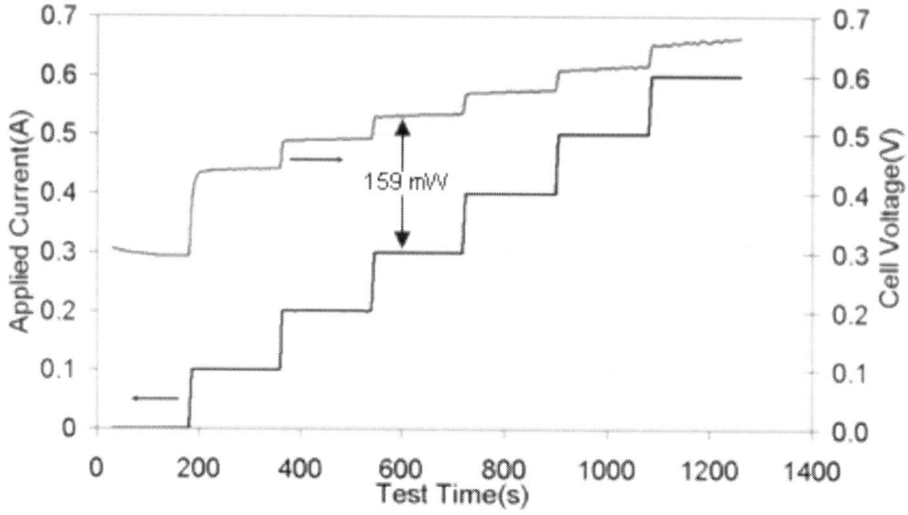

Figure 3. Galvanostatic analysis on the AEC. Potentiostat-generated data gathered by applying current to the AEC in 100 mA steps.

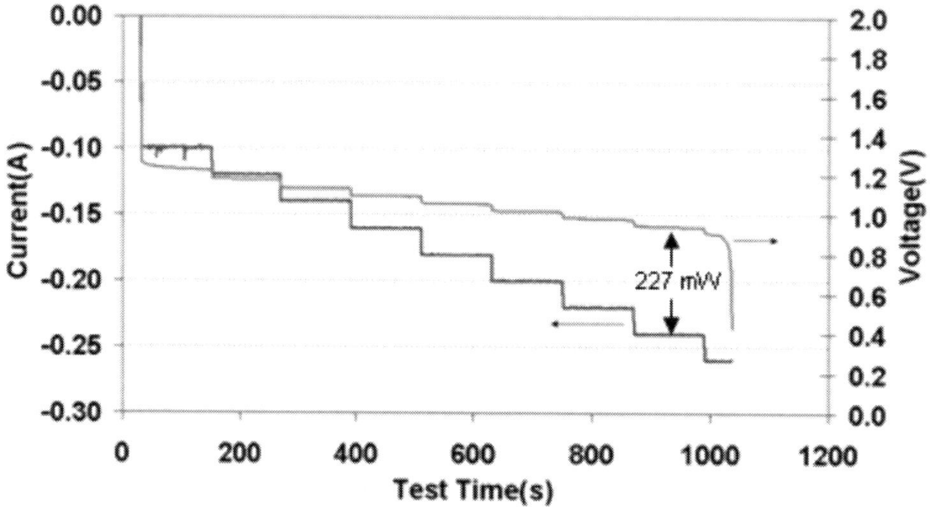

Figure 4. Galvanostatic analysis on the PEMFC. The potentiostat was used to apply amperic loads in 20 mA increments.

4. Conclusion

Based on the data collected, the amount of energy supplied by the PEMFC surpasses that which is required to operate the AEC. Assuming the power generated by the PEMFC can be separated as shown in Figure 5, there is enough net energy from the PEMFC to power a DC-geared motor (Model 3440 Hankscraft Motors). This means that a bench-scale automobile featuring an AEC-PEMFC-motor process can be self sustaining until ammonia is depleted. Fig. 6 shows the evolution of this bench-scale prototype. Currently, a circuit which can split this power is being developed.

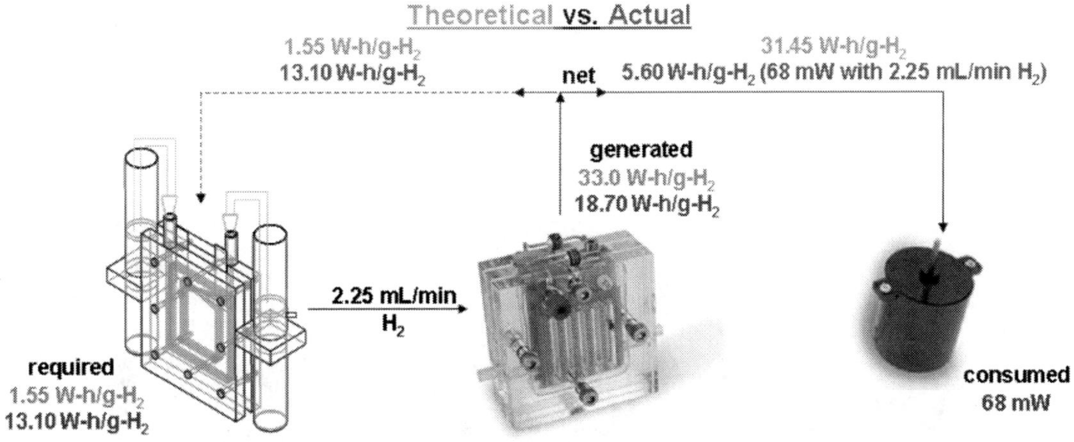

Figure 5. Theoretical energy values required to power each component, as well as those determined experimentally, are shown above.

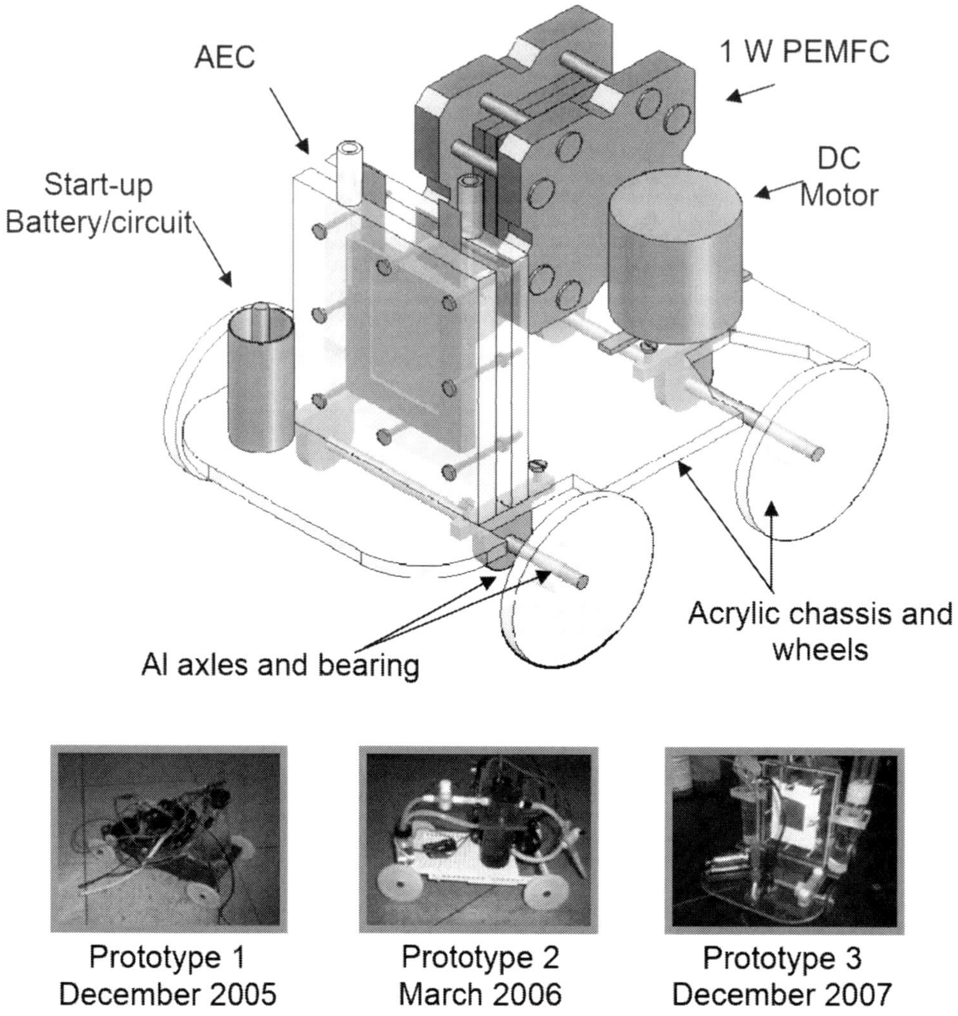

Figure 6. The evolution of the bench-scale model AEC-PEMFC automobile.

References

1. *Bouwman, A.F., et al., A global high-resolution emission inventory for ammonia. Global Biogeochemical Cycles, 1997.* ***11****(4): p. 561-587.*
2. *Dash, B.P. and S. Chaudhari, Electrochemical denitrificaton of simulated ground water. Water Research, 2005.* ***39****(17): p. 4065-4072.*
3. *Basakcilardan-Kabakci, S., A.N. Ipekoglu, and I. Talini, Recovery of ammonia from human urine by stripping and absorption. Environmental Engineering Science, 2007.* ***24****(5): p. 615-624.*
4. *Thomas, G. and G. Parks, Potential roles of ammonia in a hydrogen economy: A study of issues related to the use of ammonia for on-board vehicular hydrogen storage. U.S. Department of Energy, 2006: p. 23.*
5. *Vitse, F., M. Cooper, and G.G. Botte, On the use of ammonia electrolysis for hydrogen production. Journal of Power Sources, (2005).* ***142****(1-2): p. 18-26.*
6. *Balat, M. and N. Ozdemir, New and renewable hydrogen production processes. Energy Sources, 2005.* ***27****(13): p. 1285-1298.*
7. *Satyapal, S., et al., The US Department of Energy's National Hydrogen Storage Project: Progress towards meeting hydrogen-powered vehicle requirements. Catalysis Today, 2007.* ***120****(3-4): p. 246-256.*
8. *Granovskii, M., I. Dincer, and M.A. Rosen, Exergetic life cycle assessment of hydrogen production from renewables. Journal of Power Sources, 2007.* ***167****(2): p. 461-471.*

Direct Sequence Detection of Human H5 Influenza Viral RNA

Matthew B. Kerby, Sarah Freeman, **Kristina Prachanronarong**, Andrew W. Artenstein, Steven M. Opal, Anubhav Tripathi

Influenza, an emerging infectious disease and major public health concern, continues to threaten human health both in its seasonal form and its potential to cause pandemics. Three pandemics were documented in the 20th century; the most devastating, that of 1918, caused at least 50 million deaths worldwide. It is widely believed that a future influenza pandemic is inevitable and that such an event could result in global mortality in the range of tens to hundreds of millions if unchecked.

Structural genes coding for two viral proteins, hemagglutinin (HA) and neuraminidase (NA), specifically identify influenza subtype. Influenza viruses have the propensity to undergo genetic reassortment resulting in the presentation of altered antigenic surfaces for interaction with immune cells. When such variation results in novel subtypes, any preexisting immune recognition is lost. This phenomenon is referred to as antigenic shift and may be a precursor of a pandemic. To date there have been sixteen HA subtypes and nine NA subtypes identified. Although most have not been associated with human disease, a future pandemic may result from one of the many possible permutations of these types.

The potential health threat posed by various influenza subtypes is dependent on the intrinsic virulence of the virus and the level of type-specific immunity within a population. Epidemiologic sequelae may range from local outbreaks to a global pandemic. The specter of a recurrent pandemic has recently been raised by the occurrence of severe human disease due to subtype H5N1, a novel form of avian influenza, in Asia. While there has been a relatively low incidence of human disease due to H5N1 since 2003, a major concern is that the number of cases will rise exponentially should the virus adapt biologically, resulting in sustained human-to-human transmission. Since "humanized" influenza spreads efficiently, the ability to rapidly identify influenza subtypes will be essential in order to provide health care workers and public health professionals with key information in real-time fashion. Early detection will allow timely public health responses such as community mitigation strategies, antiviral and vaccine deployments.

Various methods have been proposed for the molecular detection and identification of a wide array of viruses, although many are labor intensive, expensive and/or sacrifice information content for rapid identification. Influenza is traditionally identified by type, subtype and strain, with each sublevel providing greater detail about the virus. Low-cost immunochromatographic test strips, similar in design to commercially available pregnancy tests, use antibodies to rapidly identify single classes of viral nucleoproteins. This qualitative assay is limited by suboptimal sensitivity, which restricts its use as a screening tool. Additionally, the results must be confirmed by growth of the virus in cell culture for definitive diagnosis. A second approach employs microsphere immunoassays for detection of antibodies to influenza virus. The method involves the use of nucleoprotein conjugated beads and a biotin–streptavidin detection system. Although this assay is associated with comparatively high sensitivity and specificity, it requires specialized equipment and is time-consuming. This and other immunoassays such as direct enzyme-linked immunosorbent assay (ELISA) or direct immunofluorescence (IF) largely suffer from the non-specific interactions between large numbers of antibodies and analytes that sometimes limit test accuracy and multiplexing capabilities.

A third approach using microarrays has been extensively investigated in biology, biodefense and infectious disease applications. Recent advances in microarray systems have demonstrated the ability to identify viral subtypes and mutations with a high degree of specificity and accuracy. However, hybridization efficiency of RNA to a surface-bound capture probe is strongly influenced by its surface transport properties and by the orientation of its secondary and tertiary structure. Viral RNA is often fragmented leading to altered secondary and tertiary structure, thus interfering with hybridization and specificity. Hence, the reproducible use of microarrays across the platforms has yet to be resolved. Moreover, microarray slide preparation, imaging and analysis require considerable labor, capital and time. A fourth approach uses reverse transcription polymerase chain reaction (RT-PCR) and real-time RT-PCR for detecting a specific subtype. Typically, a single primer pair targets a short, well conserved portion of the influenza RNA or occasionally a mutation region. However, these PCR-based tests require temperature cycling equipment, which complicates field deployment as part of a portable diagnostic device. Pourmand et al recently demonstrated a diagnostic method based on rapid pyrosequencing using 7 select H5N1 sequences. Reverse transcription PCR amplified 4 separate segments identifying HA clade, a glycosylation motif, sialic acid receptor specificity, and HA cleavage motif, each of which help predict the pathogenicity. The sequencing requirement currently restricts this otherwise informative method to a laboratory environment.

An ideal diagnostic method would incorporate the specificity derived from genetic-based analyses, the speed and simplicity of a one-step test, and the higher information content provided by an array-based or sequencing method. In this work, we describe an influenza subtype identification method based on a solution phase, sequence-specific hybridization of a dual labeled probe designed to hybridize directly with the viral RNA of interest without the need for thermal cycling. Dual labeled probes, also known as molecular beacons, are stem loop structures of single stranded DNA, where one end is labeled with a fluorophore and the other end a quencher molecule. In the absence of a target sequence, the complementary ends of the probe form a 5-7 base pair stem by hybridization, which quenches the fluorescent signal. The fluorophore unquenches and fluoresces when the 15-25 base loop hybridizes to the complementary RNA target sequence with higher affinity than the stem. Availability of a probe binding site on the RNA depends on conformation, which may change either spontaneously or induced through binding of probes. Native RNA structure may be determined by folding kinetics rather than by thermodynamics. Thermodynamically, the values for free energies for formation of nearest neighbor base pairs from single strands and for the formation of secondary structures can be altered by changing temperatures. Prediction of secondary structure *in silico* is limited by the accuracy with which the empiric parameters can be determined and by principal deficiencies, for example by the lack of energy contributions resulting from tertiary interactions. Therefore, experimental measurement of a probe response to RNA best confirms their activity. Experimental parameters include temperature, concentration and cations. The critical role of divalent cations in disruption or stabilization of influenza RNA structures is not known. RNA phosphate groups are believed to provide strong binding sites for Mg^{2+}, making secondary or tertiary structures particularly sensitive to concentration.

While dual labeled probes have been used in PCR assays and in live cell detection of RSV virus, to our knowledge, this is the *first report* detailing the use of dual labeled probes for direct detection or subtyping of influenza viral RNA. This work details the design of high information-content probes for the HA segment of the H5 influenza genome, experimentally characterizes the interaction of a dual-labeled DNA probe with H5 viral RNA, and demonstrates the detection method given the proper conditions of vRNA concentration and probe.

AMP-ACTIVATED PROTEIN KINASE (AMPK) MAY AFFECT FAT STORAGE BY CONTROLLING AN ENZYME THAT OXIDIZES FATTY ACIDS

Alisha Bloodworth[1,2] and Elmus G. Beale[3]

From the [1]Honors College, Texas Tech University, the [2]Department of Chemical Engineering, College of Engineering, Texas Tech University, and the [3]Department of Cell Biology and Biochemistry, Texas Tech University Health Sciences Center.

In mammals, AMP-activated protein kinase (AMPK) is responsible for the control of the body's fuel levels. Essentially, AMPK has multiple metabolic roles, one of which is the regulation of triglyceride metabolism. Because dysregulation of AMPK has been linked to diabetes, obesity, and cardiovascular diseases, it has become a prime target for study. It is important to develop a model species, such as the nematode *Caenorhabditis elegans*, in which genetic and environmental factors can be controlled. AMPK is conserved in *C. elegans*, which has two AMPK catalytic subunit orthologs (α1 and α2). Previous research in our lab shows that a null mutation in the α2 subunit, AMPKα2, causes a 90% reduction of the mitochondrial fatty acid oxidative enzyme, ACS-2, due to fasting. The predicted result of this defect would be a reduced ability to metabolize stored lipids during starvation. To test this prediction, Nile red, a lipophilic stain, was used to dye intracellular lipids red in *C. elegans*. To determine if image analysis is a reasonable way to compare wild-type *C. elegans* with the AMPKα2 knockout worms, L4 stage worms were stained with Nile red and then starved for 4 and 8 hours. Red fluorescence images were taken of the worms after each time period and fluorescence intensity, above background, was then quantified with software and statistically analyzed. Quantification of fluorescence levels revealed that AMPKα2 knockout worms had

significantly higher levels of fat compared to wild-type worms following starvation. Taken together, the available data suggest that one mean by which AMPKα2 regulates fuel homeostasis is by controlling expression of mitochondrial b-oxidative enzymes such as *acs-2*. Future experiments will examine whether this also occurs in mammals.

AMPK functions to control fuel levels in the body. Certain antidiabetic drugs function by activating AMPK and thus prevent metabolic syndrome. Metabolic syndrome describes the link between diabetes, obesity, and cardiovascular diseases. Stimulation of AMPK leads to energy production and inhibition of AMPK causes energy consumption. A properly regulated AMPK pathway provides the correct fuel homeostasis for the body and inhibits metabolic syndrome (Fig. 1).

Because of the intricacy of metabolic pathways and our inability to control genetic and environmental factors, a simple model species would greatly facilitate new discoveries. *C. elegans* is a powerful model species for a number of reasons. They have a quick generation time of about 3.5 days, are very hardy and easily grown in a laboratory, and their small genome is entirely sequenced and cloned. Also, according to Ashrafi, et al. significant metabolic pathways linked with human obesity are conserved in *C. elegans* (1). Two AMPK catalytic subunit orthologs are conserved in *C. elegans*, α1 and α2 (Fig. 2). In a previous study in our lab, the discovery was made that a null mutation in the α2 subunit, AMPKα2, causes a 90% reduction of the mitochondrial fatty acid oxidative enzyme, ACS-2, due to fasting (Fig. 3). The result of this null mutation should be reduced ability to use up stored lipids during periods of starvation.

Nile red is a dye that stains intracellular lipids red and can be used with *C. elegans*. In a lipid-rich environment, the stain is fluorescent so it is good for use in experiments involving image analysis. With the use of digital photography under a fluorescence microscope and software, one can quantify the intensity of the fluorescence inside an organism such as *C. elegans*.

EXPERIMENTAL PROCEDURES

Plate Preparation. Three sets of plates were made. The first consisted of nutrient agar, *E. coli* as food source, and 0.05 μg/mL Nile red as described by Ashrafi et al (1). The second set of plates contained nutrient agar with a 2% agarose layer spread across the top to prevent worms from burrowing into the agar in the absence of food. Nile red solution was put on the second set of plates as above. And the third set of plates contained only nutrient agar with the agarose layer but no Nile red.

Sample Preparation. Worms were plated on the first set of plates and left for 2 days at 20°C to ensure intake of Nile red. These worms were taken off the plates and washed three times with M9 buffer. They were then split into three groups. Two groups were moved to the second set of plates containing no food source and left at 20°C. The other group of worms was washed with water and then placed in 50 μL of 0.225 M sodium azide for about 5 minutes until all of the worms were anesthetized. The worms were then spread out on one of the plates from the third set. After 4 hours and 8 hours, this process was repeated for the two sets that were placed on the plates containing no food source.

Image Analysis. All images were captured using a Zeiss™ AxioCamMR3 at 91 ms exposure. Analysis was performed using ImageJ (2) and the 3D Color Inspector plugin (3) to determine the intensity of red fluorescence in each worm.

Statistical Analysis. All data were analyzed using Kruskal-Wallis nonparametric ANOVA with InStat™ (Graphpad Software). Post hoc data analysis was with Dunn's Multiple Range Test. $P < 0.05$ was considered a statistically significant difference.

RESULTS AND DISCUSSION

Verification of a uniform starting level for red fluorescence intensity was shown by comparing fed wild-type worms with fed AMPKα2 knockout worms. These worms showed no significant difference in red fluorescence intensity (Fig. 4). This means that both the wild-type and the knockout worms are storing the same amount of lipids when their food source is not restricted. After 4 hours of fasting, there is a significant drop in red fluorescence intensity in the wild-type worms (Fig. 4). Because the intensity of the fluorescence has decreased, this means that an amount of the dye in the worm has disappeared which indicates that some of the lipids have been used. After 8 hours of fasting, there is no further change in red fluorescence intensity from the 4-hour level, and as at 4 hours, the lipid levels were reduced only in the wild-type worms. Data indicate that the wild-type worms are capable of mobilizing their stored lipids when the food source is restricted.

The AMPKα2 knockout worms showed no significant decrease in red fluorescence intensity from the fed level at either of the 4 or 8-hour time intervals (Fig.

4). However, when comparing the AMPKα2 knockout worms with the wild-type worms after 4 hours of fasting, there is a significant difference in red fluorescence intensity with the knockout worms having a much higher intensity than the wild-type worms. The AMPKα2 knockout worms seem unable to mobilize their stored lipids as evidenced by no change in Nile red staining. This is what was predicted to happen based on the finding that a null mutation in the α2 subunit causes a 90% reduction of the mitochondrial fatty acid oxidative enzyme, ACS-2, due to fasting (Fig. 3). The wild-type worms are able to use their stored lipids as usual and the Nile red is leaving their system as the lipids are used up and the intensity is decreasing.

In summary, AMPK plays a central role in regulating fuel availability. It is a viable target for new therapies and diagnostics for metabolic syndrome. Proof of this assertion is the fact that two currently used antidiabetic drugs function by activating AMPK. It is thus important to elucidate all of the regulatory and metabolic pathways on which it acts. The present study provides additional motivation to use *C. elegans* as a model system to elucidate how AMPK controls gene expression to regulate metabolic pathways. In the case described here, AMPK participates in the mobilization of stored fat in the response to fasting. Future studies will need to assess whether this pathway also applies to mammals including humans.

REFERENCES

1. Ashrafi K, Chang FY, Watts JL, Fraser AG, Kamath RS, Ahringer J, Ruvkun G (2003) Genome-wide RNAi analysis of Caenorhabditis elegans fat regulatory genes. *Nature* 421:268-272
2. Rasband, W.S., ImageJ, U.S. National Institutes of Health, Bethesda, Maryland, USA, http://rsb.info.nih.gov/ij/, 1997-2007.

3. Barthel, K.U., 3D Color Inspector, Internationale Medieninformatik, Berlin, Germany, http://www.f4.fhtw-berlin.de/~barthel/ImageJ/ColorInspector//help.htm, 2005.

ACKNOWLEDGEMENTS

We would like to thank Dr. Jannette Dufour for allowing us to use her microscope to digitally photograph the worms. The work was supported by a Seed Grant from the TTUHSC Center for Cardiovascular Disease and Stroke, the Texas Tech University Honors College, the Allensworth Family, and the Helen Jones Foundation.

FIGURE LEGENDS

Fig. 1: AMPK functions to maintain fuel homeostasis thereby preventing metabolic syndrome.

Fig. 2: AMPK is conserved in *C. elegans*.

Fig. 3: A null mutation in the α2 subunit causes a 90% reduction of the mitochondrial fatty acid oxidative enzyme, ACS-2, due to fasting, as shown by previous research performed in our lab.

Fig. 4: The AMPKα2 knockout worms seem unable to mobilize their stored lipids, as evidenced by no change in Nile red staining, while the wild-type worms seem to be capable of mobilizing theirs over both the 4-hour and 8-hour periods of time.

FIGURE 1

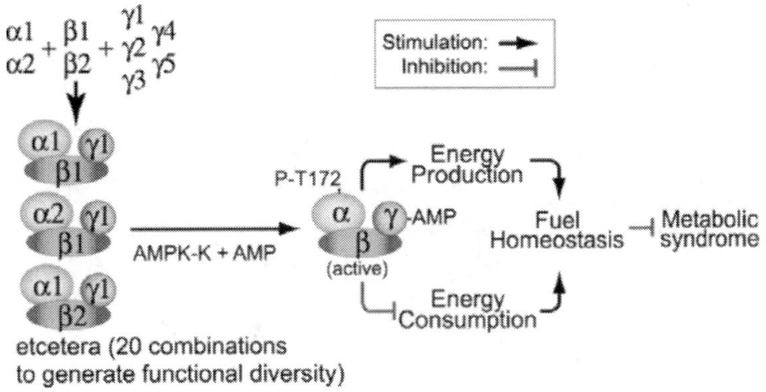

FIGURE 2

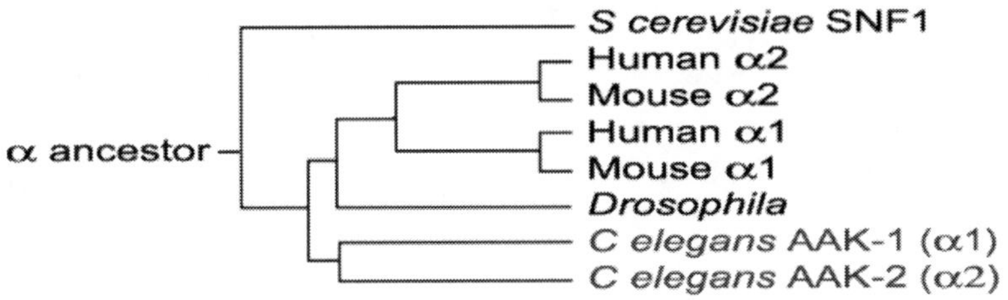

FIGURE 3

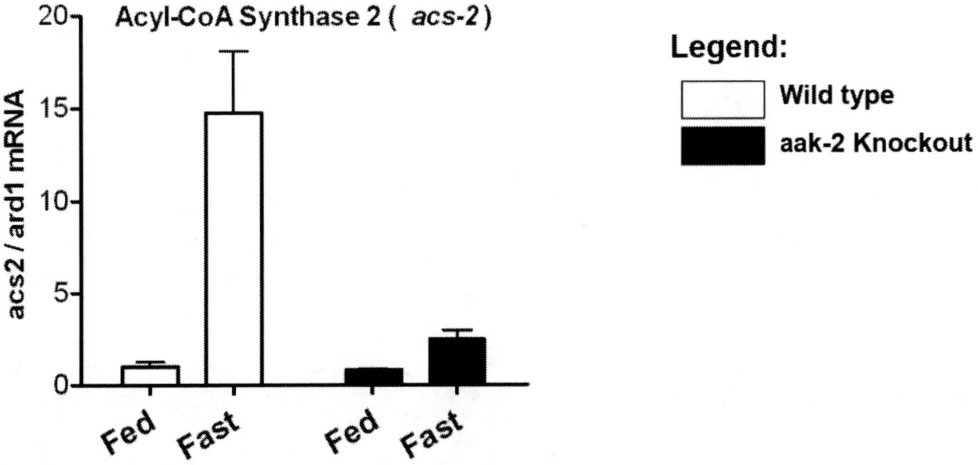

FIGURE 4

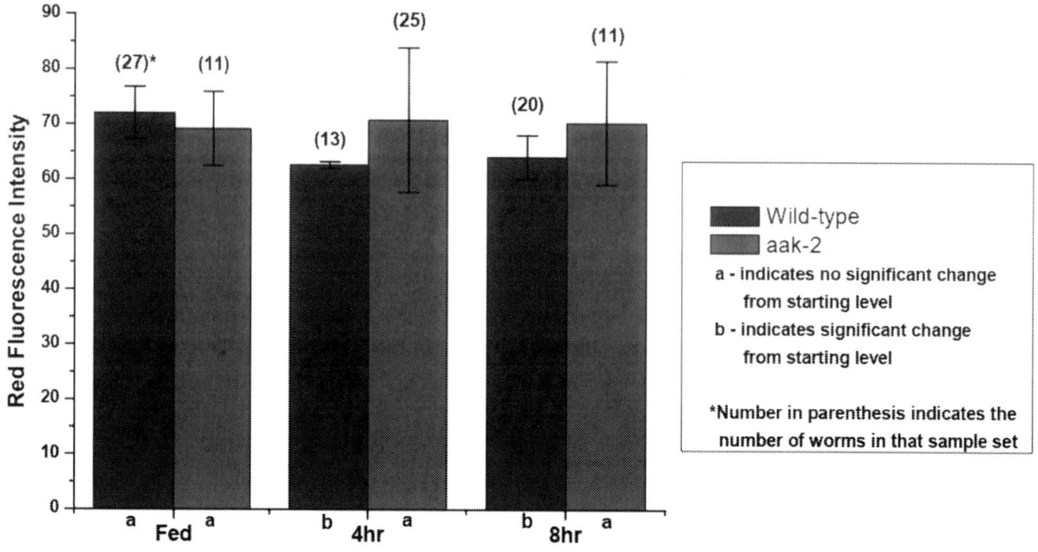

Red Fluorescence Intensity
90
80
70
60
50
40
30
20
10
0
(27)*
(11)
(13)
(25)
(20)
(11)
a
Fed
a
b
4hr
a
b
8hr
a
Wild-type
aak-2
a - indicates no significant change from starting level
b - indicates significant change from starting level
*Number in parenthesis indicates the number of worms in that sample set

Retrofit of Sour Water Networks in Oil Refineries: A Case Study

Lisa Scodari, Daniel Sujo-Nava, C. Stewart Slater, Mariano Savelski, and Kevin Dahm. Chemical Engineering, Rowan University, 2455 Weldwood Dr, Apt. 6302 , Baton Rouge, LA 70816

Refining sour crude with high NH_3 and H_2S content has increased significantly in recent years due to oil scarcity and high prices. However, refining operations are constrained economically by the difficulty of meeting environmental regulations. Managing sour water efficiently is imperative for mitigating the environmental impact of sweet oil reserves depletion. A case study is presented of the retrofit of a sour water network in a petroleum refinery that generates 581,000 m^3/yr of sour water. Flowrates and contaminant levels of ammonia and hydrogen sulfide were traced for all major water streams in the sour water network. By grouping some process units, a reuse scheme was found that only generates 280,000 m^3/yr of sour water, saving 83% of freshwater and 52% in energy and CO_2 emissions associated with pumping. Additionally, it was found that the remaining water, after treatment in the sour water stripper, could be reused elsewhere in the plant, if cyanide ions were eliminated. This case study has provided the foundation for further plant studies to validate the water profile of the proposed scheme.

Porous Monolith Structures in Microfluidic Devices

Thomas Schwei

Laminar flow in the channels of microfluidic devices causes mixing to be diffusion-limited and slow. In order to improve mixing for fluorescently labeling proteins we create a passive mixer within a microfluidic channel by patterning an array of porous structures. These polymer monoliths are patterned directly within plastic chips via ultra-violet (UV) irradiation through a mask positioned over the desired area of the channel. Through optimization of the prepolymerization solution and irradiation time we demonstrate the ability to pattern plugs of monolith in 100-micrometer segments within 5 to 6 minutes of UV irradiation. Future work will include evaluating the resolution limits and testing a small library of structures for mixing efficiency.

6a Molecular Interactions to Product and Process Design: Crystallization, Education and beyond

Ryan C. Snyder, University of California Santa Barbara, Department of Chemical Engineering, Santa Barbara, CA 93106

Chemical Engineering and product and process design are continually evolving as both our fundamental understanding of molecular interactions increases and as the areas of interest for cutting edge research have evolved from liquids to micro-structured products such as pharmaceuticals, food products, solar cells and nanowires. Our goal is to use this continually increasing knowledge of molecular interactions to develop experimentally verified predictive models that can be implemented on time scales relevant for product and process development of new materials. The focus of my research has been in the area of the crystallization of organic molecular systems such as those found in the pharmaceutical and food products industries. Two of the primary product properties desired for crystallization are the polymorph (packing structure) and habit (shape), since the physical, chemical and processing properties of the crystal are highly dependent on both of those product properties.

One of my research thrusts has been to develop methodologies to predict the growth and dissolution rates of faceted crystal surfaces. These growth and dissolution rates are the foundation for the prediction of dynamic shape evolution and solution mediated polymorphic phase transformations. The growth and dissolution model is based upon the spiral mechanism of Burton, Cabrerra and Frank[1], and it takes into account the effect of important process parameters such as solvent, temperature and supersaturation. The model has been automated by eliminating the need for arbitrary human decisions[2], and generalized to account for organic molecules of API-level complexity[3]. The predictions of this model have been validated by comparison to experimentally observed crystal shapes obtained in our laboratory and from the literature.

These methods for the prediction of growth and dissolution rates can be used to predict morphology changes given a faceted shape evolution model. Thus, I have developed such a faceted shape evolution model in the context of both growth and dissolution[4]. The relative growth and dissolution rates needed to implement this model can come from the a priori methods described above, or from previously measured experimental data. The model demonstrates that a crystal's shape evolves toward the single stable steady state shape in growth; however, it evolves away from the single unstable steady shape in dissolution. In addition, the predictions made using the methods above and the dissolution shape evolution model have been validated using video microscopy in a Peltier viewcell[5]. The growth and dissolution models also have been combined to demonstrate the feasiblity for the use of cycles of growth and dissolution for the enhancement of crystal shape[6].

Additionally, I have a strong interest in teaching all aspects of the chemical engineering curriculum. In addition to crystallization, one particular area in which I have developed a unique expertise throughout my graduate career is that of conceptual design using the hierarchical decision procedure of Douglas[7]. The systematic nature of this decision making process allows for the rapid selection of a near-optimal design, and similarly to my research in crystallization the decisions are highly dependent on the underlying fundamental physics and chemistry. This combined experience in teaching process design and research in product and process design for crystallization provides a synergistic path forward for the development of product and process design methodologies for crystallization as well as new research areas.

1. Burton, W.K., Cabrera, N., and Frank, F.C., "The Growth of Crystals and the Equilibrium Structure of their Surfaces," Phil. Trans. R. Soc. 1951; 243, 299.

2. Snyder, R. C. and Doherty, M. F., "Predicting Crystal Growth by Spiral Motion", in preparation.

3. Snyder, R. C., Sizemore, J. P. and Doherty, M. F., "Growth Prediction for Molecular Crystals of API-Complexity", in preparation.

4. Snyder, R. C. and Doherty M. F., "Crystal shape evolution during dissolution and growth", AICHE J. 2007; 53: 1337-1348.

5. Snyder, R. C., Veesler S. and Doherty M. F., "The Evolution of Crystal Shape During Dissolution: Predictions and Experiments", in preparation.

6. Snyder, R. C., Studener S. and Doherty M. F., "Manipulation of Crystal Shape by Cycles of Growth and Dissolution", AICHE J. 2007; 53: 1510-1517.

7. Douglas, J. M., "Conceptual Design of Chemical Processes." Boston, MA., McGraw Hill Inc., 1988.

6b 3-Dimensional in Vitro Model of Hepatic Tissue for Investigating Liver Physiology and Pathophysiology

Rohit Jindal, Center for Engineering in Medicine, Massachusetts General Hospital, Harvard Medical School, 51 Blossom Street, Boston, MA 02114

In vivo, cells are organized in complex tissue architecture with various extracellular matrix (ECM) and cellular signals governing cell behavior. I am interested in developing in vitro models of tissue that mimic this 3 dimensional arrangement of ECM and different cell types. These models offer the potential of capturing the complex signaling environment found in vivo and thus provide more physiologically relevant systems for investigating tissue physiology and pathophysiology. Currently, I am working on the development of coculture model of hepatic tissue that mimics the sinusoid structure found in the liver. This in vitro model is being used for evaluating the role of cell-cell signaling and cell-ECM interactions in a variety of physiologically relevant settings including maintenance of hepatocytes differentiated function, ischemia- reperfusion injury, and inflammation.

In this presentation, I will also briefly describe my doctoral work that included patterning of polymers in the microfluidic channels; and on chip electrochromatography using sol-gel immobilized stationary phase particles.

6c Advanced Nonlinear Programming Formulations and Algorithms: Expanding the Scope of Industrial Nmpc Applications

Victor M. Zavala, Department of Chemical Engineering, Carnegie Mellon University, 5000 Forbes Ave., Pittsburgh, PA 15213

During the last decades we have witnessed a widespread use of nonlinear programming (NLP) techniques and algorithms for the solution of a wide range of problems arising in engineering and science. In addition, it has also become clear that advances in optimization and the continuous increase of computational power are directly coupled to the application scope. In this work, we will present a complex problem posed by an industrial collaborator: the low-densitiy-polyethylene (LDPE) tubular reactor process. This application has served as a clear reference of the scope of current optimization capabilities. This has been useful in making a critical assessment of the potential of different optimization strategies and algorithms and, at the same time, has motivated the development of a number of novel NLP formulations and the expansion of current capabilities of NLP solvers.

In collaboration with Prof. Lorenz T. Biegler (CMU), Dr. Carl D. Laird (Texas A&M) and our industrial partners, we have expanded the current capabilities of the large-scale NLP solver (IPOPT) that has allowed for the solution of different types of optimization problems arising in Nonlinear Model Predictive Control frameworks. Some highlights of the results accomplished so far are the following:

1) Advanced Parameter Estimation Formulations: We have developed a general parameter estimation framework around IPOPT that allows for the analysis and solution very large-scale parameter estimation problems in both standard and parallel computer architectures [1][2]. We have tested our developments using an industrial low-density polyethylene (LDPE) tubular reactor model.

In this particular application, the LDPE reactor model is so complex that these estimation problems were previously thought intractable [3]. In our framework, we have been able to overcome these limitations and solve error-in-variables (EVM) parameter estimation problems with up to 32 data sets containing around 4,100 ordinary differential equations, 16,000 algebraic equations and 2,100 degrees of freedom in a distributed cluster in around 4 minutes of CPU time. The framework consists of a simultaneous collocation-based strategy and a parallel Schur complement decomposition strategy implemented IPOPT. In addition, we have established an efficient strategy to extract large-scale parameter covariance matrices from IPOPT which are essential for further statistical analysis. Significant improvements in the model predictions have been obtained and a systematic parameter inference analysis has been performed for the first time in this type of rigorous models. The rigorous LDPE reactor model has been tested against industrial plant data and an on-line parameter estimator has been developed and delivered to our industrial collaborator. Since the estimation framework uses general optimization techniques, it can also handle more general parameter estimation formulations.

2) Fast Nonlinear Model Predictive Control: When dealing with large-scale rigorous models in

optimization-based control strategies such as NMPC, it is clear that not even the fastest optimization algorithms currently available will be able to provide fast feedback to the plant. Instead, what is usually observed is that there exists a long computational delay (several minutes) between the instant in which the new state of the system is received and the instant in which the optimization problem has been solved and the control action becomes available. This computational delay has been identified as one of the main limitations associated to NMPC since it has a direct impact on the stability and performance of the controller. On the other hand, the moving horizon optimal control problems (OCPs) solved in NMPC present an interesting property that can be exploited to obtain very fast approximate solutions: the problem is parametric in the initial conditions given by the moving state of the system. While fully explicit parametric control approaches cannot be currently applied to complex industrial applications, it is possible to exploit the parametric property of OCPs by computing and updating (on-line) explicit control laws using the concept of NLP sensitivity. We have developed different computational delay-free NLP sensitivity controller variants based on a simultaneous collocation-based approach and full-space optimization solvers (such as IPOPT) and tested our developments on a simplified industrial LDPE process model [4]. In this application, it has been possible to reduce the on-line computational tasks by 2-3 orders of magnitude and we are now able to provide instantaneous feedback in around 1 second. Finally, we have studied the nominal and robust stability properties of the proposed fast NMPC controllers to analyze their interactions with numerical errors, model-mismatch and state estimation errors [5].

3) Fast Moving Horizon Estimation: The parametric property of moving horizon problems arising in NMPC can also be exploited in the moving horizon estimation (MHE) case through appropriate formulations. With this, we have developed a fast MHE strategy that is able to provide state estimates without computational delay [6]. We have tested the developments on the LDPE process model were fast and accurate state estimates have been obtained. Interestingly, the numerical properties of the KKT matrix can be related to the observability of properties of the dynamic system. In addition, we are currently studying the theoretical properties of the fast MHE algorithm and analyzing the interactions of NLP sensitivity errors on the stability properties of the estimator.

4) Parallel Solution of DAE-constrained Optimization Problems with Loosely Coupled Algebraic Components: We have found that the main challenge posed by our industrial application is the size and complexity of the LDPE tubular reactor model. On the other hand, the LDPE process presents the interesting property that the dynamics of the reactor and those of the rest of the plant occur at significantly different time-scales. Because of this, the LDPE reactor model can usually be assumed to be at quasi-steady-state. This gives rise to a special class of DAE constrained problems were the proportion of algebraic states is much larger than that of differential states. This property is not specific to this application but is also present in some other complex systems such as metabolic networks, dynamic models with complex thermodynamic calculation routines, among others. We have developed a simple and practical reformulation of this type of DAE-constrained problems that leads to a general and efficient Schur complement parallel solution strategy. This decomposition required of some minor implementation modifications to the current parallel IPOPT capabilities and can be used in a wide range of optimization problems. Finally, these developments have been used to solve complicated multi-stage dynamic optimization problems arising in our industrial LDPE process in a few minutes [7].

Having this entire computational framework, we are currently incorporating all these developments on a stand-alone NMPC framework for the LDPE process. This will enable us to increase the profitability of the process through a more systematic decision-making. While the developments are directly applicable to many different industrial processes, our challenging application has also helped us in identifying potential improvements that will further expand the scope of nonlinear programming in some other industrial applications.

[1] Zavala, V. M. and Biegler, L.T. Large-Scale Parameter Estimation in Low-Density Polyethylene Tubular Reactors. Ind. Eng. Chem. Res. 45, 7867 -7881, 2006.

[2] Zavala, V. M.; Laird, C.D. and Biegler, L.T. Interior-Point Point Decomposition Approaches for Parallel Solution of Large-Scale Nonlinear Parameter Estimation Problems. Chem. Eng. Sci. Accepted for Publication, 2007.

[3] Bokis, C. P.; et.al. Physical Properties, Reactor Modeling, and Polymerization Kinetics in the Low-Density Polyethylene Tubular Reactor Process, Ind. Eng. Chem. Res. 41, 1017-1030, 2001.

[4] Zavala, V. M.; Laird, C.D. and Biegler, L.T. Fast Implementations and Rigorous Models: Can Both be Accommodated in NMPC? Int. J. Robust Nonlinear Control. Accepted for Publication, 2007.

[5] Zavala, V. M.; and Biegler, L.T. The Advanced Step NMPC Controller.

Proceedings of the IEEE Workshop on Advanced Process Control Applications for Industry, 2007.

[6] Zavala, V. M.; Laird, C.D. and Biegler, L.T. A Fast Computational Framework for Large-Scale Moving Horizon Estimation. Proceedings of the 8th International Symposium on Dynamics and Control of Process Systems, 2007.

[7] Laird, C.D.; Zavala, V. M. and Biegler, L.T. Efficient Parallel Solution Of DAE Constrained Optimization Problems With Loosely Coupled Algebraic Components. AIChE Annual Meeting, 2007.

6d Transport Phenomena within Tissue Engineering Materials

Benjamin J. Lawrence, School of Chemical Engineering, Oklahoma State University, 423 Engineering North, Stillwater, OK 74078

My research interests include biomaterials, tissue engineering, and bioreactor design. The main thrust of my dissertation is applying engineering principles to investigate transport phenomena within tissue engineering scaffolds. Mass transfer is an important aspect of tissue engineering that is often overlooked when designing scaffold materials. Nutrient mass transfer limitations lead to a host of problems during tissue regeneration. The issue is complicated further when using composite materials reinforced with impermeable membranes. One such composite has been developed to mimic the mechanical properties of small intestinal submucosa, a natural material clinically used in regeneration. The composite is a sandwich configuration with porous natural polymers surrounding a synthetic polymer membrane [1]. I am investigating both local and global scaffold mass transport using a nanoparticle controlled release system, computational simulations, and a bioreactor setup. For local mass trasport, I am utilizing nanoparticles prepared in house.

[1]. Lawrence B, Maase EL, Madihally SV. Synthetic Scaffolds Mimicking Small Intestinal Submucosa. J. Biomedical Materials Research-Part A. (accepted pending revision). 2007.

6e Nanocomposite Properties and the Polymer Interphase

Karl Putz, Mechanical Engineering, Northwestern University, 2145 Sheridan Road, Evanston, IL 60208

The surface area per gram of a rod is inversely proportional to the radius of the rod. With the development of single walled carbon nanotubes (SWNTs), the large surface area (1200 m2/g) has been both a curse and a blessing. The attraction between individual SWNTs combined with the surface area has driven aggregation, which reduces the effectiveness of the nanotubes in polymer nanocomposites. However, as the scientific community has explored very low concentrations and developed methods of dispersing the nanotubes, better than expected results (mechanical and thermal) have been demonstrated.

The positive results originate from the interaction between the SWNTs and the polymer matrix. The polymer near the surface of the SWNT has slower kinetics and higher elastic modulus; this region of the matrix is called the interphase. The interphase is important not only because it allows us to create polymer nanocomposites with enhanced properties, but to explore polymer physics as well.

This poster will highlight my graduate work on exceptional modulus changes at ultra low concentrations of single walled carbon nanotubes. In this work, we demonstrated that attractive interactions between the nanotube and polymer system, as demonstrated by Raman spectroscopy, combined with dispersed nanotubes at concentrations below 0.35 wt. % led to higher than volume additive enhancements of the storage modulus. The results were demonstrated with pristine SWNTs and PMMA as well as functionalized SWNTs with polystyrene.

In addition, my current postdoctoral work on the effect of nanoparticle shape on the creation and properties of the interphase will be highlighted. In this work, nanoparticles of different dimensionality (0D, 1D, 2D) with the same surface modifications are used to investigate how the shape of the interface changes the properties of the interphase. The results are discussed in light of polymer physics and a method to new model the properties of the interphase is presented.

6f Effects of Nanoparticle Addition on the Bulk, Surface and Interfacial Properties of Polymers

Anish Tuteja[1], Wonjae Choi[2], Joseph M. Mabry[3], Michael E. Mackay[4], Gareth H. McKinley[2], and Robert E. Cohen[5]. (1) Department of Chemical Engineering, Massachusetts Institute of Technology, 77 Massachusetts Avenue, Bldg. NE-47, Room 583, Cambridge, MA 02139, (2) Mechanical Engineering, Massachusetts Institute of Technology, Building 3-252, 77 Massachusetts Avenue, Cambridge, MA 02139, (3) AFRL/PRSM, ERC Inc., Edwards Air Force Base, 10 East Saturn Blvd., Bldg. 8451, Edwards, CA 93524, (4) Michigan State University, Department of Chemical Engineering and Materials Science, East Lansing, MI 48824, (5) Chemical Engineering, Massachusetts Institute of Technology, 77 Massachusetts Ave., 66-554, Cambridge, MA 02139

Polymer–nanoparticle composite materials offer the unique opportunity to synergistically combine the properties of the polymer and filler on a nanoscale. My research during my Ph.D. and postdoctoral work has been aimed to discover and understand some of the unusual and unexpected enhancements in both the bulk and surface properties caused by the addition of nanoparticles to polymers.

Some of the projects that I worked on during my Ph.D. work with Prof. Michael E. Mackay at Michigan State University are described below:

Novel strategies for nanoparticle dispersion. [1] We demonstrated that the relative size of the nanoparticle with respect to the linear chain is the key factor governing miscibility of nanoparticles in polymers. Phase stability is possible in chemically dissimilar systems like polystyrene nanoparticles [2] in polymethylmethacralate or polyethylene nanoparticles in polystyrene (both of these are classical phase separating systems) as long as the nanoparticle radius is less than the radius of gyration of the linear chain; on the other hand, if the nanoparticles become bigger than the linear chain, even polystyrene nanoparticles phase separate from linear polystyrene. Using our understanding of the physics of particle dispersion, we also developed methods for producing miscible nanoparticle-polymer blends at relatively high loadings, from systems (like fullerenes C60 in polystyrene) that phase separate using traditional blend preparation methods.

Effects of nanoparticle addition on the viscosity of polymer melts and solutions. [3, 4] Traditionally the addition of particles to suspensions and melts causes an increase in their viscosity (this was first predicted by Einstein about 100 years ago). We demonstrated for the first time non-Einstein-like behavior in polymer melts by establishing that the addition of nanoparticles, even at low volume fractions (less than 5%), to polymers can reduce the melt viscosity by as much as 90%, contradicting Einstein's prediction for an increase in viscosity of solutions on the addition of particles. This work has now been extended to show that the viscosity of concentrated polymer solutions can also be reduced by nanoparticle addition.

Diffusion of nanoparticles in polymer melts. [5] One of our key predictions after the study on the viscosity decrease caused by the addition of nanoparticles to polymer melts was that the nanoparticles must diffuse faster than predicted by the Stokes-Einstein relation, to cause the observed viscosity decrease. Recent experiments using X-ray photon correlation spectroscopy (XPCS) have confirmed this prediction as we obtained order of magnitude faster diffusion coefficients than those predicted by the Stokes-Einstein relation.

Enhanced thermal stability in polymer-nanoparticle composites. We demonstrated for the first time that the addition of nanoparticles greatly improves (up to 10 fold) the thermal stability of polymers. This observation is particularly surprising in the case of polymeric nanoparticles (like polystyrene nanoparticles) where the nanoparticles and the linear polymer have a much faster degradation rate individually.

Chain expansion on the addition of nanoparticles. Recently there has been a lot of controversy in the literature debating the effects of nanoparticle addition on the radius of gyration of the linear polymer. Our extensive experiments on the ideal system of polystyrene nanoparticles in linear polystyrene using small angle neutron scattering have shown conclusively that the nanoparticle addition causes an expansion in the radius of linear chains (swelling). We were also able to provide a detailed description of the effects of changing the nanoparticle size and linear chain lengths, on the observed expansion.

Rheological characterization of intramolecularly crosslinked nanoparticles. In this work, novel polystyrene nanoparticles were synthesized by the controlled intramolecular crosslinking of linear polymer chains to produce well defined single-molecule nanoparticles of varying molecular mass, corresponding directly to the original linear precursor chain. These molecules were ideal to study the rheological/relaxation behavior of high molecular mass polymers in the absence of entanglements, as the high crosslink density present in these molecules potentially inhibits entanglements. It was found that the reptation model (or the presence of a tube for relaxation) does not explain many of the important rheological features seen for our systems. Instead, a coupling or cooperative motion between the crosslinked loops of the molecules was found to play a significant role in the relaxation process.

Effect of nanoparticle addition on the mechanical properties of polymers. We studied the effects of nanoparticle addition on the mechanical properties of polymers and established that nanoparticle addition can be used to significantly enhance (up to 5 fold) the damping properties of polymers. Quite significantly, this increase in the damping properties is achieved without affecting the elastic modulus. In fact, a slight increase in the elastic modulus was observed (coupled with a decrease in viscosity!).

Multifunctionality in nanocomposites. Many of the surprising property enhancements caused by the addition of nanoparticles to polymers, as described above, can be engineered to happen for the same volume fraction of the added nanoparticles. Our studies have shown that the addition of certain nanoparticles can simultaneously lead to better processability (reduction in viscosity of up to 80%, contrary to Einstein's prediction), higher electrical conductivity (greater than Maxwell's prediction) and improved mechanical damping and enhanced thermal stability, leading to truly multifunctional nanocomposites.

My postdoctoral work with Prof. Gareth H. McKinley and Prof. Robert E. Cohen at Massachusetts Institute of Technology has been aimed at using nanoparticles to modify the surface properties of polymer composites, and have used nanoparticles to design the first ever super-oleophobic surfaces as elaborated below:

Designing super-oleophobic surfaces. The combination of surface chemistry and roughness' on the micron and nanoscale imbues enhanced repellency to the lotus leaf surface when in contact with a high surface tension liquid such as water (surface tension $\gamma = 72.1$ mN/m). This understanding has led to the creation of a number of artificial superhydrophobic surfaces (water contact angles greater than 150°, low hysteresis). However, researchers so far have been unsuccessful in producing super-oleophobic surfaces for liquids with much lower surface tensions; for example alkanes such as decane ($\gamma = 23.8$ mN/m) or octane ($\gamma = 21.6$ mN/m). Theoretical calculations suggest that a super-oleophobic surface would need to have a surface energy < 5 mN/m, whereas the lowest solid surface energies reported to date are in the range of ~6 mN/m. In this work, we explain how a third factor, surface curvature (apart from surface chemistry and roughness), can be used to significantly enhance liquid repellency, by studying electrospun polymer fibers containing very low surface energy perfluorinated nanoparticles (FluoroPOSS). Increasing the POSS concentration in the elecrospun fibers allows us to systematically transcend from super-hydrophilic to super-hydrophobic and finally to the first ever super-oleophobic surfaces (exhibiting low hysteresis and contact angles with decane and octane greater than 150°).

References:

1. Mackay, M. E., Tuteja A. et al. General Strategies for Nanoparticle Dispersion. Science 311, 1740-1743 (2006).

2. Tuteja, A., Mackay, M. E., Hawker, C. J., Van Horn, B. & Ho, D. L. Molecular architecture and rheological characterization of novel intramolecularly crosslinked polystyrene nanoparticles. Journal Of Polymer Science Part B-Polymer Physics 44, 1930-1947 (2006).

3. Mackay, M. E., Dao T.T., Tuteja A. et al. Nanoscale effects leading to non-Einstein-like decrease in viscosity. Nature Mat. 2, 762-766 (2003).

4. Tuteja, A., Mackay, M. E., Hawker, C. J. & Horn, B. V. Effect of Ideal, Organic Nanoparticles on the Flow Properties of Linear Polymers: Non-Einstein-like Behavior. Macromolecules 38, 8000-8011 (2005).

5. Tuteja, A., Mackay, M. E., Narayanan, S., Asokan, S. & Wong, M. S. Breakdown of the continuum stokes-einstein relation for nanoparticle diffusion. Nano Lett 7, 1276-81 (2007).

web.mit.edu/atuteja/www/home.html

6g Towards Understanding the Gecko Adhesive System

Noshir S. Pesika, Chemical Engineering, University of California, Interfacial Science Lab, Engineering II, Room 3357, Santa Barbara, CA 93106

Understanding the mechanism behind the gecko adhesive system has recently received significant interest because of the potential of creating synthetic dry adhesives inspired by nature and evolution. Since van der Waals forces are responsible in the adhesion mechanism of geckos, the hierarchical structure and geometry of the system become critical as opposed to surface chemistry. This poster demostrates (1) the role of geometry (Peel Zone Model) in the enhancement of adhesion, and (2) the influence of the spacing parameters (Stacking Model) in determining the maximum achievable adhesion. Insights provided by the proposed models will potentially guide the design of future synthetic dry adhesives inspired by the gecko adhesive system.

Engineering Drug, Gene, and Cell-Based Delivery Systems for the Treatment of Human Disease

Michelle Dawson, Dan G. Duda, Rakesh K. Jain
Edwin L. Steele Laboratory, Massachusetts General Hospital and Harvard Medical School

My research will focus on the design of therapeutic drug, gene, or cell-based delivery systems that can be used for the local or systemic treatment of genetic diseases, including cancer. I believe I can best accomplish this goal in an innovative academic research environment by guiding future scientists and engineers in this endeavor.

As a doctoral candidate in the laboratory of Dr. Justin Hanes at Johns Hopkins University, my research focused on the design of gene vectors for the treatment of cystic fibrosis. My doctoral dissertation was titled "Mucosal Barriers to Non-Viral Gene Delivery in the Cystic Fibrotic Lung." This project involved three main facets including: (i.) the formulation and characterization of polymeric and liposomal gene delivery systems, (ii.) the development and characterization of in vitro models for testing the efficiency of gene delivery systems, which included mucus from cystic fibrosis patients, synthetic mucus used to mimic lung mucus, and lung cells grown in polarized monolayers, and (iii.) the development and application of sophisticated techniques for comparing the efficiency of nano-sized gene delivery systems at overcoming biological barriers, which included flow cytometry, electron microscopy, confocal and epiflourescent imaging, multiple particle tracking, rheology, light scattering, and quantitative and statistical image analysis. The work included in my doctoral dissertation resulted in 6 publications (3 in press and 3 in preparation or submitted), a book chapter, two patents, 9 podium presentations, 6 poster presentations, and 7 awards, including an NSF fellowship and prestigious drug delivery awards from the International Society of Aerosols in Medicine and Capsugel/Pfizer.

After graduation, I was eager to pursue a more translational postdoctoral training position related to the design and delivery of cancer therapeutics. Thus, I joined Dr. Rakesh Jain's laboratory at Massachusetts General Hospital, where I continue to focus on gene delivery using cell-based delivery systems. Specifically, I am designing genetically engineered mesenchymal stem cell-based therapeutics for the treatment of solid tumors. The ability to engineer cells and infuse them into the body to serve as therapeutic vehicles has become increasingly feasible in recent years. The advantage of cell-based delivery protocols resides in their ability to incorporate into tumor tissues and release locally sustained levels of therapeutic cytokines. To facilitate the production of mesenchymal stem cell-based therapeutics, my project has focused on the three primary aspects: (i.) the characterization of the phenotype of bone marrow-derived mesenchymal stem cells, (ii.) the development of quantitative homing assays for studying the migration of different bone marrow stroma-derived cell populations to tumors, and (iii.) the development of genetically-altered stem cell therapeutics, which have been tested in mouse models of neoplasia. Using multiphoton laser scanning microscopy, we have tracked intravitally the recruitment and engraftment of bone marrow derived stromal cells in tumors. I hope this training experience will help me to develop vital insight into the biological response to therapeutics, which is essential in the development of drugs that can overcome biological barriers to treat human disease.

6i Engineering a Mulitfunctional Scaffold for Spinal Cord Repair

Noelle K. Comolli, Drexel University, Department of Chemical and Biological Engineering, Philadelphia, PA 19104, Itzhak Fischer, College of Medicine, Drexel University, Philadelphia, PA 19104, Birgit Neuhuber, Nuerobiology & Anatomy, Drexel College of Medicine, 2900 Queen Lane, Philadelphia, PA 19129, and Anthony M. Lowman, Deptartment of Chemical and Biological Engineering, Drexel University, 3141 Chestnut St., Philadelphia, PA 19104.

Spinal cord injury affects nearly twenty five thousand Americans, with an additional ten thousand new injuries per year. This project proposes the use of a non-biodegradable hydrogel as a mechanical support to fill the injury site and provide a protected environment for the regenerating neurons. The hydrogel can be designed to match the mechanical stiffness of the local tissue to reduce the amount of stress shearing around the implant. The use of a thermally responsive, injectable polymer, PNIPAAm-PEG branched copolymer, is proposed as a minimally invasive surgical technique. Below its LCST, typically around 29-32C, the polymer forms a miscible solution with water, but above its LCST, it becomes hydrophobic, separating from water and forming a semi porous gel. The aqueous polymer solution can be created with cell culture media and seeded with bone marrow stromal cells in order to physically entrap cells into the scaffold when it is injected into the defect. This would allow the cells to proliferate creating an environment more attractive to growing neurons. This environment can be enhanced with the addition of biodegradable polymer microparticles loaded with trophic factors to promote both nerve growth as well as angiogenesis.

6j Effect of Phosphorus on Acid Cracking of Lipids for the Production of Green Diesel

Stephen Dufreche[1], Rafael Hernandez[1], Todd French[2], Mark G. White[1], Earl G. Alley[3], and William E. Holmes[4]. (1) Dave C. Swalm School of Chemical Engineering, Mississippi State University, 330 Swalm President's Circle Dr., Mississippi State, MS 39762, (2) Chemical Engineering, Mississippi State University, 330 Swalm Circle President Dr., Mississippi State University, MS 39762, (3) Mississippi State Chemical Laboratory, Mississippi State University, Mississippi State, MS 39759, (4) Mississippi State Chemical Laboratory, Mississipp State University, P.O. CR, Mississippi State, MS 39762

Biodiesel is one of the most widely used biofuels in the world, due in part to its simplicity of production, compatibility with existing engines, and reduction of green house gas emissions. However, technical difficulties with biodiesel include: (1) the need of highly refined oil for ASTM compliance, (2) incompatibility with the petroleum-diesel pipeline distribution system, and (3) a relatively small inventory of expensive feedstocks. Issues (1) and (2) could be overcome by the production of biofuels using chemical processes associated with petroleum refining. Catalytic lipid cracking could result in green diesel, a fuel chemically similar to conventional diesel but derived from a clean renewable feedstock. A homogeneous reaction of a model lipid with triflic acid (100 times more acidic than sulfuric acid) resulted in products indicative of cracking. The final phase of the project expands on these results with the cracking of model lipids over a solid acid catalyst zeolite (ZSM-5). Since lipid feedstocks contain small amounts of phospholipids knowledge of the interactions between phospholipids and zeolites is crucial to a system-wide understanding of the lipid cracking process. Phosphorus-containing compounds were used to poison ZSM-5 in order to simulate the cracking of phospholipids. Model lipid compounds were then cracked over the poisoned zeolite, with differences in product distribution and kinetics based on phosphorus loading recorded.

Publications

• Dufreche, S., Hernandez R., French, T., White, M., Alley, E. and Holmes, W., (2007) “Superacid Cracking of Methyl Palmitate with Trifluoromethanesulfonic Acid for Production of Green Diesel,” Journal of Molecular Catalysis A, Anticipated Submission 05/07.

• Dufreche, S., Hernandez R., French, T., Sparks, D., Zappi, M., and Alley, E., (2007) “Extraction of Lipids from Municipal Wastewater Plant Microorganisms for Production of Biodiesel,” Journal of the American Oil Chemists' Society, Online 2006

• Zappi, P. A., Dufreche, S. T., and Galjour, J. R. , 2004, “Managing Lift-Station-Derived Odors with Biofiltration,” Water Environment & Technology 16, 57-61 2004.

6l Proteins of Novel Composition

Tae Hyeon Yoo, Division of Chemistry and Chemical Engineering, California Institute of Technology, 1200 E. California Blvd., MC 210-41, Pasadena, CA 91125

The properties of proteins are strikingly susceptible to engineering via rational and combinatorial strategies. In chemical terms, however, the engineering strategies employed to date are remarkably conservative, in that they employ a common set of twenty "canonical" amino acid constituents. Recent work from several laboratories has shown that a substantially expanded repertoire of non-canonical amino acids can be incorporated into recombinant proteins, either by amino acid replacement or by nonsense or frameshift suppression. This work raises intriguing questions about the prospects for creating novel proteins by recruitment of new amino acid constituents. In particular, global replacement of one of the canonical amino acids by a non-canonical analog would be expected to cause marked changes in protein structure, dynamics, and function. The subject of my thesis research (conducted in the laboratory of David Tirrell) has focused on three specific topics in the area of protein engineering via global incorporation of noncanonical amino acids: 1) Green fluorescent protein was engineered to tolerate replacement of Leu by 5,5,5-trifluoroleucine (Tfl) by using a combination of global incorporation of Tfl and directed evolution; 2) A high-throughput method of screening aminoacyl-tRNA synthetase libraries has been developed, and the strategy was demonstrated by identifying an Escherichia coli methionyl-tRNA synthetase variant that activates 6,6,6-trifluoronorleucine; 3) The effects of Tfl on hydration dynamics in a coiled-coil protein (A1 protein) are currently being examined in collaboration with the Zewail group at the California Institute of Technology.

6m Enhanced Polymeric Nanoparticles for Gene Delivery

Jordan J. Green, Biological Engineering, Massachusetts Institute of Technology, 77 Massachusetts Ave., Room E25-342, Cambridge, MA 02476

Viral gene therapy has high efficacy, but is plagued by serious safety risks, production and manufacturing challenges, and other limitations including nucleic acid cargo capacity. To address these challenges, we have developed poly(beta-amino ester)s, easy-to-synthesize biodegradable polymers, that are able to deliver DNA to primary human umbilical vein endothelial cells (HUVECs) at levels comparable to adenovirus at a Multiplicity of Infection (MOI) between 100 and 500, and two orders of magnitude better than the commonly used non-viral polymeric vector, polyethylenimine (PEI). Interestingly, small structural changes to these polymers were found to have dramatic effects on gene delivery including tightening of the polymer/DNA binding constants, altered biophysical properties of the DNA-encapsulating nanoparticles, and enhanced subsequent protein expression. In vivo, these polymer modifications dramatically enhance DNA delivery to ovarian tumors. We also demonstrate a general method of coating these polymer/DNA nanoparticles for targeted delivery. At appropriate formulation conditions including near-neutral charge ratio, the coated nanoparticles enable effective ligand-specific gene delivery. We believe the development of this nanoparticle drug delivery system with virus-like efficacy, ligand-based specificity, biodegradability, and low cytotoxicity, could set a new benchmark in non-viral transfection capability.

6n Processing and Phase Behavior: Tools for Creating Nanoscale Polymer Structures

Christopher J. Ellison, Dept. of Chemical Engineering and Materials Science, University of Minnesota, 421 Washington Ave SE, Minneapolis, MN 55455

In this poster I will highlight two projects that I have been involved with during my postdoctoral work as a member of the Bates group at the University of Minnesota. Following the conclusion of my postdoctoral work, I would like to begin a tenure track faculty position in a chemical engineering department. As a faculty member, the overarching theme of my research program will be focused on nanostructured polymeric materials including block copolymers, nanocomposites, thin films and membranes. One of the cornerstones of my research program will be polymer synthesis. The current palette of synthetic tools available to polymer scientists has facilitated nearly unbridled control of molecular scale architecture. I view synthetic capability to be paramount in designing and studying novel polymeric materials. The projects that I am currently involved with can be summarized as follows:

Polydispersity Effects in the Phase Behavior of Poly(isoprene-b-styrene-b-ethylene oxide) (ISO) Triblock Copolymers

The phase behavior of block copolymers has been studied intensely for several decades, both theoretically and experimentally. Despite this long history, relatively little attention has been focused on understanding the impact of polydispersity. Presumably, this has been a consequence of the living synthetic methods commonly employed for block copolymer synthesis that naturally produce polydispersities < 1.1. Nevertheless, the importance of polydispersity was noted as early as 1980 (1) when Leibler stated: "Much more important than fluctuation effects are those effects resulting from the polydispersity of the copolymer blocks…it seems that even a small polydispersity of the sample may be crucial." Since 1980, there have been only a few experimental studies which have aimed to address this issue. We have recently studied (2) functional organolithiums that can be used to prepare α-hydroxy polystyrene with polydispersity indices ranging from ~1.1-1.6. Here we used 3-triisopropylsilyloxy-1-propyllithium (TIPSOPrLi) to prepare monomodal α-hydroxy poly(styrene-b-isoprene) diblocks containing a polystyrene block with M_w/M_n = 1.31 or 1.44. More than 15 ISO triblocks were synthesized from these two parent diblocks along the $f_I = f_S$ isopleth. These studies (3) of polydispersity effects are unique in that they address the impact of polydispersity in the S block which is forced to bridge between the I and O domains. Surprisingly, we did not observe the orthorhombic network morphology which has been observed in monodisperse analogs. These results clearly illustrate the important role that polydispersity can play in the phase behavior of block copolymers and are a demonstration of the delicate nature of the free energy balance governing morphological behavior.

(1) Leibler, L. Macromolecules 1980, 13, 1602-1617. (2) Meuler, A.J.; Mahanthappa, M.K.; Hillmyer, M.A.; Bates, F.S. Macromolecules 2007, 40, 760-762. (3) Meuler, A.J.; Ellison, C.J.; Evans, C.M.;

Hillmyer, M.A.; Bates, F.S. Macromolecules, submitted.

Nanofibers Produced by Melt Blowing

Melt blowing is a more than 50 year old polymer processing method for generating polymer fibers often as nonwovens. It employs a jet of hot air which creates an extensional force on an extruded polymer filament resulting in attenuation of its diameter by more than 1000 times in some cases. Although historically this method has been limited to the production of fibers exceeding 1-2 microns, we have recently demonstrated the ability to produce long, defect free fibers several hundred nanometers in average diameter. Traditionally, electrospinning has been the only processing method available for producing fibers on this size scale. However, electrospinning is inherently less desirable than melt blowing due to its low production rate, solvent requirement, and high operating voltage. In this study (1), nanofibers were produced from commercial melt blowing grades of polypropylene and poly (butylene terephthalate) using a laboratory melt blowing device (both single- and multi-orifice) which was designed after commercial equipment. This demonstration of melt blowing capabilities is a step towards closing the gap between electrospinning and melt blowing in terms of the ability to produce nanoscale fibers. In addition, it highlights the potential for increasing the number of applications served by melt blowing. Finally, we have attempted to understand the ultimate limitations of this process by exploring the processability of a series of low and high molecular weight polystyrenes and their blends. Surprisingly, conditions can be identified where exceedingly low molecular weight polystyrene (M_n = 2.6 kg/mol and PDI = 1.09) can be readily melt blown into defect free fibers with an average diameter less than 1 micron. This result illustrates how much is yet to be learned about this process.

(1) Ellison, C.J.; Phatak, A.; Giles, D.W.; Macosko, C.W.; Bates, F.S. Polymer 2007, 48, 3306-3316.

6o Novel Methods for Microfabrication of Cellular Interactions and Detection of Cell Phenotype Expression

Ji Youn Lee, Department of Biomedical Engineering, University of California at Davis, 451 East Health Sciences Dr., Davis, CA 95616

My overall research interests lie in development of microtechnologies for cell cultivation and analysis. The research work employs principles of tissue engineering, microfabrication, biomaterials and biotechnology.

Currently I am working as a postdoctoral researcher at the University of California, Davis. The research focus is in creating in vitro hepatic microenvironment which mimics complexity of the in vivo interactions to define the microenvironment niche required for hepatic phenotype expression in fetal hepatocytes and stem cells. I have developed a micropatterned co-culture method based on microarraying and photoresist lithography, and also demonstrated local cellular analysis method by laser microdissection and gene expression analysis. Most recently, I have expanded the complexity of the cell culture system to create micropatterned triple cell cultures. These tri-cultures are currently being investigated as a microenvironment conducive to stem cell differentiation toward liver cells.

During my graduate study at Seoul National University, I have performed interdisciplinary research in a field of DNA computing with colleagues in Departments of Computer Science and Biochemistry. I have introduced new molecular algorithm and experimental implementation to solve a graphical problem. Also, I have worked on projects discriminating acute leukemia based on gene expression profiling and developing lab-on-a chip system to automate biological operators for DNA computing.

The future research plans are to merge the study during Ph.D and postdoctoral research toward integrated cell-chip for clinical and toxicological applications and also fundamental studies. One project will be long-term cultivation of functional hepatocytes by rational design of microenvironment and perfusion culture to develop cell-based analysis platform. Second project will involve the study of stem cell differentiation toward specific lineage (e.g. hepatocytes). Rapid, rational and systematic screening of diverse inducers leading to differentiation will be performed in a microfabricated platform.

6p Quantifying Cellular Physiology Using Metabolic Models and Isotope Labeling

Patrick F. Suthers, Department of Chemical Engineering, The Pennsylvania State University, 147 Fenske Lab, University Park, PA 16802

One of the key descriptors of cellular physiology is the set of flux values that flow through the cell's metabolic pathways. With the rapid generation of genomic data, creating such metabolic models in a systematic framework becomes crucial. Once constructed, these models can be used to propose experimentally testable hypotheses and guide modifications of the organism, e.g., to improve product yield. In this poster, I present examples in two areas: constructing metabolic models and quantifying fluxes in a large-scale model using isotope labeling.

First, with a genome size of ~580 kb and approximately 480 protein coding regions, *Mycoplasma genitalium* is the smallest known free-living organism. The reduced genomic content of *M. genitalium* has led researchers to suggest that the molecular assembly represented by this organism may be a close approximation to the minimal set of genes required for bacterial growth. One challenge to its study, though, has been that it has extremely fastidious nutrient requirements. Here, we present a flux balance analysis (FBA) approach to examine the substrate uptake requirements for *M. genitalium*. This approach involves constructing a metabolic model that encompasses the reactions carried out by the enzymes encoded in the genome.

Once a metabolic model was constructed, it was validated by comparing FBA simulation predictions with experimental data for growth of various mutant strains. For instance, these include classifications of genes as essential or nonessential for growth. The validated model is employed to predict mathematically viable uptake environments for *M. genitalium*. Specifically, this process is used to systematically guide the construction of chemically defined medium based on the underlying metabolic pathways.

The second topic concerns an important tool that can be used to quantify the flux values in the network of a metabolic model: isotope labeling. Here, we present a computational framework that combines a constraint-based modeling framework with ^{13}C isotopic label tracing. This *Escherichia coli* model includes 393 fluxes, 214 metabolites, and balances on cofactors such as ATP and NADH as well as the electron transport chain, full amino acid biosynthesis and degradation, and a detailed biomass equation. By including pathways such as cofactor balancing and the electron transport chain, we can ensure that the results are biologically relevant.

Two of the important considerations for analysis of the isotope model are labeled substrate choice and how isotope measurements impact the elucidation of fluxes in large-scale metabolic reconstructions. By introducing a degrees of freedom based optimization method it is possible to exhaustively identify all combinations of isotope labeling experiments and flux measurement that completely resolve all flux values in the network. These measurements, consisting of both partial or full isotope state determination, are assigned relative costs that allow the experimentalist to select the measurements that will be both sufficient and economical. Additional measurements can be taken in order to provide information about the confidence intervals of the fluxes.

Experimental results of flux elucidation are presented for an *Escherichia coli* strain engineered to produce amorphadiene, a precursor to the anti-malarial drug artemisinin. These include a statistical analysis of fluxes determined for the system such as the minimal and maximal values of the fluxes given measurement noise. We also introduce a degree of resolution calculation that quantifies how the isotope data restrict these fluxes within the ranges allowed by overall stoichiometry.

Once these flux ranges are determined, they can guide engineering of the system to improve product yields. For instance, if we wish to increase the production rate of a target compound, we can identify other fluxes that need to be changed beyond the ranges determined by the isotope data. These fluxes then become potential targets of manipulation. This approach is applied to the production of amorphadiene and other bio-compounds of industrial interest.

6q Molecular Simulation of Heterogeneous Polymer Systems: From Biomimetic Materials to Energy Research

Vikram K. Kuppa, Chemical Engineering, Massachusetts Institute of Technology, 77 Mass Ave, Building 66 Room No-453, Cambridge, MA 02139

The study of polymers in solutions as well as bulk melts is a well established field that has been the topic of much research over the past few decades. In recent years, the spotlight has shifted in favor of multi-component polymer blends and composites, and is specifically directed at tailoring their behavior by controlling interfacial characteristics. The properties of such heterogeneous polymer systems are taken advantage of in myriad uses ranging from high mechanical strength and barrier property applications to bio-inspired and biomimetic materials, and considerable interest has been generated in industry as well as in the scientific community. In particular, designed polymeric and colloidal nanostructures constitute a majority of the currently viable applications of nanotechnology (official funding by the US Govt. exceeding $1 billion for 2007), with examples that include the use of nanocomposites in car parts, stain resistant garments and light-weight beer bottles. However, to completely fulfill the exciting promise of novel hybrid materials, a fundamental comprehension of nanoscale behavior is essential. The focus of my proposed research is on employing computer simulations to provide a molecular understanding of the nanometer level attributes that dictate macroscopically observed properties in heterogeneous polymer systems. Using molecular modeling and statistical mechanics, the specific-interaction driven behavior of macromolecules will be investigated, especially focusing on the structure and dynamics of multi-phase and multi-component polymer systems. Different kinds of heterogeneities will be probed, including dissimilarities between constituent segments of the polymer chains, and variations in morphological features such as spatial restrictions that arise in the form of non-bonded attraction to adjacent surfaces and the presence of multiple phases (polymer/surface, polymer/solvent or polymer/polymer blocks or blends). The aims of the molecular modeling approach are twofold: first, to contribute quantitative and qualitative predictions that complement empirically motivated experiments and synthesis protocols; second, to provide a nanoscale perspective of structure-property relationships in the inhomogeneous materials that are increasingly important in today's world. The research will synergistically impact the fields of chemical engineering and materials science (polymer surfaces and interfaces), energy research (electrolytes for fuel cells) and biological engineering (paradigms for membrane proteins, molecular design of biomimetic heteropolymers). The ultimate goal is to elucidate and derive more insightful approaches to polymer bio- and nanotechnology.

6r Analysis of Recognitive Polymer Systems: Focusing on the Tailorablity of the Macromolecular Structure in Order to Enhance Binding Characteristics

Asa D. Vaughan, Biomimetic & Biohybrid Materials, Biomedical Devices, and Drug Delivery Laboratories, Department of Chemical Engineering, Auburn University, Auburn, AL 36849-5127

Research interest are biomimetic materials for sensor materials and drug delivery carriers. Examples of this research would be crosslinked polymer systems such as contact lenses that deliver drug to a specific location for an extended period of time. Teaching interests include thermodynamics, polymer thermodynamics, environmental transport, and chemical engineering thermodynamics. Current Research Novel recognitive polymers are macromolecular structures that retain their binding characteristics in a wide range of pH and temperature environments. These bio-compatible materials show great promise for robust and inexpensive sensor coatings and point of care devices. At the molecular level, binding characteristics are a direct resultant of functional interactions between the incorporated functional monomer and the target molecule. At the intra-macromolecular level, the macromolecular structure has formed size specific pockets containing matching functional moieties to that of the target molecule thus providing multiple functional complexation points. Both functional interactions and macromolecular structure play a determining role in the resultant binding characteristics shown by the recognitive polymer for the target molecule. Biomimetic acrylate and methacrylate based co-polymer networks which compromise the majority of recognitive polymer networks have been "trained" to have a macromolecular memory for the target molecule. However, characterization of the polymer network in order to determine relationships on how to tailor feed compositions in order to optimize the binding characteristics of the resultant polymer network has been largely neglected. This work reflects on what effect compositional parameters have upon resulting binding characteristics. Reaction analysis was done on polymerization reactions with the focus on determination how the feed composition affects double bond conversions, kinetics, and resultant binding characteristics. Results show that increasing the double bond conversion have effects upon binding characteristics by modest increases in loading capacity while maintaining relatively equivalent binding affinity. This work also focuses on the macromolecular architectural structure and the resultant binding characteristics. The use of living/controlled polymerization has shown in linear polymer systems to give low polydispersities and the unique ablity to form specific block copolymers. Exploration of living/controlled polymerization It is hypothesized that a more homogenous architectural structure would lead to more optimally formed binding sites that would essentially increase the affinity, selectivity, and loading capacity of these recognitive polymer networks. Living/controlled polymerization techniques have shown increases the loading capacity while retaining selectivity for the target molecule. Examination of mesh size for each type recognitive network will give insight as to what living/controlled polymerization techniques have upon the resulting mesh size for these recognitive polymer networks.

6s Dynamics of DNA and Swimming Microorganisms Using Theory and Coarse-Grained Simulations

Patrick T. Underhill, Chemical Engineering, University of Wisconsin-Madison, Madison, WI 53706

Mesoscopic computer simulations of polymers and colloids have provided an important bridge between atomistic and continuum models. These models have been used as alternatives to constitutive equations or to develop constitutive equations. They also allow for calculations of microstructure and how flow influences that microstructure. One example of these models is bead-spring chain models of polymers. They have been used extensively to understand the behavior of polymers, in particular the dynamics of DNA in microfluidic devices. My thesis work under Patrick Doyle at MIT focused on developing and analyzing the response of new bead-spring chain models. As microfluidic devices get increasingly small and approach the regime of nanofluidics, the previously used bead-spring chain models are no longer accurate coarse-grained representations, while more detailed micromechanical models are too computationally expensive. Using statistical mechanics and Brownian dynamics simulations, we developed new models which bridge the gap between the micromechanical and coarse-grained models. These intermediate level models capture the necessary physics, are computationally tractable, and are needed when analyzing new devices for DNA separations and gene mapping.

My postdoctoral work under Michael Graham at the University of Wisconsin-Madison has focused on simulating suspensions of self-propelled particles. Recently large collections of swimming microorganisms have been observed producing collective motions on a scale much larger than the scale of a single organism. The study of swimming microorganisms has a long history in the fluid mechanics literature. The linearity of the hydrodynamic equations at small Reynolds number require that microorganisms swim differently than macroscopic organisms. This has been studied previously by looking at the swimming of a \emph{single} organism. In contrast, large collections of swimming organisms have been analyzed using continuum theories written as differential equations for field variables.

We have approached the problem by directly simulating a large collection of coarse-grained representations of the swimming organisms. Inspired by the coarse-grained models of polymers and colloids, we model them as a series of beads connected by rods. The model is coarse enough that we can directly follow the dynamics of large populations while still including the far-field multi-body hydrodynamics. The model also natural extends to include other interactions between organisms.

6t Improved Polymer Properties for Use in Nanolithography

Robert A. Riggleman, Department of Chemical and Biological Engineering, University of Wisconsin, Madison, 1415 Engineering Drive, Madison, WI 53706 and Juan De Pablo, Department of Chemical and Biological Engineering, University of Wisconsin-Madison, Madison, WI 53706.

As the semiconductor industry strives to create devices with features smaller than 40 nm in the next five to ten years, a number of technical challenges must be overcome. The key step in the production of such devices requires manufacturing stable polymer nanostructures with large aspect ratios and nanoscopic lateral dimensions. One of the major challenges that must be overcome is that the properties of polymer materials change from their bulk values when they are confined to such small geometries. It is well-established in the literature that confinement can reduce the glass transition temperature, leading to a softening of the material and collapse during normal semiconductor processing. Here, we use molecular simulations to provide insight into how confinement affects the properties of polymers, and how additives can be used to create more stable nanostructures. We have developed a coarse-grained model that exhibits antiplasticization, a phenomenon where the addition of low molecular weight solvent molecules can increase the density and stiffness of a polymer glass while reducing the glass transition temperature. This is in contrast to the more widely-studied plastification of a material, where the elastic constants, density, and glass transition temperature decrease. Using our model, we have shown how antiplasticizer molecules can reduce the effects of confinement on the properties of polymers by reducing the size of the cooperative rearrangements. This effect is explained through the fragility of glass-formation, where antiplasticization is associated with reducing the fragility of our glass-forming polymer. A reduction in fragility is commonly associated with a decrease in the size of cooperative rearrangements within the framework of the Adam-Gibbs theory of glass-formation. In addition, our results are in excellent agreement with experimental results published in the literature.

6u Nucleation Behavior of Nanoparticles and Crystals

Venkateswarlu Bhamidi, Department of Chemical and Biomolecular Engineering, University of Illinois at Urbana-Champaign, 600 S. Mathews Ave., Urbana, IL 61801

Nucleation of crystals and nano-particles from solution plays an important role in several areas of science and technology like manufacture of photonic band gap materials, formation of protein crystals for structure determination, and industrial production of pharmaceutical compounds and semiconductor materials etc. At present, limited understanding exists towards the underlying processes of particle formation from solution. My research is geared toward acquiring that fundamental understanding about the governing processes of nano particle formation and developing innovative methods and techniques to apply this knowledge in solving problems of great interest to the industry and academia.

My current research focus is on studying the particulate processes in proteins and pharmaceutical systems. The insights obtained through my studies will assist not only the practice of industrial crystallization, but also the crystallization of proteins for structure determination, and rational drug design methodology. A few specific issues that I will address in this poster are:

a). Understanding Metastability in Nucleation from Solution

While thermodynamics requires the phase transition to occur as soon as one crosses the equilibrium boundary (solubility), well known observations indicate that nucleation occurs at a relatively high supersaturation. Crystallizing solutions can sustain significant metastability. This phenomenon is often attributed to the so called unclear 'kinetic effects'. The key questions then become, is there any limit to the metastability that a system can endure and if so what are the factors that determine such a limit? What will be the influence of 'kinetic effects' on nucleation when one generates the supersaturation infinitesimally slow such that the system is always at equilibrium?

The above questions are addressed here by studying the nucleation behavior of various compounds under extremely slow rates of supersaturation. Such slow rates of supersaturation are achieved by crystallizing the solutions in an evaporation based micro-device. Experimental evidence to the existence of a lower boundary on the MZW is presented and this observation is explained in terms of our current understanding of kinetic / thermodynamic processes underlying nucleation. Thus, this boundary on MZW can be viewed as a 'kinetic limit' of nucleation. We attempt to link this 'kinetic limit' to the intermolecular interactions and suggest methods to predict nucleation behavior of compounds.

b). Influencing Polymorph Selectivity by Regulation of Diffusive Mixing in Microfluidic Devices

Rapid polymorph screening techniques that reduce the time, effort and material consumed are an essential requirement in pharmaceutical industry. Microfluidic technology offers an effective approach in developing innovative methods and techniques to address this issue. Here, a microfluidic mixer design that facilitates polymorph screening is discussed. This design effects anti-solvent crystallization utilizing the fact that mixing at the micro-scale is by diffusion only (no turbulence). Also, the rate of mixing and the composition of the liquid phase can be varied by changing the flow rates and concentrations, thus influencing the polymorph selectivity. The results of our experiments on some model systems are presented and the applicability of the concept towards the development of a high throughput microfluidic device for polymorph screening is discussed.

6v Isotopically Nonstationary Metabolic Flux Analysis

Jamey D. Young, Chemical Engineering, Massachusetts Institute of Technology, 77 Massachusetts Ave. 56-439, Cambridge, MA 02139

Metabolic flux analysis (MFA) is a powerful platform for studying biochemical reaction networks. Results are typically obtained by (1) introducing an isotopically labeled substrate into a cell culture at metabolic steady state, (2) allowing the system to reach isotopic steady state, (3) measuring relative labeling in metabolic intermediates and byproducts, and (4) computationally processing these measurements to estimate metabolic fluxes. An emerging area of interest is in the application of MFA to address fundamental physiological questions in higher organisms, especially plant and animal systems of specific agricultural or clinical significance. Because MFA has been traditionally applied to microbial systems, however, there are several features that limit its straightforward extension to higher organisms. In the case of animal studies, it is often difficult or impossible to maintain primary cells or whole animals in a fixed metabolic state long enough to achieve steady isotopic labeling. Plant systems growing under photoautotrophic conditions, on the other hand, only assimilate carbon from CO_2 and therefore produce a uniform steady-state ^{13}C-labeling pattern in all metabolites, irrespective of flux distribution. In response to these and other drawbacks of the traditional MFA framework, we have developed an approach that involves repeated measurements of intracellular label enrichment during the transient period preceding isotopic steady state. Nonstationary Metabolic Flux Analysis (NMFA) is subsequently applied to leverage these measurements into quantitative flux estimates that characterize the metabolic phenotype. Recent advances by Antoniewicz et al. (2007) and Young et al. (2007) have made the NMFA computational problem tractable for systems of realistic size and complexity.

As an illustration of the NMFA approach, we report results from two experimental systems. First, NMFA has been applied to characterize time-dependent phenotypes in a rat hepatoma cell line (H4IIEC3) following exposure to varying levels of palmitate, oleate and amino acids. Preliminary studies have shown that elevated palmitate levels lead to apoptosis, while combined exposure to palmitate and oleate leads to excess lipid accumulation (steatosis) but does not trigger apoptosis. It is also possible to modulate the divergence between apoptotic and steatotic outcomes by varying the amino acid composition of the culture medium. Because the phenotypes are fully developed within 12-24 hours following fatty acid exposure, one cannot estimate fluxes during the important interval when the cells are transitioning from a normal to a diseased state using the traditional MFA approach. However, we have been able to examine this transition using NMFA and are beginning to uncover the mechanisms by which metabolic perturbations lead to phenotypic switching in this system. Second, we have applied NMFA to estimate fluxes in the prokaryotic cyanobacterium *Synechocystis* sp. PCC 6803 under photoautotrophic conditions. This represents the first time that photoautotrophic fluxes have been estimated using a ^{13}C labeling experiment. We show that, even though the steady-state ^{13}C distribution is insensitive to fluxes, transient measurements of isotope incorporation following a step change from unlabeled to labeled CO_2 can be used to estimate fluxes successfully.

References

Antoniewicz, M. R., Kelleher, J. K., Stephanopoulos, G., 2007. Elementary metabolite units (EMU): a novel framework for modeling isotopic distributions. Metab. Eng. 9, 68-86.

Young, J. D., Walther, J. L., Antoniewicz, M. R., Yoo, H., Stephanopoulos, G., 2007. An elementary metabolite unit (EMU) based method of isotopically nonstationary flux analysis. Biotechnol. Bioeng., accepted.

6w Colloidal Suspensions: Fundamental Physics and Engineering Applications

Jacinta C. Conrad, Materials Science and Engineering, University of Illinois at Urbana-Champaign, 1304 W. Green St., Urbana, IL 61801

Colloidal suspensions are ubiquitous in nature and serve as excellent model systems for a variety of complex fluids. Furthermore, such suspensions are found in many industrial and technological applications. The microstructure and dynamics of colloidal suspensions directly influence their macroscopic behavior; thus, engineering functional materials demands a thorough knowledge of the relationship between the microscopic properties of these suspensions and their bulk physical properties.

My graduate work at Harvard University focused on the phase behavior, structure, and rheology of model non-equilibrium colloidal suspensions, including both glasses and gels.

1.1. *Colloidal microstructure and dynamics near the colloidal glass transition* (with F. W. Starr, P. P. Dhillon, E. R. Weeks, D. R. Reichman, and D. A. Weitz): We use confocal microscopy and simulation to investigate the structure and dynamics of model colloidal suspensions near the glass transition. First, we investigate the relationship of local static structure to dynamics in suspensions of nearly-hard-sphere poly-(methyl methacrylate) colloids. We calculate the Voronoi volume for all particles and show that this quantity is only weakly correlated to local particle displacements; we obtain qualitatively similar results in a simulation of a polymer melt. These results suggest that static local structure alone does not predict dynamical behavior in experimental colloidal suspensions. Second, we investigate the behavior of the slowly-relaxing particles. These particles form spatially correlated clusters that evolve over time; furthermore, the structure of these clusters is correlated to the macroscopic properties of the suspension. In supercooled fluids, the largest cluster spans the system on short time scales but breaks up on longer time scales; in glasses, a percolating cluster exists on all accessible time scales. Using molecular dynamics simulation, we show that these clusters make the dominant contribution to the bulk elasticity of the sample.

1.2. *Phase behavior, structure, and rheology of weakly attractive colloidal suspensions* (with S. Manley, H. M. Wyss, P. J. Lu, K. Miyazaki, V. Trappe, L. J. Kaufman, A. B. Schofield, D. R. Reichman, and D. A. Weitz): We investigate the phase behavior, structure, dynamics, and rheology of colloidal suspensions with weak attractions. We induce a weak depletion attraction between poly-(methyl methacrylate) colloids by adding linear polystyrene. We investigate the phase behavior of these suspensions as a function of colloid volume fraction and polymer radius of gyration and concentration, and show that we can obtain interconnected gels, stable fluids of clusters, and colloidal fluids. We thoroughly investigate one region of the phase space, with intermediate volume fraction ($\phi = 0.15 - 0.35$) and a fixed range of the interparticle attraction. In this region colloid-polymer mixtures can undergo spinodal decomposition into colloid-rich and colloid-poor regions. Gelation results when the interconnected colloid-rich regions solidify; we show that this occurs when these regions undergo a local attractive glass transition. This arrest scenario bridges the gap between colloidal gelation and the colloidal glass transition; the resulting structures display hallmarks of both glasses and gels. The resultant macroscopic rheology of these structures is strikingly different than that observed for either colloidal glasses or gels; while the relaxation time depends strongly on colloidal volume fraction, the storage modulus is relatively insensitive to changes in volume fraction.

My postdoctoral work at the University of Illinois focuses on studying colloidal suspensions under flow. We investigate the basic properties of flowing suspensions, and exploit the flow properties to assemble novel colloidal granules and to pattern colloidal films.

2.1 *Flow of attractive colloidal suspensions in microchannels* (with R. F. Shepherd, S. K. Rhodes, and J. A. Lewis): We use confocal microscopy to investigate the behavior of suspensions of attractive colloids at a constriction. We flow suspensions of attractive silica colloids through microchannels and directly image both their structure and dynamics during flow. We investigate the flow properties as a function of applied pressure, microchannel geometry, and colloid volume fraction. We also explore the non-equilibrium clogging when the system experiences jamming within the microchannels.

2.2 *Microfluidic assembly of homogeneous and Janus colloid-filled hydrogel granules and shells* (with R. F. Shepherd, S. M. Menke, S. K. Rhodes, D. R. Link, M. Marquez, D. A. Weitz, and J. A. Lewis): We use microfluidic techniques to assemble novel colloid-hydrogel granules and shells. We suspend silica microspheres in an aqueous acrylamide suspension and use microfluidic devices to emulsify this suspension in a continuous oil phase. By varying the microchannel geometry, we can form spheres, discoids, and shells; by varying the input stream, we can form homogeneous and chemically heterogeneous (Janus) surfaces. This approach offers a facile route for assembling granules and shells of controlled shape and composition.

2.3 *Patterning colloidal films via evaporative lithography* (with D. J. Harris, H. Hu, and J. A. Lewis): We investigate evaporative lithography as a route for patterning colloidal films. Films are dried beneath a mask that induces periodic variations between regions of free and hindered evaporation; fluid and entrained particles migrate towards the regions of highest evaporative flux, leading to lateral segregation of particles within the film. By tuning the colloidal suspension composition, separation distance between the mask and underlying film, and mask geometry, we can control the pattern formation within the dried films.

6x Molecular Mechanisms for the Aggregation of Proteins and Therapeutic Antibodies

Naresh Chennamsetty, Chemical Engineering, Massachusetts Institute of Technology, 77 Massachusetts Ave, E19-528, Cambridge, MA 02139

My overall research goal is to understand the mechanisms behind protein and therapeutic antibody aggregation. Antibody aggregation leads to immunogenicity and loss of activity in the pharmaceutical formulations used in the treatment of numerous cancers, chronic inflammatory diseases, and infectious diseases. Furthermore, protein aggregation *in vivo* is associated with a number of neurodegenerative disorders such as Alzheimer's and Creutzfeldt-Jakob diseases, and type II diabetes. We study these aggregation mechanisms using multi-scale modeling and simulation techniques in close collaboration with experiments.

I. Mechanisms for the Aggregation of Therapeutic Antibodies

In my current post-doctoral position with Prof. Bernhardt L. Trout at the Massachusetts Institute of Technology, we study the mechanism behind the aggregation of therapeutic antibodies. Antibodies constitute the most rapidly growing class of human therapeutics. These antibodies are, however, thermodynamically unstable under conditions required for storage and administration and degrade due to aggregation. We use molecular simulation tools in collaboration with experimental techniques to identify the regions of the antibody most prone to aggregation and devise ways to prevent aggregation.

II. Development of Multi-scale Modeling Techniques for Soft Condensed Matter Systems

During my PhD work with Prof. Keith E Gubbins at North Carolina State University, we developed a multi-scale methodology connecting the atomistic and meso-scale models useful in understanding the long length and time scale phenomena involved in surfactant self-assembly, polymers, colloids and protein solutions. In addition, we have developed a novel technique to determine the rigorous effective pair potentials for a simple binary A/B mixture by integrating out the degrees of freedom of solvent B using Widom's particle insertion method. We have also explored the effect of cosolvents and cosurfactants such as alcohols on surfactant self-assembly and solubilization capacity in supercritical carbon dioxide with applications in separation processes in the food industry such as extraction of water-soluble vitamins and proteins, dry cleaning and polymerization.

6y Surfactant Adsorption at Fluid Interfaces

Alissa J. Prosser, Oregon Health & Science University, Portland, OR 97239

The aim of this study is to further the understanding of pulmonary surfactant adsorption and the resultant film mechanics. Pulmonary surfactant is a complex mixture of lipids and proteins that forms a film within the terminal airspaces and prevents the airspaces from collapsing during exhalation. The proper function of pulmonary surfactant is essential for normal respiration. Premature infants that lack sufficient surfactant suffer from Respiratory Distress Syndrome, the treatment for which is the instillation of an exogenous surfactant replacement. The exogenous surfactant is generally pulmonary surfactant obtained from bovine or porcine sources. Dysfunction of the surfactant layer occurs in Acute Respiratory Distress Syndrome. Although surfactant dysfunction is a consequence rather than a cause of this syndrome, ongoing clinical trials are investigating the efficacy of exogenous surfactant replacement therapy. However, surfactant therapy is expensive and the supply is limited. There is a clear need for a mechanistic understanding of the function of pulmonary surfactant and the components thereof in the context of respiratory cycling to aid in the development of inexpensive, readily available synthetic surfactant replacements. To function effectively, surfactant must adsorb quickly and remain at the interface during respiratory cycling, forming films of near zero surface tension during exhalation. Preliminary data on a simple model system of one component of pulmonary surfactant have shown that the pathway of film formation has an effect on the properties of the adsorbed film. This study has tested the following hypotheses. First, that factors that affect transport but do not affect the microstructure of the surface-active aggregate do not affect the mechanics of the resultant film. Second, factors that affect the microstructure of the surface-active aggregate affect the mechanics of the resultant film. The primary goal of this study is to emphasize that the dynamics of surfactant adsorption and the mechanical response of the resultant film are intrinsically coupled.

Reactions of Alcohols over H^+/ZSM-5

Mark G. White[a*], *Amit C. Gujar*[a], *Vamshi Krishna Guda*[b], *Qiangu Yan*[a], *Hossein Toghiani*[a],
[a]Dave C. Swalm School of Chemical Engineering, Mississippi State University, MS-39762
[b]Department of Chemistry, Mississippi State University, MS-39762

Extended abstract

Introduction- Converting syngas to gasoline is an attractive proposition given the current price of oil and an ever-increasing demand for gasoline. Syngas can be obtained from non-renewable resources by steam reforming of petroleum hydrocarbons or by coal gasification, or from renewable resources by gasification of biomass. Gasoline is an important fraction of petroleum, which is widely used as a transportation fuel. Syngas can be converted directly to gasoline-range compounds by Fischer-Tropsch chemistry using Co or Fe catalysts. An alternative route to product gasoline from syngas is to convert syngas to alcohols over catalysts like Cu/Co/Cr alcohol synthesis catalyst and then convert the alcohols to gasoline (ATG) over H^+/ZSM-5.

The separate reactions of the homologous series of alcohols (methanol, ethanol, 1-propanol, 2-propanol, 1-butanol, 2-butanol, 2-methyl-1-propanol and 2-methyl-2-propanol) over H^+/ZSM-5 were completed in an, isothermal, constant-volume, batch reactor. The effects of various process parameters such as pressure, temperature, reaction time, etc., on the methanol to gasoline reaction were examined first. Subsequently, the other alcohols were studied for the changes in their product distribution when operating under similar conditions and compared to the results for reacting methanol

Experimental Details- Various reactions involving different alcohols as the reactant were carried out in a 450 cc batch reactor (Parr Instrument Co., Moline, IL) equipped with a catalyst basket. In a typical reaction, the reactor was loaded with 1 g of H^+/ZSM-5 (prepared from NH_4^+/ZSM-5, SiO_2/Al_2O_3=23, obtained from Zeolyst International) and charged with 46.64 g of methanol. The reactor was heated in 50 °C increments to 350 °C and held at 350 °C for 2 hours. The stirrer speed was held constant at 200 rpm. The reactor was then allowed to cool to room temperature and the gas and liquid fractions were collected and analyzed using GC-MS (Agilent 6890N, Agilent Technologies)

Results and Discussion- Process parameters including reaction time, initial catalyst charge, initial methanol charge, initial He pressure and presence of H_2 were varied to determine their effect on gasoline yields. The effect of reaction time on gasoline yield with 48 g methanol charge, 1 g of H^+ZSM-5 and near atmospheric initial He pressure is shown in Figure 1.

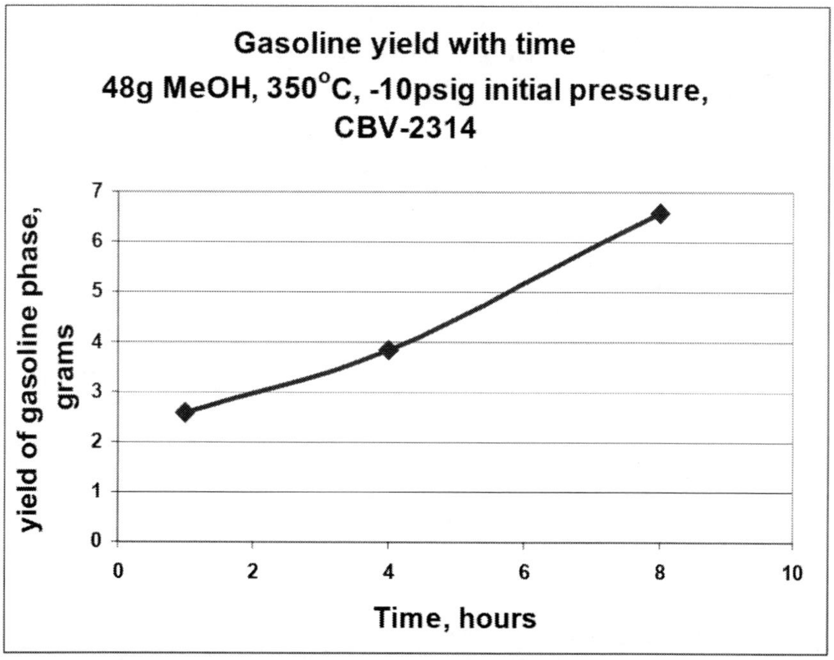

Figure 1. Gasoline yield with varying time over H^+/ZSM-5

As shown in Figure 1, gasoline yield increases with increasing reaction time. This shows that the methanol to gasoline reaction is kinetically limited. It is expected that, at long enough times, the yield of gasoline will stabilize due to the coke formation inside the pores of the zeolite, thus, leading to cessation of the gasoline formation reactions.

The effect of catalyst (H^+/ZSM-5) amount on gasoline yield is shown in Figure 2. As expected, the gasoline yield increased with increasing catalyst amount due to the kinetic control of the reaction.

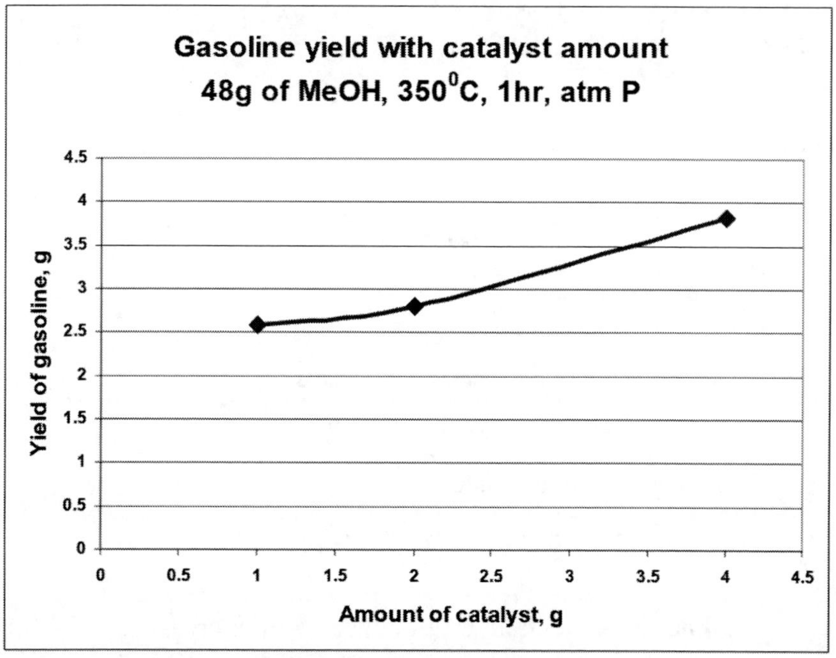

Figure 2. Gasoline yield with varying catalyst amounts.

The effect of initial methanol charge on gasoline yield is shown in Figure 3.

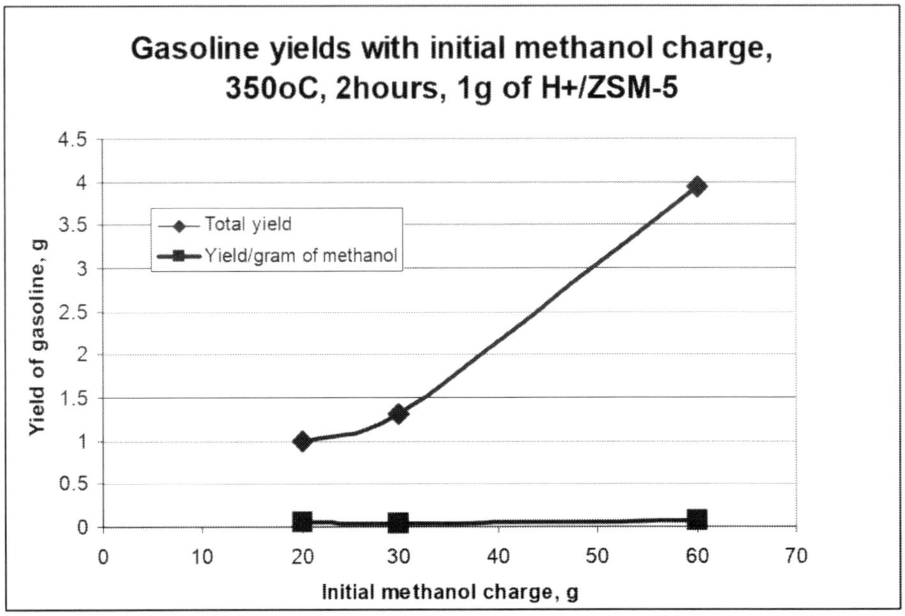

Figure 3. Gasoline yield with varying initial amounts of methanol.

Figure 3 shows that the gasoline yield increased with increasing methanol yield, but the yield of gasoline per gram of methanol remained constant. This demonstrates that the MTG reaction kinetics are not affected by the initial reactant charge.

The effect of initial He pressure on gasoline yield and the final reactor pressure are shown in Figure 4. Also included are data demonstrating the effect of initial H_2 pressure on the gasoline yield.

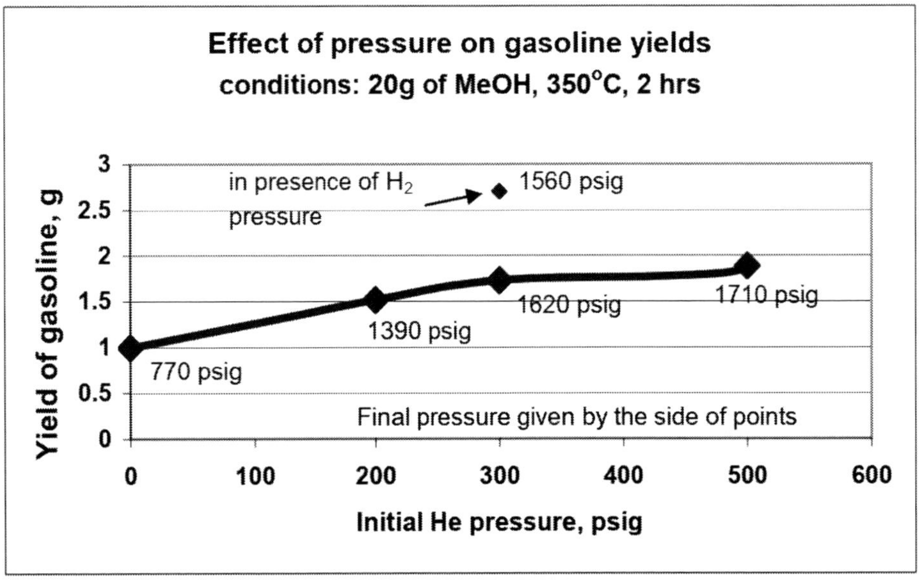

Figure 4. Gasoline yield with varying initial He pressures of methanol.

As evident in Figure 4, gasoline yield tended to increase with increasing initial He pressure. This could be due to the fact that at higher pressures the reactant and products are in supercritical state leading to easy ingress and egress from the zeolite pores. It is interesting to note that H_2 has a significantly positive effect of gasoline yield. It was also observed that the catalyst was visually less coked and that the color of gasoline obtained was much lighter, leading to the conclusion that formation of higher aromatic compounds is reduced. This is probably due to hydrogenation of reactive olefins and dienes, which are the major cause of coking in the MTG reaction.

To test the effect of higher alcohols over H^+/ZSM-5, various C_2 to C_4 alcohols were reacted over the zeolite in the batch reactor. All the alcohols produced gasoline with varying yields. Figure 5 shows the gasoline yield for the various alcohols tested at constant moles of carbon.

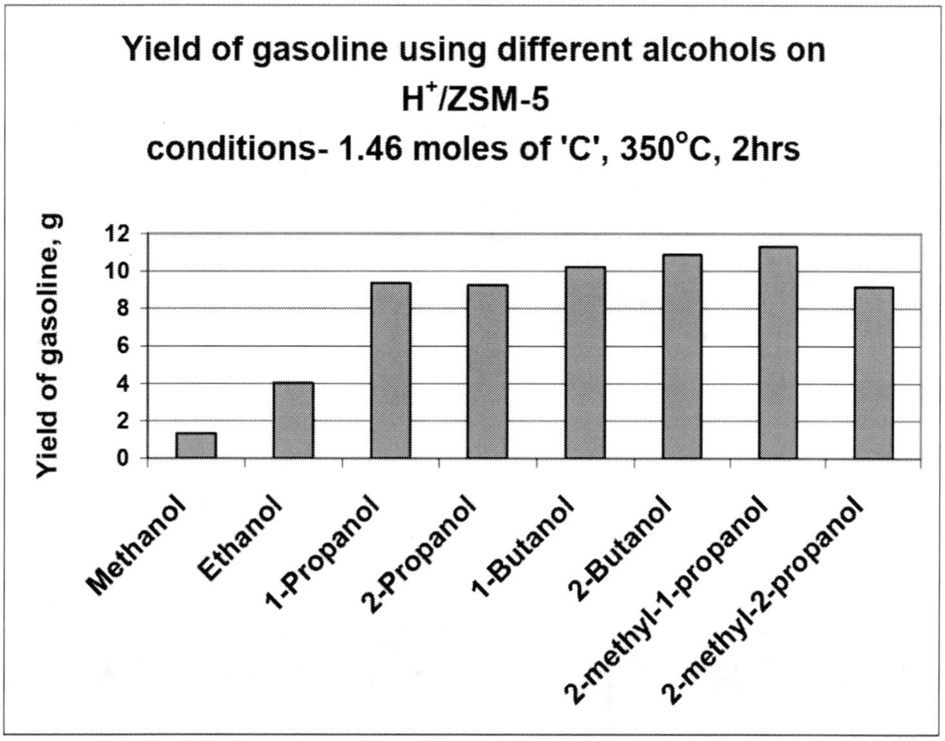

Figure 5. Gasoline yield for different alcohols at constant moles of carbon.

As evident in Figure 5, gasoline yield tends to increase with increasing length of carbon chain. This is probably due to the fact that it is easier to form olefins from higher alcohols than from methanol. In case of methanol, methanol dehydrates to dimethyl ether, which in turn is converted to ethene and propene, the primary intermediates in aromatics formation. As the carbon chain increases, the tendency to form ethers seems to decrease and dehydration of alcohols tends to form olefins. The observed yields were compared to the theoretical yields of C_8 species based on number of carbon atoms to give the percentage of theoretical yield, Table 1. These results may be interpreted to show that longer chain alcohols produce more C8-type hydrocarbons than what is observed for shorter chain alcohols. The significance of these results is that the alcohol-forming function in the composite catalyst should be optimized to convert syngas into C_3-C_4 alcohols so as to maximize gasoline yields.

Table 1: Yields of C_8 based on number of carbon atoms for different alcohols.

Alcohol	Yield of C_8 based on carbon
Methanol, 46.6432g	4.67%
Ethanol, 33.53g	14.33%
1-Propanol, 29.3483g	32.81%
2-Propanol, 29.43g	32.27%
1-Butanol, 26.63g	37.30%
2-Butanol, 26.55g	40.02%
2-methyl-1-propanol, 26.57g	41.50%
2-methyl-2-propanol, 26.41g	33.95%

The octane rating for the gasoline produced by these alcohols was determined using an NDIR instrument (LT Industries Complete Octane Monitoring System, COMS). This instrument determines the research octane number, R, and the motor octane number, M. The octane ratings, (R+M)/2, are shown in Figure 6. Gasoline produced from higher alcohols show higher octane ratings than those from lower ones.

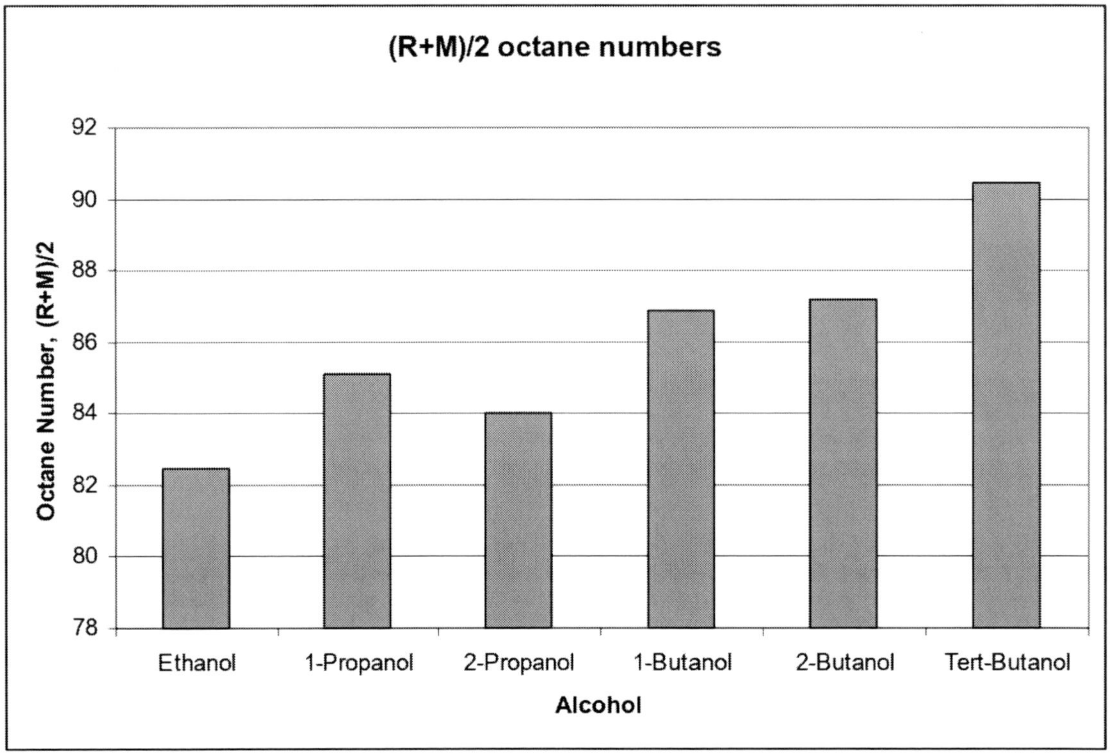

Figure 6. Octane ratings (R+M)/2 for gasoline produced from various alcohols.

Another interesting observation from these higher alcohols to gasoline tests were that the tendency to form aromatics followed the trend such that-

Tertiary > Secondary > Primary

This trend was especially evident at lower temperatures where the primary alcohols produced negligible amounts of aromatics. Table 2 shows the aromatics content for various gasoline produced by different alcohols at 350 °C.

Table 2: Aromatics yields for gasoline produced by different alcohols at 350 °C.

Alcohol	Aromatics yield, wt%
1-PrOH,	8.027
2-PrOH,	13.36
1-BtOH,	3.747
2-BtOH,	7.375
tert-BtOH,	14.53

The higher aromatics yield for tertiary alcohols could be explained by the fact that they have a tendency to form carbocations more readily than their primary alcohol counterparts. These carbocations are suspected to be the key intermediates that form aromatics.

Conclusions: The following conclusions can be made from the above experiments-

- The methanol to gasoline reaction is kinetically limited;
- Increasing He pressures tend to give higher yields of gasoline;
- The presence of H_2 has a positive effect on gasoline yield;
- Higher alcohols tend to give significantly higher gasoline yield;
- The tendency for aromatics formation is greater for branched alcohols than for straight chain alcohols.

Acknowledgements: We acknowledge the support of this work under the US DOE contract award DE-FG36-06GO86025.

6aa Multi-Compartment Drug Carriers

Guohui Wu, Department of Chemical Engineering, University of California, Santa Barbara, Santa Barbara, CA 93106 and Joseph A. Zasadzinski, Department of Chemical Engineering, University of California, Santa Barbara, CA 93106.

Vesosome, a multi-compartment structure consisting of drug-loaded liposomes encapsulated within another bilayer, is a promising drug carrier with better retention and stability compared with unilamellar liposomes. The vesosome preparation takes advantage of the interdigitated phase of saturated lipids, which causes lipid bilayers to form flat, open sheets at low temperature, that close to form large unilamellar vesicles at higher temperatures. Adding poloxamer 188 (P188) to the Dipalmitoylphosphocholine (DPPC) interdigitated sheets prior to closure leads to ~ 0.6 μm diameter vesosomes. Applying osmotic pressure by diluting the vesosomes with buffer of increased salt concentration can further decrease the vesosome diameter by 10%. Freeze-fracture TEM and light scattering were used to evaluate the size distribution and polydispersity of the vesosomes formed. Nearly monodisperse hollow gold nanospheres with absorbance wavelength in the visible and infrared range have been synthesized and loaded inside vesosome to be externally triggered via near Infra-Red light for controlled drug release.

6ab Cellular Bioengineering: Towards in Vitro Models for Disease Pathophysiology and Toxicology

Anand K. Ramasubramanian, UC - Berkeley, Dept of Chemical Engineering, Berkeley, CA 94720 - 1462

Cellular engineering is the understanding and control of cellular behavior for developing novel diagnostic medical tools and therapeutics, or harnessing cellular function for commercial applications. In this talk, I shall discuss our research in this burgeoning area aimed at developing in vitro models for arterial thrombosis and breast cancer, and also a commercially viable cytotoxicity screen. First, our results on the kinetics of interaction between platelet receptor and a subendothelial protein that mediate arterial thrombosis and hemostasis will be presented. Our dynamic studies indicate that this interaction is transient and rapid, and may partly explain the pathophysiology of thrombus formation. Next, I will discuss our on-going attempts at delineating the pathways that lead to the selectivity of anti-cancer drugs on cancerous and non-cancerous human mammary epithelial cells. Finally, I shall demonstrate the application of a high-throughput mammalian cell-based assay, which can provide a rapid, inexpensive and convenient evaluation of drug toxicity.

6ac At the Interface of Neuroscience and Cell & Biomolecular Engineering

David Colby, Institute for Neurodegenerative Diseases, UCSF, Box 0518, San Francisco, CA 94143

In science, the most exciting discoveries often occur at the interface between two disciplines. In both my doctoral and postdoctoral work, I have been applying my training in chemical engineering to problems in neurodegeneration. During my graduate studies, I worked with Dane Wittrup at MIT to engineer antibody fragments to counteract disease in cellular models of Huntington's Disease. Since completing my doctorate, I have been working with Stanley Prusiner at UCSF, engineering "synthetic prion" strains, which are comprised of recombinant prion protein refolded into an infectious form *in vitro*. In the next phase of my career, I would like to establish a lab at the interface of engineering and neuroscience, applying cell and biomolecular engineering approaches to the most pressing problems in neurobiology.

6ad Drug Delivery through Epithelial Tissues

Harvinder S. Gill, Biomedical Engineering, Chemical and Biomolecular Engineering, Georgia Institute of Technology, 315 Ferst Drive, Atlanta, GA 30332

Majority of the biopharmaceutical entities are administered into the human body using hypodermic needles. However, hypodermic needles cause pain, needle phobia, and pose safety risks by causing transmission of blood borne pathogens like HIV and hepatitis B viruses. The oral route is not available for biopharmaceutical delivery due to degradation and poor absorption of biopharmaceuticals in the gastrointestinal tract. Therefore, novel drug delivery methods that are safer and minimally painful are required to promote delivery of biopharmaceuticals.

The major transport barrier to the entry of biopharmaceuticals is provided by the protective epithelial tissues lining the organs, for example, the skin protects the exterior body surface and the epithelial tissue protects the inside of hollow cavities like the the gastrointestinal tract. These barrier tissues naturally have low permeabilities to prevent entry of pathogens and other harmful substances into the body. This same protective property also prevents transport of biopharmaceuticals like proteins and DNA.

The overarching theme of my proposed research is to characterize and develop novel physical methods to permeabilize epithelial tissues for biopharmaceutical delivery. My doctoral research involved permeabilization of skin using two physical methods, i.e., microneedles and microdermabrasion, Because there is structural similarity between the different epithelial tissues lining the various body organs, my knowledge of skin permeabilization will enable me to develop methods to permeabilize other epithelial tissues.

I will present my research proposal to permeabilize the skin and the epithelial tissue of the oral cavity and the urinary bladder for vaccine delivery and gene therapy.

6ae Ordering of Spherical-Domain Block Copolymers in Monolayers and Multilayers

Gila E. Stein, Center for Nanoscale Science and Technology, National Institute of Standards and Technology, 100 Bureau Drive, Stop 8412, Gaithersburg, MD 20899-8412 and Edward J. Kramer, Materials and Chemical Engineering, University of California, Santa Barbara, CA 93106.

Thin films of block copolymers are a simple, inexpensive route for patterning nanoscale features, and are widely investigated in the semiconductor industry as an alternative (or complement) to established lithographic techniques. While the physics of bulk systems are well understood, the ordering of domains confined to a thin film can differ due to surface interactions and geometric frustration. We have identified some remarkably complex behavior in monolayers and multilayers of a model spherical-domain diblock copolymer using grazing-incidence small-angle x-ray scattering (GISAXS), which provides a quantitative measure of the structure over macroscopic areas. Below the melting transition, the hexagonally-packed monolayer has long-range orientational order and quasi-long-range translational order (algebraic decay), consistent with theoretical predictions for two-dimensional crystals. As the number of layers of spheres is increased from one to 23, competition between the packing preferred in the bulk with that at the interfaces drives a transition from a hexagonal to a body-centered cubic lattice. The hexagonal symmetry of the monolayer persists through four layers, and at five layers the symmetry breaks to form a face-centered orthorhombic lattice. As the number of layers is further increased from five to 23, the face-centered orthorhombic unit cell deforms to approach asymptotically a structure similar to the body-centered cubic lattice with the closest-packed (110) plane oriented parallel to the substrate. GISAXS results are supported by real-space imaging of the lattice with scanning force microscopy (both high resolution and under-sampled) and transmission electron microscopy.

Publications:

7. J. Bang, B.J. Kim, G.E. Stein, T.P. Russell, X. Li, J. Wang, E.J. Kramer, C.J. Hawker, "Effect of humidity on the ordering of PEO-based copolymer thin films," submitted for publication 2007.

6. G.E. Stein, E.J. Kramer, X. Li, J. Wang, "Thickness dependent ordering in laterally confined monolayers of spherical-domain block copolymers," Macromolecules, In Press, 2007.

5. G.E. Stein, E.W. Cochran, K. Katsov, G.H. Fredrickson, E.J. Kramer, X. Li, J. Wang, "Symmetry Breaking of In-Plane Order in Confined Copolymer Mesophases," Physical Review Letters, vol. 98, no. 158302, 2007.

4. G.E. Stein, E.J. Kramer, X. Li, J. Wang, "Layering transitions in thin films of spherical-domain block copolymers," Macromolecules, vol. 40, pp. 2453-2460, 2007.

3. G.E. Stein, E.J. Kramer, X. Li, J. Wang, "Single-crystal diffraction from two-dimensional block

copolymer arrays," Physical Review Letters, vol. 98, no. 086101, 2007.

2. V. Khanna, E.W. Cochran, A. Hexemer, G.E. Stein, G.H. Fredrickson, E.J. Kramer, X. Li, J. Wang, S. F. Hahn, "Effect of chain architecture and surface energies on the ordering behavior of lamellar and cylinder forming block copolymers," Macromolecules, vol. 39, pp. 9346-9356, 2006.

1. A. Hexemer, G.E. Stein, E.J. Kramer, S. Magonov, "Block copolymer monolayer structure measured with scanning force microscopy moiré patterns," Macromolecules, vol. 38, pp. 7083-7089, 2005.

6af Systems Biology Approach to Endocrine Signaling

Pamela K. Kreeger, Biological Engineering, Massachusetts Institute of Technology, 77 Massachusetts Ave, 56-389, Cambridge, MA 02139

Advances in understanding the molecular basis of normal physiology and disease states are critical to developing improved diagnostic tests and therapeutic interventions for human health. A new avenue toward this objective has arisen from the advent of biomimetic tissue culture, increasing the relevance of experimental results, and of systems biology, comprehending molecular mechanisms in terms of integrated networks rather than as mere isolated components. Combined, these two technological revolutions offer the promise of more powerful studies of cell behavior in a physiologically appropriate microenvironment that examines the integration of external cues (*e.g.*, extracellular matrix and soluble factors) and intrinsic cues (*e.g.*, gene expression levels).

Developing a 3D Microenvironment for Ovarian Follicle Development

My graduate work, under Dr. Lonnie Shea and Dr. Teresa Woodruff, focused on the importance of matching the microenvironment to the needs of a tissue for appropriate development. We applied this conceptual question to the development of the ovarian follicle, in order to generate an alternative method of producing mature eggs for fertility treatments. In my research, I developed three-dimensional synthetic matrices that promoted ovarian follicle development and produced meiotically competent oocytes for fertilization. In this system, ovarian follicles were isolated from immature mice, encapsulated, and cultured within alginate matrices to maintain cell-cell connections which are critical for growth and maturation of the enclosed oocyte. Through modification of the scaffold material and culture conditions, I examined the regulation of follicle development by the extracellular matrix and follicle stimulating hormone (FSH). These distinct signals integrated to regulate the quality of the oocyte. As an example, earlier stages of follicle development were promoted by the addition of collagen I, while FSH was critical for growth of later stage follicles and oocyte maturation.

Systems Biology Analysis of the Effects of RAS Mutations in Apoptosis Signaling

In my postdoctoral research, with Dr. Douglas Lauffenburger, I have been developing computational models of a 'cue – signal – response' form, where the cues to the cell are both external (*e.g.*, cytokine treatments) and internal (*e.g.*, genetic mutations). In particular, in order to examine the role of RAS mutations, which are identified in many human cancers, we are developing a partial least-squares regression model for apoptosis in an isogenic panel of colon carcinoma cells lines with varying RAS mutations. Our hypothesis is that cell fate decisions are regulated through a quantitative interplay between the intracellular signaling pathways downstream of apoptotic stimuli and RAS, and that elucidation of this function will enable us to discern the role of mutant RAS in apoptosis. To develop this model, we are constructing an extensive dataset of apoptotic responses to a combination of extracellular cytokine stimuli and RAS genotype variations, along with the corresponding intracellular signal measurements downstream of RAS and tumor necrosis factor (TNF) receptor. Our results indicate that the K-RAS mutation sensitizes cells to TNF stimulated apoptosis, increasing cell death nearly 33% by 48 hours. Our results indicate that in addition to the phenotypic differences with varying RAS state, there are distinct signaling profiles in response to treatment with TNF. Future work will extend this system biology approach to animal models that conditionally express these mutations in the intestinal epithelium.

Combined Application of these Approaches to Understanding Fundamental Biological Questions of Importance in Women's Health

Integrating my doctoral and postdoctoral research experiences, my research program will investigate fundamental biological questions using a systems biology/engineering approach and disease models of importance in women's health. I will extend a systems biology approach to incorporate endocrine signaling, which provides a master level of biological regulation through organ to organ communication and is critically related to a variety of disease states in women. Here, I outline initial studies which will incorporate both experimental and modeling approaches to improve understanding of:

• The impact of immediate versus delayed signals in response to cellular cues - we will develop a 'cue-signal-response' model of a panel of estrogen mimicking compounds in breast cancer cells, examining the multiple modes of estrogen-induced signaling.

• How cells interpret multiple synergistic and antagonistic cues - we will examine the extensive network of cross-talk between the EGF family receptors and the estrogen receptors.

• How mutations such as those found in cancer rewire the cell network response to physiological inputs - we will examine mutations implicated in the progression of ovarian cancer, and the cellular response to endogenous paracrine and endocrine ligands.

6ag Reversible Addition-Fragmentation Chain Transfer in Microemulsion Polymerizations

Jennifer O'Donnell, School of Chemistry, University of Sydney, Sydney, Australia and Eric W. Kaler, Center for Molecular and Engineering Thermodynamics, Department of Chemical Engineering, University of Delaware, 102 P.S. DuPont Hall, Newark, DE 19716.

Over 20 million tons per year of polymer latex is produced worldwide by free radical emulsion polymerization for applications such as coatings, adhesives and rheological modifiers. Unfortunately, these polymerizations often suffer from poor control of molecular weight and yield high polydispersity polymers. Therefore, there is strong interest in incorporating the controlled polymerization mechanism of reversible addition-fragmentation chain transfer (RAFT) into heterogeneous polymerizations. RAFT controls the bulk free radical polymerization of many monomers over a wide range of reaction conditions and allows for the production of polymers with predetermined molecular weight, but frequently results in high polydispersity polymers and loss of colloidal stability in heterogeneous polymerizations. Additionally, rate retardation is often observed in RAFT polymerizations, and the source of this rate retardation is not fully understood.

The goals of this work were twofold: (1) to incorporate RAFT into microemulsion polymerizations to produce stable latex nanoparticles composed of low polydispersity polymers of predetermined molecular weight and (2) to identify the key parameters for achieving control in RAFT microemulsion polymerizations. Microemulsion polymerization is an attractive alternative to emulsion polymerization because of the absence of large monomer droplets and the elimination of biradical termination reactions. These characteristics were expected to increase the colloidal stability of the resulting latex and enhance the control of the polymerization. In fact, RAFT microemulsion polymerizations of acrylates with xanthate chain transfer agents produced stable latex nanoparticles (20-30 nm in diameter) composed of low polydispersity polymers of predetermined molecular weight. The RAFT microemulsion polymerization kinetics, polymer molecular weights and polydispersities, and latex particle sizes identified the key parameters for achieving control as the chain transfer agent per micelle ratio, and the water solubilities of the monomer and the chain transfer agent. With this information, a simple kinetic model was developed to further examine the RAFT microemulsion polymerization mechanism. The model demonstrates the significance of the rate of fragmentation of the intermediate radical formed during the equilibrium transfer reaction and the rate of diffusion of the chain transfer agent to the locus of polymerization. The model is also useful for the selection of appropriate chain transfer agents for different microemulsion polymerizations.

6ah Nanoengineering Systems for Targeted Drug Delivery, Cell-Based Therapy, and Microfluidic Biosensors/chips

Hongyan He, Department of Chemical and Biomolecular Engineering, The Ohio State University, 140 WEST 19TH AVE, Columbus, OH 43210

Many characteristics of polymers and ease of processing make them attractive for various biomedical applications. Proper combination of functional polymers and a number of manufacturing modules can offer tailored properties for sophisticated technologies used in targeted drug delivery, cell-based therapy, and microfluidic biosensors/chips. 1.Multifunctional delivery systems for oral and implant administrations Many protein- and DNA-based drugs need to be properly protected and precisely controlled during administration. A robust manufacturing protocol has been developed to produce particulate-like polymeric devices based on functional hydrogels to achieve multi-functionalities, such as drug protection, self-regulated oscillatory release, enhanced mucoadhesion and targeted unidirectional release. The delivery devices will be of great benefit to the advancement of oral administration of existing therapeutic agents. A nanoporous miniature device made of biodegradable poly(lactide-co-glycolide acid) (PLGA) has also been designed based on the integration of a number of micro-manufacturing modules, such as phase inversion, a sacrificial template nano-imprinting, and carbon dioxide (CO_2) -assisted bonding. The biocompatibility, biodegradation, and release performance of this device have been evaluated in animal studies. Such devices are able to enhance the bioavailability of existing therapeutic and provide a continuous, pre-designed release profile within the local tumor environment, potentially without the need for surgical implantation. 2.Cell-based delivery systems by micro-/nanofabrication Cell-based delivery devices holds great promise for applications requiring site-specific and sustainable drug delivery of cell-synthesized molecules. On the other hand, embryonic stem (ES) cell therapies have been proposed for regenerative medicine and tissue replacement after injury or disease. An integration of ES cells within immunoprotective and biodegradable devices will provide great potential for novel delivery platforms. Our recent work focus on the development of nanoporous PLGA-based chambers for immunoisolation of embryonic stem cells. A series of advanced techniques such as composite membrane preparation by phase inversion, surface coating by plasma polymerization, and nitrogen (N_2) -assisted bonding, have been combined to improve the device functionalities. Studies described herein also include the interactions of PLGA-based substrates with differentiated ES cells in terms of viability, proliferation, and functionality.

6ai Computer Simulations of Complex Fluids and Materials

O. Berk Usta, Chemical and Petroleum Engineering, University of Pittsburgh, 1249 Benedum Hall, 3700 Ohara St., Pittsburgh, PA 15261

Colloidal and polymeric systems and suspensions of elastic materials are often classified as "complex fluids" owing to the nonlinear relationship between the applied forces and resulting dynamics. This complexity arises from the wide range of time and length scales that govern their motion. The motion of such systems depend on a variety of different factors such as particle interactions and large scale collective motion related to structural relaxations. The simplest class of interactions are the direct inter-particle interactions such as electrostatic interactions and hard-sphere collisions which also exist in simple fluids. In addition to these, sub-micronsolute particles in a colloidal suspension are coupled in their motion through the fluid. This coupling is called "hydrodynamic interaction" and constitutes the focus of my work. These hydrodynamic interactions are long ranged and many-body, therefore they create a great deal of complexity for theoretical calculations as well as for traditional computational methods.

In my research, I investigate the dynamics of complex fluids with the lattice-Boltzmann method for fluid flow and a variety of mechanical models for suspended solid and deformable particles. Such an approach leads to the solution of Navier-Stokes equations on large time and length scales. The lattice-Boltzmann method is closely related to and is a descendent of the lattice-gas cellular automata; it has proven to be a computationally competitive, accurate, and flexible simulation method for complex fluids. In the lattice-Boltzmann method, hydrodynamic interactions, between solid objects, arise naturally as a result of local momentum conservation and exchange between solid and fluid phase. The flexibility is due to the ease of incorporating boundary conditions for a wide variety of different shapes of objects.

In this session, I will present the current research on both the method itself and applied problems in sorting of both polymer solutions and microcapsules. I will also discuss the future research ideas where this approach can make a significant impact.

6aj Estimating the Emission of Greenhouse Gases and Volatile Organic Compounds in Aqueous Electrolyte Solutions

James B. Falabella, Chemical and Biomolecular Engineering, Georgia Institute of Technology, 311 Ferst Dr. NW, Atlanta, GA 30332-0100, Xin-Sheng Chai, Institute of Paper Science and Technology, Georgia Institute of Technology, 311 Ferst Drive NW, Atlanta, GA 30332-0100, and Amyn S. Teja, Georgia Institute of Technology, 311 Ferst Drive NW, Atlanta, GA 30332-0100.

The determination of accurate air-water partition coefficients for volatile organic compounds (VOC) and greenhouse gases (GHG) is essential in pollution and global climate modeling. In the atmosphere, oceans, as well as in groundwater, the concentration of dissolved salts is sufficiently high to significantly affect the air-water partitioning behavior of the VOC or GHG. There is therefore a great need for partitioning data for common VOCs and GHGs in aqueous environments containing salts. Unfortunately, such data are not readily available, and even when available, there are large disagreements between different sets of data. Henry's constants of several VOCs in aqueous salt solutions were therefore measured using a new headspace gas chromatography (HTHSGC) method to create a set of internally consistent air-water partition coefficients for modeling. The VOCs studied included the homologous series of 1-alkanols, 2-ketones, and organic sulfides, as well as components of gasoline including toluene, ethylbenzene, o-xylene, methyl tertbutyl ether, and ethyl tertbutyl ether. A model with temperature-independent parameters based on dilute solution theory was developed using the library of data to resolve disagreements between literature sources. The model correlated Henry's constants of methane over temperature ranges as wide as 300 K, salt concentrations up to 4 molal, and pressures up to 1000 bar. Extrapolations of up to 50 K, 1 molal salt and 100 bar pressure can also be performed with the model to eliminate the need for additional experiments. The temperature-independent salt effect parameter was found to be directly proportional to the critical volume of the VOC. Also, all homologous VOCs followed the same linear trend with size, allowing prediction of the salt effect.

6ak Determination of Ozone Uptake in Human Lungs: Study of the Effects of Smoking

Tim Brenza[1], Melissa Lowe Bates[2], Aziz Ben-Jebria[1], and James Ultman[1]. (1) Chemical Engineering, The Pennsylvania State University, University Park, PA 16802, (2) Physiology, The Pennsylvania State University, University Park, PA 16802

In previous studies exposure to ozone (O_3), a common outdoor air pollutant, is shown to cause a decrement in spirometry parameters. This decrement has been shown to be attenuated in smokers. We hypothesize that hypersecretion of respiratory mucus due to smoking reduces the absorbance of O_3 in the conducting airways of smokers, resulting in decreased responsiveness to exposure to O_3. Due to the reduced amount of O_3 absorbed in the conducting airways, the amount of O_3 absorbed in the deeper respiratory airways of the lung is expected to be increased in smoking subjects when compared to non-smoking subjects.

In this study, subjects participate in three research sessions: bolus, air and ozone. In the bolus research session, measurements of flow and O_3 concentration are used in the determination of longitudinal distribution of O_3 uptake in human lungs. During the air and ozone research sessions the subject will exercise on a cycle ergometer for one hour while breathing either room air or ozone mixed with room air. In these sessions measurements of flow, ozone concentration and carbon dioxide (CO_2) concentration are recorded.

Ozone bolus measurements have been preformed in previous studies to determine flow effects and subject variability on non-smoking subjects. The objective of the present study is to determine uptake using the bolus inhalation apparatus on a population of smoking and non-smoking subjects, allowing for a comparison between these two populations. Fractional uptake (Λ) is expressed as the amount of O_3 absorbed during a single breath relative to the amount of O_3 in the inhaled bolus. Uptake is related to the penetration volume (V_P) of the bolus into the respiratory tract.

From the flow and ozone concentration data, the Λ-V_P distribution will be determined from a series of 60-80 test breaths that are collected at a fixed flow of 60 L/min for each subject using the bolus apparatus. Comparing the Λ-V_P profiles of the smoking and non-smoking populations shows a small distal shift of the absorption profile for the smoking subjects. However, this result may be due to differences in population respiratory system sizes. To minimize the effects of differences in lung size among the individuals in this study, the V_P is normalized by the anatomical dead space (V_D) obtained from carbon dioxide expirometry. V_D is a measurement of the volume of the lung that does not take part in gas exchange, and can therefore also be used as an estimate of the conducting airway volume of each subject. Comparing the Λ-V_P/V_D profiles show that there are no significant differences between the two populations in terms of ozone absorption.

For the air and ozone sessions changes were made to the equipment and data collection process from previous studies for the inclusion of CO_2 monitoring. The goal of the inclusion of the CO_2 data in the air and ozone sessions is to show the dynamic changes in lung function during ozone exposure. Data from CO_2 expirography has been used to determine anatomic information non-invasively. The CO_2 expirogram can be divided into three regions, the pure dead space, the transition region, and the alveolar plateau. V_D can be determined from the pure dead space and transition regions of the expirogram, while the normalized slope (S_N) can be computed from the alveolar plateau.

The following results are for thirty healthy non-smoking subjects (17 men, 13 women) that participated in these two research sessions (air and ozone). The value of V_D decreases during the O_3 exposure session at a rate of 0.067 ± 0.035 mL/min ($P = 0.06$). The value of S_N increases during the O_3 exposure session at a rate of 0.57 ± 0.094 E^{-3} $L^{-1}min^{-1}$ ($P < 0.001$). During the course of the air exposure, the value of V_D does not change significantly with time ($P = 0.72$). However, the value of S_N is seen to also increase with air exposure at a rate of 0.29 ± 0.11 E^{-3} $L^{-1}min^{-1}$ ($P = 0.011$). The results of the air session suggest that exercise alone can lead to an increase in S_N. The larger and more significant increase in S_N during O_3 exposure indicates there is a loss of gas exchange efficiency in addition to the exercise-induced effect. However, there was a systematic change in tidal volume observed during the one-hour exposure period. The changes in breathing pattern may have contributed to changes in S_N during the dynamic measurements.

Future work in this study will focus in the following area: 1) Establish a model that will enable parameters determined from the absorption data in the bolus session to partition the absorbed ozone during the ozone session into distinct respiratory compartments; 2) Develop methods in which anatomic information can be obtained independent of exhaled tidal volume from CO_2 measurements during the air and ozone sessions; and 3) Determine if uptake of ozone in the respiratory system can account for difference in responsiveness to zone exposure between smoking and non-smoking subjects.

6al Protein Engineering Strategies for the Creation of Fluorescent Biosensors

Tej Pavoor and Eric Shusta. Chemical and Biological Engineering, University of Wisconsin, 1415 Engineering Drive, Madison, WI 53706

Fluorescent protein-based biosensors have considerable application in cell biology, proteomics, and immunochemistry. The creation of such biosensors requires a stable fluorescent protein scaffold capable of accommodating loops that confer binding. This has been attempted by several laboratories but with limited success. In our approach, we selected green fluorescent protein (GFP) as the protein scaffold. We studied the ability of the GFP scaffold to accommodate amino acid insertions at two different positions on GFP. Since the GFP scaffold with binding loops inserted at both locations was not expressed by yeast, we successfully utilized directed evolution using yeast surface display to evolve the scaffold to be well expressed and fluorescent. Our studies led us to examine protein folding and chromophore stability and we utilized both aspects as selection criteria for directed evolution. We subsequently randomized the inserted amino acids to create a library of GFP-based biosensors. This study provides novel insight into the engineering strategies of intrinsically fluorescent proteins and provides a platform for the development of the next generation of sensors.

6am Modeling Extracellular Mass Transport of Nutrients and Byproducts around Metabolizing Bacteria

Michael R. Benoit, Microbiology & Immunology, Stanford University, 299 Campus Dr., West, Stanford, CA 94305-5124

The effect of hydrodynamic mixing in bacterial populations due to bacterial chemotaxis is a well described phenomenon known as bioconvection. The classic scenario for bioconvection entails bacterial swimming upward along an oxygen gradient, creating a dense layer of cells near the air/liquid interface, followed by sedimentation of bacteria (due to gravity). This often leads to stable cells of circular flow and creates convective mixing throughout the culture. Under certain conditions, we observed buoyant plumes that were created independently of bacterial motility. We propose that these buoyant hydrodynamic flows were caused by solute gradients, resulting from bacterial metabolism. Cultures of metabolically active non-motile E. coli consistently formed plumes visible with the naked eye, leading to complete mixing of the bacterial culture, whereas cultures of stationary phase bacteria never formed plumes, presumably due to a lack of rapid metabolism. The phenomenon of solute-induced buoyancy-driven convection has previously been characterized for growing protein crystals and soft tissues, however, both are much larger than individual bacterial cells. We developed numerical models to study the effect of nutrient consumption and by-product excretion on extracellular fluid density and the formation of buoyant plumes. The numerical solutions compared well with analytical calculations and experimental results. Furthermore, an analysis with dimensionless numbers was used to characterize the size-dependent nature of observable buoyant plumes, and to investigate the effect of altered levels of gravity, such as the microgravity conditions encountered during space flight.

6an Engineered Polymer Vesicles (Polymersomes) for Targeted Adhesion, Bioimaging, and Controlled Delivery

Anthony J. Kim and Daniel A. Hammer. Chemical and Biomolecular Engineering, University of Pennsylvania, 210 South 33rd St, Philadelphia, PA 19104

(1) Current Research: Targeted emissive polymersomes are biocompatible nanocarriers that have the enormous potential for in vivo imaging, diagnostics, and drug delivery applications. For example, imaging of cancer or inflammatory cells can be dramatically improved by attaching chemical compounds to emissive polymersomes that selectively bind to chemical markers on surfaces of these cells. In this work, we use NIR emissive polymersomes comprised of self-assembled amphiphilic diblock copolymers and hydrophobic porphyrin-based NIR fluorophores (NIRFs). These polymersomes have several distinguishing characteristics; (1) both the polymer and porphyrin-based NIRF are fully biocompatible; (2) polyethylene-oxide (PEO) head groups result in long in vivo circulation times and also available for coupling to biomolecules; (3) thickness (9-22nm) and mechanical strength of membrane can be adjusted to incorporate a wide range of both hydrophobic and hydrophilic compounds; (4) vesicle sizes are tunable from 50 nm to 10 um in diameter; (5) porphyrin-based NIRFs possess highly attractive and sensitive emissive spectra (700-950nm), where light is not absorbed by tissue and can penetrate deep into it; and (6) poly(ethylene oxide-b-caprolactone) based polymersomes hydrolyze over time for controlled release of molecules.

(2) Doctoral Research (with John C. Crocker, University of Pennsylvania): My doctoral research was focused on assembling novel colloidal crystal structures using DNA-mediated interaction. In order to expand the kind of structures that can be formed with simple spheres, specific interactions between different species in a binary mixture have been predicted to promote novel crystal formation. For instance, by surface functionalizing colloids with engineered DNA sequences we can control the range and magnitude of interparticle interaction and lead them to assemble in ways we want. We have developed a novel swelling/deswelling method for modifying the surface of colloids with a polymer brush and functional groups to make sterically stable DNA-grafted colloids that sucessfully assemble into close-packed colloidal crystal structures. In addition, we have studied crystal growth in a simple colloidal alloy: a binary solid solution, in which a minority species is randomly inserted into a lattice of the majority species as a substitutional impurity. A competition between DNA single-base mismatches was used to create a very small interaction strength mismatch that provided us with a view of the growing crystals' interfacial thermodynamics. Depending on the free energy difference, we found either complete exclusion of the minotrity species or a finite segregation coefficient.

6ao Investigating Membrane Surface Interactions with Lipid-Coated Particles

Esther W. Gomez, Chemical Engineering, University of California, Berkeley, 106 Lewis Hall, Berkeley, CA 94720 and Jay T. Groves, Chemistry, University of California, Berkeley, 105 Lewis Hall, Berkeley, CA 94720.

Assembly of lipid bilayers onto the surface of silica particles allows for the surface composition and charge of the particles to be tuned continuously over a wide range. The collective behavior of these systems depends strongly on the properties of the coating membrane with subtle changes in the surface chemistry resulting in macroscopic changes in the order within the system. Using a combination of optical imaging techniques we are able to monitor both the colloid structure and the surface potential in electrostatically levitated colloidal monolayers composed of lipid-coated particles. We observe that interactions can be tuned, from repulsive to long-range attractive. The collective behavior is asymmetric with respect to the sign of the surface potential. Incorporation of membrane-associated receptors to the lipid coating allows for biofunctional studies. We observe that binding of protein to the membrane receptors results in macroscopic changes in monolayer ordering thereby providing a label-free method for monitoring molecular interactions on membrane surfaces.

6ap Acid Catalysis of Lipids to Produce Green Fuels: Advancing Biofuels in a Fossil Fuel World

Tracy J. Benson, Chemical Engineering, Mississippi State University, Mississippi State, MS 39759

Development of renewable fuels is essential to global sustainability as Earth's fossil fuels are depleted, resulting in decreased supplies for transportation fuels and energy generation. Scientists have identified new renewable resources capable of displacing significant quantities of petroleum, the chief feedstock for transportation fuels and commodity chemicals. For example, lipids can be extracted from plants, animals, and microorganisms and used for fuel and specialty chemical production. Our research group has estimated that approximately 10 billion gallons of lipids could be generated and converted into fuels using current agricultural practices, wastewater treatment, and refining infrastructure.

During my master's work, I explored various bio-adsorbents for the removal of water from ethanol/water mixtures (1). Using a temperature swing adsorber, mixtures of 90, 95, and 97 wt% ethanol were volatilized and passed through a column bed packed with either hardwood sawdust, kenaf core, or bleached wood pulp. A glass adsorption column with an inside diameter of 2.54 cm was used to generate breakthrough curves to determine the effectiveness of the adsorbents and to allow comparability with starch based adsorbents that are currently used within the ethanol industry. Experimental results indicated that water was preferentially adsorbed allowing for complete dehydration of ethanol. Bleached wood pulp was a superior adsorbent over the kenaf core and hardwood sawdust in adsorbing water due to its large percentage of hydroxyl groups stemming from a high concentration of xylans. The bleached wood pulp was the only material that demonstrated similar water adsorbing capacities as compared to starch-based adsorbents such as corn grits.

My doctoral thesis has been devoted to the advancement of science in the cracking of lipid molecules into diesel-like and gasoline-like fuels. First, oleic acid, an unsaturated free fatty acid in many lipid sources, was reacted with triflic acid, a liquid superacid catalyst at 0°C (2). Instrumentation employed for sample analyzes included 1H-NMR, 13C-NMR, FTIR, and GC/MS (both EI and CI) to identify reaction products and to hypothesize a reaction pathway. The resultant products were saturated fatty acids with carbon lengths of C9-C14, C16, and C18. This work concluded that a Bronsted acid protonates the double bond causing methyl and hydride shifts that succumb to β-scissions resulting in linear and branched chain isomers of smaller carbon lengths than the original starting material. The cracking of oleic acid was in stark contrast to palmitic acid, a saturated fatty acid, which yielded no discernable products when reacted with triflic acid at temperatures up to 100°C.

The second step in this research was to evaluate the heterogeneous catalytic cracking chemistry of model lipid compounds, chiefly oleic acid, mono-, di-, and tri-olein. For this phase, a reactor/analyzer was constructed using a Varian 3600 GC which was reengineered to allow for reactions with online mass spectrometry analysis (3). This unit, which is capable of accepting a gas, liquid, or solid reactant, was equipped with a Saturn quadrupole ion trap mass spectrometer for mass spectral analysis in real-time as lipids were passed over a bed of highly acidic ZSM-5 catalyst. As this was an integrated reactor/analyzer, the Rxi-1ms GC column was exposed in the inlet to temperatures in excess of 400°C. This temperature is beyond the degradation point of the column, and so selective ion storage RF waveform was used to remove the siloxane masses m/z (207, 281, and 355) from the spectra. This produced lower detection limits and full scan data for identification. Electron impact and selected ejection chemical ionization were used for product identification and molecular weight confirmation. Hexane was reacted over H-ZSM-5 catalyst for instrument validation. This produced a series of alkanes and alkenes with distributions consistent with that reported for the cracking of hexane. The heterogeneous cracking of model compounds resulted in mixtures that were within the ranges of diesel and gasoline organics. Oleic acid was found to form chiefly BTEX and trimethylbenzene as initial cracking products. Decarboxylation is expected to follow a radical reaction pathway, yielding CO2. We plan to also publish these results in the Journal of Molecular Catalysis A: Chemical (4).

Along with research, I had the opportunity to teach a chemical engineering core course (CHE 3213 - Fluid Flow Operations) for two semesters. My average student evaluations (4.23/5.00) were above the average for the College of Engineering. I also had numerous opportunities to teach laboratory practices and experimental procedures to undergraduate and high school students who have worked with me to develop this research. My goal is to contribute to the education of future chemical engineers through research, teaching, and service to the community.

Refereed Publications

(1) T Benson and C George (2004), "Cellulose Based Adsorbent Materials for the Dehydration of Ethanol Using Thermal Swing Adsorption," Adsorption Journal Vol.11.

(2) T Benson, R Hernandez, T French, E Alley, and W Holmes, "Reactions of Fatty Acids in Superacid Media: Identification of Equilibrium Products" Journal of Molecular Catalysis A: Chemical (Accepted for publication, May 2007).

(3) T Benson, W Holmes, R Hernandez, T French, and E Alley, "Heterogeneous Catalytic Cracking Reactor Utilizing Online Mass Spectrometry Analysis" Journal of the American Society for Mass Spectrometry (Expected submission: June 2007)

(4) T Benson, R Hernandez, T French, E Alley, and W Holmes, "Heterogeneous Reaction of Model Glycerides over H-ZSM5 Catalyst: Elucidation of Cracking Mechanisms" Journal of Molecular Catalysis A: Chemical (Expected submission: Summer 2007)

6aq Computational and Theoretical Studies of Soft Materials and Biological Systems

Arthi Jayaraman, Department of Material Science and Engineering, University of Illinois at Urbana-Champaign, 1304 W. Green St., Urbana, IL 61801, Kenneth S. Schweizer, Material Science and Engineering, University of Illinois at Urbana-Champaign, 1304 W. Green St., Urbana, IL 61801, Carol K. Hall, North Carolina State University, Box 7905, Raleigh, NC 27695, and Jan Genzer, Chemical and Biomolecular Engineering, North Carolina State University, 911 Partners Way, Raleigh, NC 27695.

My research is aimed at understanding molecular phenomena at the interface between materials science and biology using computational and theoretical methods.

During my doctoral work with Prof. Carol Hall and Prof. Jan Genzer at North Carolina State University, I used computer simulations to study the thermodynamics of molecular recognition in bio-mimetic polymer systems, specifically, in the following two areas:

1.1. **Designing pattern recognition surfaces for selective adsorption of copolymer sequences**: We developed a novel simulation method to design surfaces for recognizing specific monomer sequences in copolymers [1]. The order parameter characterizing the surface pattern, and the surface selectivity were determined as a function of polymer-surface and polymer-polymer interactions, and the blockiness of the monomer sequence. The highlight of this work was that the designed surfaces recognize specific monomer sequences with higher selectivity than the standard checker board surfaces of sizes commensurate with the monomer sequence.

1.2. **Molecular recognition in model DNA microarrays** : We studied the thermodynamics and the kinetics of hybridization of single stranded "target" genes in solution with complimentary immobilized "probe" DNA molecules on a microarray surface using a coarse-grained lattice model [2]. We used Monte Carlo simulations to examine how various parameters affect the extent of hybridization and the kinetics of the hybridization process. This work provided a set of general guidelines for maximizing microarray sensitivity and specificity.

In my post-doctoral work with Prof. Kenneth Schweizer at University of Illinois, Urbana- Champaign, I use theoretical methods to study soft materials. Currently, I am investigating:

2.1 **Structure and phase behavior of melts and dense solutions of polymer tethered nanoparticles**: We have applied microscopic Polymer Reference Interaction Site Model (PRISM) theory to study dense solutions and melts of polymer tethered spherical nanoparticles. The complex interplay of entropy (translational, conformational and packing) and enthalpy (particle-particle attraction) leads to different spatial arrangements with distinctive scattering signatures. The tethered particles prefer to organize themselves in a specific geometry that is dictated by particle radius, length of the tethered polymer, number and positions of the tethered polymers, total fluid packing fraction and interfacial attraction strengths.

My future research plan is the following: (a) apply theory and simulation to study thermodynamics and kinetics in multi-component self-assembling mixtures, including polymers, colloids, and nanoparticles; (b) apply multi-scale modeling and simulation to study molecular recognition in bio-inspired systems: In particular, I plan to use multi-scale modeling and simulations to study recognition and repair of DNA damage, and investigate the role of damage repair in the development of resistance against anti-cancer drugs.

References:

1. A. Jayaraman, C. K. Hall, and J. Genzer, 'Designing pattern-recognition surfaces for selective adsorption of copolymer sequences using lattice Monte Carlo simulation' , *Physical Review Letters*, 94, 078103 (2005)

2. A. Jayaraman, C. K. Hall, and J. Genzer, 'Computer simulation study of molecular recognition in model DNA microarrays', *Biophys. J.*, 91, 2227 (2006)

6ar Designing for Sustainability with CO_2-Tunable Solvents

Jackson W. Ford, Chemical & Biomolecular Engineering, Georgia Institute of Technology, 311 Ferst Dr, Atlanta, GA 30332-0100, Charles L. Liotta, Chemistry and Biochemistry, Georgia Institute of Technology, PO Box 330834 Georgia Tech Station, Atlanta, GA 30332, and Charles A. Eckert, Chemical and Biomolecular Engineering, Georgia Tech, 311 Ferst Drive NW, Atlanta, GA 30332-0100.

Developing greener, more efficient, and less energy-intensive processes will lead the chemical industry into a more sustainable future. The utilization of alternative solvents to displace or reduce the use of traditional organic solvents is an important part of this effort. Gas-expanded liquids (GXLs) form a unique class of environmentally benign and tunable solvents that can be used in a variety of applications. GXLs reduce the need for organic solvents, operate at more moderate pressures than supercritical fluids, and facilitate less energy-intensive separations through depressurization to recover products and catalysts. We have recently employed GXLs in the recovery of fine chemicals from renewable resources, and in biomass pretreatment. A more complete understanding of the interactions between the gas, the organic liquid, and solutes at the molecular level will enable the full exploitation of GXLs.

We have used two kinetic probes, the Diels-Alder reaction of anthracene with 4-phenyl-1,2,4-triazoline-3,5-dione and the Menschutkin reaction of tributylamine with methyl p-nitrobenzenesulfonate to examine the local solvent structure of CO_2-expanded acetonitrile. We have correlated our kinetic results with solvatochromic probes including the Kamlet-Taft parameters and Kosower's Z to develop comprehensive models explaining the intermolecular phenomena involving the GXL and the solutes. The results of this research can be used to guide future applications of GXLs as green reaction solvents.

6as Multiscale Modeling of Structure and Transport Properties in Polymeric Materials

Xiaoyan Wang[1], Benny D. Freeman[2], and Isaac C. Sanchez[1]. (1) Chemical Engineering, University of Texas at Austin, Mail Code C0400, Austin, TX 78712, (2) Chemical Engineering, The University of Texas at Austin, 10100 Burnet Rd. Bldg 133, Center for Energy and Environmental Resources, Austin, TX 78758

Molecular modeling techniques are applied to study the cavity size distribution and transport properties of two very permeable membranes, poly (1-trimethylsilyl-1-propyne) (PTMSP) and a random copolymer of tetrafluoroethylene and 2,2-bis(trifluoromethyl)-4,5-difluoro-1,3-dioxole (TFE/BDD), which have very similar and large fractional free volumes, but very different permeabilities. Using atomistic models, cavity size (free volume) distributions determined by a combination of molecular dynamic and Monte Carlo methods are consistent with the observation that PTMSP is more permeable than TFE/BDD. The average spherical cavity size in PTMSP is 11.2 Å whereas it is only 8.2 Å in TFE/BDD. These cavity size distributions determined by simulation are also consistent with free volume distributions determined by positron annihilation lifetime spectroscopy (PALS). The diffusivity, solubility and permeability of CO_2 in these two high free volume membranes were also obtained through molecular simulation. The diffusivity and permeability of CO_2 in PTMSP are higher than TFE/BDD. Good agreement is observed between the simulation and experimental data.

Web Page: htt://www.che.utexas.edu/~xywang/index.html

6at Surfactants and Polyelectrolytes as Building Blocks for Soft Materials

Yakov Lapitsky[1], Molly S. Shoichet[1], and Eric W. Kaler[2]. (1) Donnelly Centre for Cellular and Biomolecular Research, Department of Chemical Engineering & Applied Chemistry, University of Toronto, 160 College Street, Toronto, ON M5S 3E1, Canada, (2) Center for Molecular and Engineering Thermodynamics, Department of Chemical Engineering, University of Delaware, 102 P.S. DuPont Hall, Newark, DE 19716

Many commercial products and academic research projects use "smart" materials that respond to external stimuli such as changes in pH, ionic strength, or temperature, and mechanical or electromagnetic perturbations. Surfactants and polyelectrolytes both provide acutely sensitive building blocks for these materials. Consequently, they are used together in a broad range of applications, which include household products (e.g., foods, personal care products, cleaning solutions, paints, and cosmetics), pharmaceutical formulations, oil recovery, and templates for the synthesis of novel materials. However, many frontiers of their physicochemical properties and applications remain unexplored. To this end, we have investigated the interactions, properties, and applications of surfactant-polyelectrolyte gels and solutions on both the molecular and macroscopic levels. This poster summarizes our work on the formation, characterization, and applications of surfactant-polyelectrolyte gels.

I. Formation, Structure, and Stability of Surfactant-Polyelectrolyte Gels

Interactions between oppositely charged surfactants and polyelectrolytes in aqueous solutions can lead to associative phase separation, where the concentrated phase is a viscous liquid, gel, or precipitate. We have exploited this phenomenon to form gel particles (ranging from microns to millimeters in diameter), fibers, and coatings from mixtures of cationic polysaccharides with oppositely charged surfactants via phase inversion. Their stability is governed by the equilibrium phase behavior of the surfactant-polyelectrolyte mixture, and can be enhanced by either varying the molecular structures of their constituents or through covalent crosslinking of the polymer chains. Conversely, the internal morphologies of these materials (which are either solid or hollow) are dictated by their gelation kinetics and vary with the gel dimensions and the solution compositions used in their preparation. This work illustrates that phase separation in surfactant-polyelectrolyte mixtures can be exploited to form soft materials with controlled phase stabilities and morphologies on a variety of lengthscales.

II. Applications of Surfactant-Polyelectrolyte Gels

We have tested the performance of these structures in the encapsulation and release of hydrophobic compounds and preparation of novel biomaterials. Aromatic oil (cymene) was encapsulated inside surfactant-polyelectrolyte gel particles and released into both aqueous and organic receiving mediums. The release rate is governed by the effective diffusivity and solubility of the oil in the aqueous gel matrix, and can vary from a few hours to a few days, depending on the receiving solution. Likewise, we have developed cell-adhesive and cytocompatible gel fibers with controlled degradation rates from mixtures of fatty acid salts (sodium caprate, laurate, and myristate) and a water-soluble chitosan derivative, N-[(2-hydroxy-3-trimethylammonium)propyl]chitosan chloride (HTCC). The dissolution times of these fibers depend on surfactant hydrophobicity, are modulated through the selection of the fatty acid molecule, and range from minutes (for HTCC-caprate), to hours (for HTCC-laurate), to days (for HTCC-myristate). These studies suggest that surfactant-polyelectrolyte gels can serve as controlled release agents and offer an attractive alternative to conventional hydrolytic and enzymatic mechanisms for controlling the degradation rates of biomaterials.

6au Hydrophilic Biopolymer Mediated Enhancement of Lung Surfactant Adsorption

Patrick C. Stenger and Joseph A. Zasadzinski. Department of Chemical Engineering, University of California, Santa Barbara, CA 93106

Lung surfactant (LS) is a unique mixture of lipids and proteins that lines the alveoli and lowers the surface tension in the lungs, thereby insuring a negligible work of breathing. The surface tension control imposed by LS is compromised during Acute Respiratory Distress Syndrome (ARDS), which afflicts 150,000 with a 40% mortality rate in the US. In a related syndrome, the absence of LS due to prematurity leads to neonatal Respiratory Distress Syndrome (NRDS), which is routinely treated with animal-derived replacement LS. However, replacement LS loses its ability to reduce surface tension and is said to be "inactivated" when used to treat ARDS. A common factor among ARDS patients is elevated levels of serum proteins in the alveolar fluid. In vitro, LS mixed with serum proteins shows an ARDS-like depression of LS activity; serum inactivation is one reason why treatment of ARDS with replacement LS is unsuccessful. Several hydrophilic polymers have recently been shown to enhance the ability of replacement LS to resist serum inactivation both in vitro and in vivo. The use of polymer additives to enhance LS performance might offer a scalable and cost effective treatment for ARDS. This work aims to provide the physical and chemical principles explaining both serum inactivation and inactivation reversal by hydrophilic polymers. This will help medical professionals understand the origins of ARDS and enhance efforts to identify biocompatible polymers to formulate replacement LS resistant to inactivation. From the materials side, this work will contribute to a fundamental understanding of competitive adsorption at liquid-vapor and solid-liquid interfaces. We have developed an in vitro model that approximates LS behavior in the alveoli during respiration using the compression/ expansion cycles of a Langmuir trough, with albumin as the prototypical surface active serum protein. Utilizing fluorescence and confocal microscopy, we have simultaneously visualized the competitive adsorption of LS and albumin to the air-water interface. Our results show healthy LS is inactivated by the addition of albumin; images of the interface show that an albumin film adsorbed to the interface creates a barrier to LS adsorption. This competitive adsorption of albumin to the alveolar air-liquid interface can be modeled as an energy barrier to LS adsorption, which can be quantitatively analyzed using a variation of the classical Smolukowski description of colloidal stability.

LS adsorption can be enhanced by the addition of polyethylene glycol (PEG) via a depletion attraction between the LS aggregates and the interface. The depletion attraction effectively pushes LS aggregates toward the interface due to increased polymer entropy induced by the elimination of the "excluded volumes" of the LS aggregates and the interface. Addition of ~1% wt. 10 kDa PEG is sufficient to reverse inactivation and restore LS to the interface. LS adsorption increases exponentially with PEG concentration as predicted by the Asakura and Oosawa model of depletion attraction. The PEG response is molecular weight dependent, with PEG 6-35 kDa yielding optimal inactivation reversal while the depletion attraction lacks sufficient range for smaller polymers. At the concentrations necessary for reversing inactivation, PEG greater than 35 kDa reaches overlap concentration resulting in decreasing performance and a breakdown of the simple Asakura and Oosawa model of depletion attraction. Fluorescence images detail the transition from an albumin covered interface to a LS covered interface during successful inhibition reversal. After LS breaks through the albumin film, the interface shows a coexistence of extended albumin and LS domains and finally the albumin domains forced from the interface at sufficiently high surface pressure.

Freeze-fracture transmission electron microscopy images show that PEG and albumin do not adsorb to the surfactant aggregates, nor do these macromolecules penetrate the interior water compartments of the surfactant aggregates. This results in an osmotic pressure difference that dehydrates the bilayer aggregates, causing a decrease in the bilayer spacing as shown by small angle x-ray scattering and an increase in the ordering of the bilayers as shown by freeze-fracture electron microscopy. Small angle x-ray diffraction shows that the relationship between the bilayer spacing and the imposed osmotic pressure for replacement LS is a screened electrostatic interaction with a Debye length consistent with the ionic strength of the solution.

Utilizing in situ grazing incidence x-ray diffraction (GIXD) and x-ray reflectivity (XR), we have examined the surface ordering of clinical replacement LS at the air-liquid interface and the effect of albumin and PEG. XR measurements confirm that albumin imposes a steric barrier to LS adsorption, inhibiting the LS characteristic GIXD peaks. In the LS free system, scattering experiments show no evidence of PEG surface ordering while LS on a PEG subphase shows only a subtle lateral condensation of the LS. However, the addition of PEG to albumin inhibited LS restores the LS characteristic XR and GIXD peaks and progressive cycling shows the LS replacing the albumin on the interface. These scattering results are consistent with fluorescence images of the interface which show a coexistence of ~1000 μm albumin and LS regions until all albumin regions are eventually expelled from the interface.

In addition to the PEG generated osmotic pressure, the bilayer perturbing properties of chitosan, a biocompatible cationic polysaccharide derived from deacetylated chitin can also enhance LS adsorption to the air-liquid interface. Recent results demonstrate that chitosan reverses albumin induced LS inhibition at significantly lower concentrations (0.01 mg/mL) than PEG (10 mg/mL) suggesting another promising therapy for ARDS. LS is ~70 wt % dipalmitoylphosphatidylcholine (DPPC) and freeze fracture transmission electron microscopy images of DPPC vesicles treated with chitosan show distinct morphological changes. Untreated DPPC vesicles are ~50 nm in diameter and uniformly distributed through the solution while a low concentration of chitosan (0.005 mg/mL) causes aggregation of vesicles. At higher chitosan concentrations (0.5 mg/mL), images show numerous ~50 nm vesicles entrapped inside a larger vesicle, similar to "vesosomes" used for drug delivery. These morphological changes induced by chitosan suggest that it enhances LS adsorption by facilitating the bilayer to monolayer transition at the interface. These results demonstrate the utility of applying classic materials science principles to competitive adsorption problems of biomedical interest. More broadly, this materials science approach will benefit a broad class of technological problems involving competitive adsorption to interfaces.

6av Engineering Protein Folding and Function Using Native Escherichia Coli Processes

Adam C. Fisher, Cornell University, 304 Olin Hall, Ithaca, NY 14853 and Matthew DeLisa, Chemical Engineering, Cornell University, 304 Olin Hall, Ithaca, NY 14853.

Since the widespread expanse of heterologous protein expression enabled by genetic cloning, the inability to produce active recombinant protein has been the main bottleneck in the biotechnology enterprise. Similarly, we have been severely hindered in our ability to obtain high-quality, purified protein in order to perform X-ray crystallography, NMR spectroscopy, or CD spectroscopy for structural biology. In addition, protein folding is an important prerequisite for the production of efficacious protein-based therapeutics and a crucial concern in the design and generation of de novo proteins. Despite such widespread importance, it is not possible to ad hoc identify the sequence determinants of physiochemical structure for a particular amino acid polymer. To overcome these limitations, our research has focused on the exploitation of native cellular processes in bacteria to engineer the folding properties of desired proteins or peptides through laboratory evolution. By extending and exploiting the intrinsic folding quality control capacities of the protein trafficking pathways in Escherichia coli we have developed a suite of genetic assays for isolating folding-enhanced proteins from recombinant libraries. These techniques have been used for: (1) antibody engineering, (2) fluorescent protein evolution, and (3) the isolation of difficult-to-fold recombinant proteins. Our progress suggests that protein engineering coupled with high-throughput approaches to screening or selecting from recombinant libraries should help to alleviate many of the confines associated with de novo protein design, structural biology, and biotechnology.

6aw Growth, Characterization, and Material Property Control of Silicon Carbide Thin Films for Micro- and Nanosystems

Christopher S. Roper, Chemical Engineering, University of California, Berkeley, 201 Gilman Hall, Berkeley, CA 94709

Silicon carbide is an emerging material for fabrication of micro- and nanosystems. Its high melting point, high hardness, wear resistance, and chemical resistance make it well suited for micro- and nanoscale sensors and actuators designed to operate in harsh environments. In addition, its high modulus to density ratio makes silicon carbide attractive for micro- and nanoscale resonators.

Deposition of polycrystalline silicon carbide thin films on 100 and 150mm diameter silicon wafers is performed at temperatures as low as 700°C in a low pressure chemical vapor deposition reactor from the precursors 1,3-disilabutane and dichlorosilane. By controlling wafer-boat geometry, highly uniform films can be deposited and the origin of non-uniform growth is elucidated.

Mechanical property tuning is achieved by controlling the precursor flow rates. Films with low residual stress and low strain gradient, optimal for released micro- and nanosystems, are achieved. Electrical resistivity is tuned by controlling the flow rate of a dopant precursor, ammonia, as well as by varying post-deposition annealing conditions. X-ray diffraction, transmission electron microscopy, X-ray photoelectron spectroscopy, and secondary ion mass spectroscopy are used to characterize the films. By correlating the variation in mechanical properties with film microstructure and elemental composition, the mechanisms behind the material property tuning can be explained.

These SiC thin films are implemented as wear-resistant coatings and high-performance structural layers in applications ranging from microsystems sensors to optical filters.

6ax Affinity Adsorption of Viruses

Caryn L. Heldt[1], Patrick V. Gurgel[2], Lee-Ann Jaykus[3], and Ruben Carbonell[1]. (1) Chemical and Biomolecular Engineering, North Carolina State University, Raleigh, NC 27695, (2) ProMetic Life Sciences, 8168 Montview Rd, MONT-ROYAL, QC H4P 2L7, Canada, (3) Department of Food Science, North Carolina State University, Raleigh, NC 27695

The capture of viruses from complex sample matrices is required by the FDA for therapeutics originating from human, animal, or cell-culture origins. This removal of viruses can be done by different processes, including nanofiltration, chemical deactivation, or less often by affinity chromatography. The most commonly used affinity ligand is antibodies that are specific to the target virus. Antibodies are difficult to produce, expensive, and require a tertiary structure to capture the target species. Small peptide sequences as short as three amino acids in length have been identified from a combinatorial library for the capture of porcine parvovirus (PPV). These peptides have been used in chromatographic columns for removal of PPV from complex mixtures containing 7.5% human blood plasma. This approach is a promising alternative to virus capture using antibodies because it eliminates the risk of viral contamination that may be found in antibody preparations. Furthermore, peptides are inexpensive to produce and more stable than antibodies. This stability allows the adsorption surface to be used repeatedly for its designed use.

This current research focuses on using small peptides for the removal of viruses from complex mixtures, but in the future, different biopolymers, including RNA riboswitches and other olignionucleotide constructs could be used to create specific recognition surfaces that capture viruses or other environmental toxins. These surfaces could then be used for the removal of viruses from process streams, concentration of viral vectors for gene therapy, or molecular surface detection in a sensor. Which use the discovered biopolymers are best suited for will be determined by the ease of elution that is needed for each application.

6ay From the Synthesis and Characterization of Electronic and Optical Oxide Nanostructured Materials towards Device Applications

Yuanbing Mao, Department of Chemical and Biomolecular Engineering, University of California at Los Angeles, Los Angeles, CA 90095

My overall research interests lie in the development of new nanostructured materials and devices of technological importance and scientific interest by employing liquid- and/or gas-phase materials synthetic techniques. The ability to synthesize and engineer materials at the nanometer length scale has become a powerful driving force in the global scientific and technological community. Successful manipulation of various nanostructured materials suggests two issues: (1) an increasingly important role for chemistry to play in the rational design of functional materials and devices, and (2) the ability to exert chemical control to levels of precision familiar to advanced engineering technology. As a well-trained inorganic materials chemist and chemical engineer, I plan to pursue research projects in nanoscale materials science and engineering. These projects will ultimately lead to the rational design of novel materials and their related devices that possess useful and advanced properties, partly due to their nanostructured features.

PhD Graduate Research: **Synthesis and assembly of transition metal oxide nanomaterials and investigation of their novel physical properties** (Advisor: Professor Stanislaus S. Wong, Department of Chemistry, State University of New York at Stony Brook). Except working as research assistant and teaching assistant at SUNY Stony Brook, I also had the opportunity to work at Brookhaven National Laboratory due to my advisor's joint appointments. The research experiences have resulted in more than fifteen (nine first author) publications in various prestigious journals, including *J. Am. Chem. Soc.*, *Adv. Mater.*, *J. Mater. Chem.*, and *J. Phys. Chem. B.*, in addition to one issued patent and three patent applications. Specifically, the projects were as follows:

i) The full potential of nanoscience and nanotechnologies will only be realized when nanomaterials could be (1) synthesized in large quantilities with reproducible size, shape, structure, crystallinity, and composition, and (2) prepared and assembled using simplistic, low-cost, and environmentally-friendly methodologies. Hence, I explored different synthetic approaches to synthesize various advanced transition metal oxide and fluoride nanomaterials. The synthetic methodologies include molten salt syntheses, template-directed approach, sol-gel process, and hydrothermal method. The oxide materials are ferroelectric ABO_3-type oxides, magnetic $Bi_2Fe_4O_9$ and α-Fe_2O_3, catalytic ABO_4-type oxides and TiO_2 as well as layered titanate oxides. Furthermore, a template-directed approach had been extended to fabricate alkaline earth metal fluoride and perovskite fluoride nanowires. These nanomaterials had been extensively characterized by spectroscopic, microscopic, diffraction and electrochemical techniques.

ii) Assembly of nanoscale components is a key step towards building functional devices, which are important for applications including nanoscale electronics and molecular sensing. In my graduate study, layered titanate and TiO_2 one-dimensional nanostructures and ABO_4-type nanorods were assembled into sea urchin-like aggregates and arrays, respectively, without sophisticated instruments.

iii) It would be conceptually easier to control desired target materials at the nanometer scale if one were able to start with nanoscale precursors. One exciting strategy is associated with the use of localized solid-state chemical transformations via the insertion, exchange, or deletion of individual atoms. As a model system to demonstrate this idea, I investigated the size- and shape-dependent morphological transformation of hydrogen titanate nanostructures to their anatase TiO_2 counterparts by a hydrothermal method. The photocatalytic properties of as-prepared titania nanostructures had also been studied.

To extend my expertise in synthesis, characterization, and assembly of nanomaterials to functional nanostructure-based devices, the knowledge and training in device fabrication I am currently pursuing during my postdoctoral research are crucial. At the same time, I am strengthening my engineering background.

Postdoctoral Research: **Fabrication of functional nano-based devices by utilizing as-synthesized metal oxide nanostructured materials** (Mentor: Professor Jane P. Chang, Department of Chemical and Biomolecular Engineering, University of California at Los Angeles). The following projects are mainly funded by NSF and FENA (Center on Functional Engineered and Nano Architectonics):

i) Fundamental and practical advances in ceramics, electronic materials, catalysts, optical and magnetic materials have enriched, and been enriched by, inorganic solid-state chemistry. For the advanced electronic, optoelectronic, and chemical sensing applications, metal oxide nanostructured materials (e.g.: rare-earth (RE) doped Y_2O_3 and $RE_2Zr_2O_7$) are prepared by liquid-phase and/or gas-phase synthetic routes.

ii) The connections between the synthesis of new materials, the structure of the materials, and the properties of the materials have always been emphasized in the field of materials science and engineering. Therefore, the micro-structure, morphology, composition, physical properties and formation process of the as-synthesized metal oxide nanostructures are investigated by various spectroscopic and microscopic techniques. Some of these aspects have been fulfilled through collaborations with scientists at UCLA, University of Texas at Austin, and Sandia National Laboratories. In addition, synchrotron techniques (EXAFS, XRD and UPS) have been employed at Stanford Synchrotron Radiation Laboratory.

iii) To validate the correspondingly improved device functionality due to the reduced dimension of these materials, simple and novel nanostructure-based device structures (e.g.: nano-floating gate memory, optical switch, and photodetector) are fabricated upon the as-prepared oxide nanomaterials. These devices are fabricated by using the equipments at the Nanoelectronics Research Facility of UCLA.

Research Plans: To be an assistant professor, I plan to continue to perform research on the synthesis and functionalization of nanomaterials and the fabrication of relevant nano-based devices. The major objectives are to develop simple and novel synthetic methodologies for nanosized multifunctional complex oxide materials (e.g., multiferroic, mixed ionic-electronic conducting and smart electroceramic perovskites), to modify the surface of these nanomaterials and prepare nanocomposites for processability enhancement, and to fabricate novel memory devices, fuel cell electrodes and optical amplifiers by using the as-prepared and functionalized nanomaterials. To work these projects out successfully and efficiently, critical and independent thinking, hard working, and excellent collaboration are three main components of my professional belief.

Teaching: I not only worked as a teaching assistant. Very recently, I also have had the opportunity to both prepare lecture notes and teach classes at UCLA for the course of Surface and Interface Engineering. Through the multidiscipline research and academic training, I am able to teach some of the chemical engineering core courses, on both undergraduate and graduate levels, such as Transport Phenomenon, Thermodynamics, Separation Processes, etc. I truly enjoy teaching and also would like to create an introductory class that focuses on nanoscience and nanotechnology to cover synthetic techniques for nanomaterials, their characterization tools, nano-based devices, challenges, and possible solutions.

6az Fundamental Mechanisms of Biomaterial Interfaces: An Integrated Simulation and Experimental Approach

Jason C. Hower and Shaoyi Jiang. University of Washington, Chemical Engineering, Seattle, WA 98195

The success or failure of biomaterials is decided at the interface between the foreign material and the tissue or contacting body fluid. Therefore, it is vital to control the properties of the material and the interactions occurring at this complex interface. However, the biointerface can be difficult to study as the surface of interest is buried in a solution of potentially unstable and variable chemistry that directly impacts the very nature of the interactions possible. My research focuses on studying the physical and thermodynamic mechanisms of processes at the biological interface using an integrated approach of molecular simulations and fundamental experimental techniques.

Motivated primarily by the need for robust and stable surface coatings that prevent non-specific protein adsorption and cell adhesion for the marine, sensing, and biomedical industries, my research has linked interfacial properties like surface hydration structure and dynamics to repulsive forces generated on probe proteins. My research has also investigated calculating the change in free energy during protein adsorption from simulations as a predictor of surface non-fouling ability.

My future research goals are to apply the integrated simulation and experimental approach to elucidate molecular-level mechanism of biological interfacial phenomena. Significant amounts of biological interfacial research focuses on characterization and description and offers little insight into the fundamental mechanisms. However, by pairing molecular simulations and basic experimentation techniques, the mechanisms of interactions can be discovered which would lead to rational material design and property-targeted progress.

6ba Integrating Multiscale Models, Experiments, Dynamics and Control: Applications in Energy Generation and Systems Biology

Vinay Prasad, Dept. of Chemical Engineering, University of Delaware, 150 Academy Street, Newark, DE 19716

Catalytic chemical systems are truly multiscale in nature, and have the additional complexity of being composed of heterogeneous materials. Biological systems are also multiscale, and are often of larger dimension. The complexity and interconnected nature of these systems means that modeling purely from first principles is not practical, and optimally designed experimentation is required to capture and predict the behavior of these systems. As with most other multiscale systems, experiments provide information about variables at the coarser or macro scales, while many important variables that are to be inferred and controlled operate at the micro and nano scales. A hierarchy of models that connect across length and time scales is necessary to effectively understand the behavior of these systems. The multiscale models are used to generate optimal experimental designs (input sequences and operating conditions), and the experiments are used to identify and estimate the parameters of the multiscale models.

My current research focus is the optimal model-based design of experiments for catalytic systems, specifically those used for energy generation (e.g., ammonia decomposition for hydrogen production). The energetics of the components of a catalytic reaction mechanism are obtained from first principles quantum calculations (density functional theory) or semi-empirical techniques (UBI-QEP). The effects of catalyst structure are captured using kinetic Monte Carlo models, and microkinetic models are used at the macroscale to connect with experiments. The multiscale models are used with a variety of model based experimental design techniques to identify the maximum number of system parameters. In particular, dynamic information from optimally designed transient experiments is used to maximize system identifiability, and to design and control input sequences and operating conditions.

Another area of my current research is studying the effect of variations at the genetic level on the phenotypic level. Once again, models are used in a hierarchy – stochastic models for genetic transcription, continuum metabolic models and simple cell growth and division models are used along with experimental information at the cellular and protein distribution level. The models are used to understand the mechanisms of regulation in the GAL genetic switch in saccharomyces cerevisae, and the contributions of intrinsic and extrinsic noise to the phenotypic behavior of the system.

The ultimate goal is the effective design and control of these systems based on a combination of multiscale models and experiments – this translates to selecting optimal catalysts and operating conditions (to maximize selectivity and conversion, for example) in catalytic systems. In systems biology, it means controlling the phenotypic response by manipulations at the molecular and genetic level. My future research will focus on attempting to bridge the materials gap and understand, predict and improve the behavior and performance of real catalysts through modeling and experimentation. In systems biology, I will focus my attention on the elucidation of the links between intracellular events (e.

g., signaling networks implicated in cancer) and the system or population-level behavior of cells in a tissue or organ. Specifically, I am to study the evolution of cancer progression, which runs the gamut from intracellular events including mutations to tissue invasion and metastasis. Another area of focus is the discrimination between molecular and genetic level mechanisms from population distribution data, which is the normal type of experimental data available at the protein and cellular level. This is a difficult problem for which no standard methods exist at present, but is of enormous importance in systems biology.

6bb Dynamics of Colloidal Dispersions at Equilibrium and Under Flow

Samartha G. Anekal, Chemical Engineering, University of Wisconsin-Madison, Madison, WI 53706

Understanding dynamics of colloidal dispersions is important for several applications ranging from paints to building templates for photonic bandgap materials. The general theme of my research is using simulations to investigate dynamic properties of colloidal dispersions under equilibrium and non-equilibrium conditions.

During my PhD (under Dr. Michael Bevan, Chemical Engineering, Texas A&M University), I used Stokesian Dynamics simulations to study the effect of inter-particle forces on the dynamics of colloidal dispersions. Specific examples include:

1. Quantitatively explaining the observation of anomalous attraction between like-charged colloids. 2. Developing a new criterion and rheological definition for percolation in attractive colloidal fluids, and 3. Validating the use of analytical expressions for quantifying self-diffusion in interfacial colloidal fluids.

In my current research as a post-doc (under Dr. Michael Graham, Chemical Engineering, University of Wisconsin-Madison), I am investigating dynamic properties of colloidal dispersions under flow. I am currently involved in the development of an O(N) method for calculating hydrodynamic interactions of particles in unbounded and confined geometries. This simulation method is being used to investigate:

1. Shear-induced diffusion of particles for enhancing mixing in microfluidic devices. 2. Effect of polymers on distribution of red blood cells under channel flow.

In addition to simulations, I am also currently involved in microfluidic experiments for studying these problems.

6bc "Nanoions": Fundamental Properties and Applications of Charged Nanoparticles

Kyle J. M. Bishop, Chemical & Biological Engineering, Northwestern University, 2145 Sheridan Rd/TECH E136, Evanston, IL 60208 and Bartosz A. Grzybowski, Chemical Engineering, Northwestern University, 2145 Sheridan Rd/TECH E136, Evanston, IL 60208.

What happens when solutions containing objects bearing charges of opposite polarities are mixed? It depends on the scale and is thought to be fundamentally different in the molecular and the colloidal regimes. Whereas oppositely charged microparticles precipitate continuously over a wide range of relative particle concentrations, mixtures of oppositely-charged nanoparticles (NPs) precipitate sharply only at the point of NP electroneutrality. This behavior – reminiscent of threshold precipitation of inorganic ions – is specific to the nanoscale and is due to the formation of like-charged, "core-and-shell" clusters stabilized in solution by mutual electrostatic repulsions. The formation of these charged clusters is rationalized by deriving the electrostatic interaction potentials between NPs accounting for the adsorption/desorption of counterions from the NPs' surface. Interestingly, this model predicts that the attractive energy between oppositely charged NPs at contact is nearly twice the repulsive energy of like-charged NPs at the same distance. Monte Carlo simulations employing these potentials confirm the formation of "core-and-shell" aggregates and sharp precipitation at electroneutrality. Finally, I show how nanoparticle titrations based on these phenomena provide a uniquely accurate tool for determining charges tethered onto nanoscopic objects and for studying the thermodynamics of surface reactions at the nanoscale.

6bd Mechanistic Study of Methanol Synthesis Via CO2 Hydrogenation on Cu (111)

Lars C. Grabow, Chemical and Biological Engineering, University of Wisconsin-Madison, 1415 Engineering Drive, Madison, WI 53706

In light of the continuous rise of atmospheric CO_2 concentrations, the utilization of CO_2 has become an important global issue with many barriers such as CO_2 capture, purification, separation, energy requirements for chemical CO_2 fixation, economical considerations and many others. In this work we focus on the catalytic methanol synthesis from CO_2/H_2 mixtures as one of many possibilities for CO_2 utilization. Despite the fact that methanol is produced on industrial scale from syngas ($CO/CO_2/H_2$) at elevated pressures and temperatures over Cu/ZnO-based catalysts, the reaction mechanism is not well understood. We use periodic, self-consistent density functional theory (DFT) calculations to characterize a large number of well-known and novel surface intermediates on the Cu(111) surface in terms of their binding structures and energetics, and vibrational frequencies. Further, we use the nudged elastic band method to find transitions states for a large number of possible elementary steps occurring during methanol synthesis. The DFT results are subsequently used to derive all necessary parameters in order to develop a comprehensive microkinetic model that allows us to investigate preferred reaction pathways and relevance of surface intermediates and oxygenated byproducts under realistic reaction conditions. The first principles derived parameters for the microkinetic model (enthalpies, entropies, activation barriers, pre-exponential factors) were adjusted to experimental data where necessary. The interplay between theory, experiment and microkineitc modeling of rate data allows for the identification of major routes/intermediates to MeOH synthesis from CO_2 hydrogenation.

6be System Engineering: Applications for Space Missions, Energy Analysis and Healthcare

Selen Aydogan-Cremaschi, e-Enterprise Center, Discovery Park, Purdue University, 1201 West State Street, West Lafayette, IN 47907

The development of new products and processes is one of the most important aspects that industries need to address, in particular during their growth and expansion. This becomes a difficult task due to uncertainties in raw material availabilities, product demand profiles, energy availability, environmental sustainability, and safety. At the early stages of the design, the system is also subject to a myriad of uncertainties, such as estimates of the parameters (recovery rates, process efficiencies, or process reliabilities), because most of the technologies involved are still under development. My Ph.D. work focuses on developing novel algorithms to provide input into decision making on design under uncertainty, for a challenging application, i.e., for designing a life-support-system for survival and well-being of the crew during space missions. The algorithms utilize a simulation-based optimization (SIMOPT) approach that combines a stochastic discrete-event simulation with a deterministic mathematical programming approach to generate multiple, unique realizations of the controlled evolution of the system. The realizations are then analyzed using time-series data-mining techniques and statistical tools to determine the necessary technologies, their deployment schedules (retrofitting) and processing capacities, and the necessary inventory levels to support crew life and activities for the specified mission duration.

6bf Engineering Model Catalysts towards Efficient Energy Conversion

Weiwei Gao[1], Eric I. Altman[2], and Cynthia M. Friend[1]. (1) Chemistry and Chemical Biology, Harvard University, Cambridge, MA 02138, (2) Chemical Eng, Yale University, New Haven, CT 06511

Concerning global energy sustainability, model catalysts bridge fundamental understandings of catalytic reactions and the improvement of the efficiency of industrial energetic processes. This poster summarizes a research background in this regard and the outline of prospective continuing investigations. The following aspects are emphasized to achieve this goal:

1) Novel catalytic processes via clean and efficient reaction routes;

Our effort has been focused on gold catalysts, which show extraordinary low temperature catalytic capability including CO oxidation, olefin partial oxidation and functionalization. We also discovered halogen promotion on gold which dramatically reduces olefin partial oxidation temperature and improves selectivity.

2) Thin film growth and characterizations of model transition metal catalysts;

We have been pioneering molecular beam epitaxy (MBE) technique to grow transition metal oxide films as platform for model catalysts (oxides and oxide supported materials). MBE provides accurate thickness and composition control and oxides that are yet commercially available. These advantages open a wide field for oxides research.

3) New catalysts towards clean and efficient energy conversion;

Using MBE we have been able to produce high quality anatase TiO_2 thin films. Our effort has been focused on tuning various dopants and doping concentrations. Through the interplay between dopants and anatase lattice, our goal is to achieve low band-gap and stable TiO_2-based catalysts towards efficient photocatalysis.

4) Interrelationship between materials structure and chemical reactivity and selectivity;

Towards understanding the origin of reactivity and selectivity of gold catalysts, we have been studying nanoscale restructuring and interaction between oxygen and chlorine on gold. Halogen induced oxygen dispersal and redistribution have been established to explain halogen promotion on olefin partial oxidation.

Detailed research proposal is within the above scope. More information available upon request.

6bg Simulations of Polymer Self-Assembly Using Field Theoretic Techniques

Erin Lennon, Chemical Engineering, University of California, Santa Barbara, Engr. II, Room 3357, UCSB, Santa Barbara, CA 93106-5080 and Glenn H. Fredrickson, Chemical Engineering and Materials, University of California, Santa Barbara, Engr. II, Room 3357, UCSB, Santa Barbara, CA 93106-5080.

Modeling the self-assembly of polymeric systems is vital to understanding their properties and creating novel materials. For example, by varying the molecular weight and composition of various block copolymers, one can form a host of structures that exhibit features on multiple scales---domains on the micron scale and interfaces of only a few nanometers.

One method that has proven useful for modeling these systems is a field-theoretic approach. This technique has been used successfully to study many models in which an additional mean-field approximation is imposed. However, to study fluctuation-dependent phenomena, one must use the full field theory which has thus far proven computationally intractable. The work presented here demonstrates how the use of advanced numerical techniques and appropriate thermodynamic integration methods can allow the practical study of complex polymeric systems.

6bh Directed Evolution of Enzymes and Biosynthetic Pathways

Tyler Johannes and Huimin Zhao. Chemical and Biomolecular Engineering, University of Illinois, 600 S Mathews, Box C-3 RAL 221 MC 712, Urbana, IL 61801

Biocatalysts are increasingly used in industrial synthetic chemistry, particularly in cases in which chemical routes are difficult to implement. For instance, NAD(P)H-dependent oxidoreductases are valuable tools for synthesis of chiral compounds. However, the expensive cost of the cofactors requires *in situ* cofactor regeneration for preparative applications. To help solve this problem, we have developed an enzymatic system based on phosphite dehydrogenase (PTDH) to regenerate the reduced nicotinamide cofactors NADH and NADPH. We used directed evolution to address one of the main limitations with the wild-type PTDH enzyme, its low stability. Using random mutagenesis and high throughput screening we were able to create a mutant with a half-life of thermal inactivation at 45°C that is >7000-fold greater than that of the parent enzyme. Another area in which biocatalysts are increasingly being used is in the production of small secondary metabolites, particularly antibiotics. Antibiotics containing P-C bonds (phosphonates and phoshinates) are a small but growing class of compounds with important biological activities. Phosphonates have been shown to display a wide range of antibacterial, antiviral, pesticidal, and anti-cancer properties. The phosphonic acid antibiotics fosmidomycin and FR-900098 represent a new class of antimalarial compounds that can be used to inhibit the nonmevalonate pathway for isoprenoid biosynthesis in the malaria-causing parasite *Plasmodium falciparum*. Although no biosynthetic pathway for fosmidomycin has been elucidated as of yet, the biosynthetic pathway for FR-900098 has been recently cloned from *Streptomyces rubellomurinus* and heterologously expressed in *Streptomyces lividans*. We have worked to develop an *E. coli* strain capable of producing these antimalarial drugs. Metabolic engineering strategies aimed at optimizing gene expression and increasing cellular substrate concentrations are also being explored. In parallel to our work in *E. coli*, we also explore the possibility of improving the production of FR-900098 in *Streptomyces lividans* using metabolic engineering.

6bi Developing Predictive Statistical Models to Understand the Dynamics of Inflammatory Cell Signals

Arthur C. Goldsipe[1], Christopher W. Espelin[2], Peter K. Sorger[3], and Douglas A. Lauffenburger[1]. (1) Biological Engineering, Massachusetts Institute of Technology, 77 Massachusetts Ave. 56-353, Cambridge, MA 02139, (2) Pfizer Research Technology Center, Cambridge, MA 02139, (3) Systems Biology, Harvard Medical School, 200 Longwood Ave, WAB 438, Boston, MA 02115

Inflammation involves coordinated communication and response by the vascular and immune systems. When properly regulated, inflammation assists an organism in dealing with injury or pathogenic invasion. Unregulated inflammation is also associated with diseases such as rheumatoid arthritis, which is an autoimmune disorder in which the joints become inflamed. Although the details of signaling cascades of inflammation are not completely known, many of the key players have been identified. In particular, both c-Jun N-terminal kinases (JNKs) and p38 are mitogen-activated protein kinases (MAPKs) with critical roles in the mammalian response to stress.

In order to better understand the signaling events surrounding the inflammatory response, high-throughput techniques have been used to measure the dynamic phospho-protein levels and cytokine release from U937 cells treated with a various inflammatory stimuli. The U937 cell line serves as a model for monocytes and macrophages, both of which are present in inflamed tissue. The cells were treated with lipopolysaccharide (LPS) and either pro-inflammatory cytokines or small-molecule inhibitors.

The experimental data have been modeled by a class of predictive statistical techniques known as partial least-squares regression (PLSR). For such biological data sets, PLSR offers several advantages, including the ability to handle highly collinear data, robustness to missing data and biological noise, and the ability to summarize complex data. Because of the underlying experimental data is multidimensional in nature, multi-block and multi-way variations of PLSR were also investigated. Various applications of these PLSR models will be illustrated, such as the prediction of cytokine release, the identification of the most informative experimental measurements, and the elucidation of the potential for crosstalk and autocrine feedback.

More generally, high-throughput biological data are being generated at an increasing rate. Therefore, it is also important to automate and simplify repetitive and error-prone aspects of data analysis and model development. To this end, we are developing modules for PLSR analysis that will be incorporated into a general data workflow for systems biology.

6bj Field-Mediated Control of Materials Synthesis and Biological Applications

William D. Ristenpart, Division of Engineering and Applied Science, Harvard University, 402 Pierce Hall, 29 Oxford St., Cambridge, MA 02138

My research focuses on the use of externally applied fields for guiding colloidal particles, fluids, chemical species and biological cells into configurations useful for materials synthesis and biological applications. Gradients in electric potential, temperature, concentration, and shear stress are examples of driving forces that are externally tunable. Frequently these driving forces have non-intuitive consequences on systems with boundaries which themselves contain gradients either imposed or modulated by the external field. The goal of my research is to elucidate the physics underlying the interactions between externally applied and internally mediated fields in technologically relevant systems in materials science and bioengineering, using a combination of experiments, theory and numerical modeling. This poster summarizes six ongoing research projects:

[1] Electrohydrodynamic Aggregation and Segregation of Unilamellar Vesicles

[2] Enzymatic Reactions in Microfluidic Devices

[3] Influence of Shear Stress on ATP Release from Red Blood Cells

[4] Effects of Substrate Thermal Conductivity on Flow in Evaporating Droplets

[5] Non-coalescence of Oppositely Charged Droplets

[6] Electrically Driven Flow on Inhomogeneous Electrodes

...

[1] Electrohydrodynamic Aggregation and Size Segregation of Unilamellar Vesicles (Collaborators: S. Lecuyer, O. Vincent & H.A. Stone)

We demonstrate that electrohydrodynamic (EHD) flow induces polydisperse suspensions of unilamellar vesicles to aggregate laterally along electrodes. Upon aggregation, smaller vesicles move underneath the larger vesicles, gradually lifting the larger vesicles off of the electrode entirely. We demonstrate that this phenomenon provides a method for gently separating giant unilamellar vesicles from polydisperse suspensions.

...

[2] Enzymatic Reactions in Microfluidic Devices (Collaborators: J. Wan & H.A. Stone)

We establish simple scaling laws for enzymatic reactions in microfluidic devices, and we demonstrate that kinetic parameters obtained conventionally using multiple stop-flow experiments may instead be extracted from a single microfluidic experiment. Introduction of an enzyme and substrate species in different arms of a Y-shaped channel allows the two species to diffuse across parallel streamlines and to begin reacting. For reactions that follow standard Michaelis-Menten kinetics, the power law takes the form $P \sim (V_{max} / K_m)\, x^{5/2}$. Numerical simulations and experiments using the reaction between luciferase and ATP as a model system are both shown to accord with the model.

...

[3] Influence of Shear Stress on ATP Release from Red Blood Cells (Collaborators: J. Wan, C. Best, R. Horton, G. Guidotti, E. H. Abraham & H. A. Stone)

To understand the influence of shear stress on the kinetics of adenosine triphosphate (ATP) release from red blood cells (RBCs), we mimic arterial constrictions using a series of channels in microfluidic devices. We show that the amount of released ATP increases roughly exponentially with the magnitude of the shear stress, but that there is a critical duration of stress (>2 ms) required for RBCs to release significant amounts of ATP. The results suggest that RBCs are sensitive not only to the diameter of arterial constrictions but to their length, an effect with important physiological and medical implications.

...

[4] Effects of Substrate Thermal Conductivity on Flow in Evaporating Droplets (Collaborators: P.G. Kim, C. Domingues, J. Wan & H. A Stone)

Non-uniform evaporation from sessile droplets gives rise to radial convection within the drop, yielding the well-known coffee stain pattern. Any non-uniformity in the evaporative flux, however, also induces a gradient in temperature and consequently a gradient in surface tension, yielding a Marangoni flow. Here we report the existence of a critical substrate thermal conductivity below which the direction of the Marangoni flow reverses, thereby affecting the particle deposition pattern.

...

[5] Non-coalescence of Oppositely Charged Droplets (Collaborators: A. Belmonte, F. Dollar, D. A. Weitz & H. A. Stone)

We demonstrate the existence of a critical electric field strength above which oppositely charged drops do not coalesce. Application of an external electric field causes appropriately positioned and oppositely charged drops to migrate toward one another. Upon contact, surface tension acts to pull the drops together. For low applied field strengths, the oppositely charged drops coalesce, but at higher field strengths the drops are repelled from one another after contact. Qualitatively, the drops appear to 'bounce' off one another. We derive a critical field strength for bouncing based on a competition between the time scales for charge transfer and the action of surface tension. The results have broad implications for applications where charged drops are manipulated by electric fields, including microfluidics, atmospheric science, and electrospray ionization.

...

[6] Electrically Driven Flow on Inhomogeneous Electrodes (Collaborators: P. Jiang, M. Slowik, I. A. Aksay, D. A. Saville)

Current density inhomogeneities on electrodes – of physical, chemical, or optical origin – induce long-range electrohydrodynamic (EHD) fluid motion directed toward the regions of higher current density. Here we analyze the flow and its implications for the orderly arrangement of colloidal particles on patterned electrodes. A scaling analysis indicates that the EHD flow velocity is proportional to the product of the applied voltage and the difference in current density between adjacent regions on the electrode. Exact analytical solutions for the EHD streamlines are derived for the case of a spatially oscillatory perturbation in current density along the electrode. We demonstrate that proper placement of scratches on an electrode yields arbitrary patterns of colloidal particles.

6bk Micro/nanofluidic Devices for Sensing and Reaction Engineering

Adarsh D. Radadia, Chemical and Biomolecular Engineering, University of Illinois, 600 S. Mathews St., Urbana, IL 61801

Following two topics summarize the works accomplished in my Ph.D research.

(I) Separation of organophosphate species on a microfabricated GC column

My main contribution to the microGC research was to provide design rules on how to build microGC columns including effects of channel configurations and turn geometries, passivation and coating methods. The research work presented at TRANSDUCERS '07 investigated dispersion effects applicable to lab-on-chip devices involving separation channels.

(II) Three Dimensional Microfluidic Laminates (3DMFL)

Dry Film Resist (DFR) has been used previously for making fluid handling microcomponents. My contribution is to develop a method to use DFR for fabricating devices with multilayer of microfluidic channels. A split-split-join type serial lamination mixer was demonstrated at MEMS'07.

6bl Design of Multifunctional Polymer-Polymer Nanocomposites

Aflal M. Rahmathullah and Giuseppe R. Palmese. Department of Chemical and Biological Engineering, Drexel University, 3141 Chestnut St., Bossone 514, Philadelphia, PA 19104

Functional materials, such as hydrogels and ionomers, are finding wide ranging applications such as in chemical valves, drug delivery systems, self healing materials, proton exchange membranes and artificial muscles. A setback with these materials is their weak mechanical properties, and while most of these functional materials are hydrophilic in nature, structurally stable materials are most often hydrophobic. This apparent mismatch in structural properties is interesting for the fact that it may be possible to bring into intimate contact materials that are thermodynamically dissimilar while retaining the properties that make them intriguing. An approach to carry this out is the stabilization of these functional moieties within a hydrophobic network, resulting in the formation of stable, functional composites. My work will discuss how this strategy of combining such hydrophobic (non-ionic) and hydrophilic (ionic) polymers can range in a wide variety of useful applications. The goals have been on developing such polymer-polymer nanocomposites for applications in proton exchance membranes, self-healing materials and permselective barrier membranes. Results have shown that it is possible to develop strategies to rationally design novel structures based on composites of hydrophobic and hydrophilic polymers and these will be presented during the session as well as discussions on thoughts for future research.

6bm Nanostructured Oxide for Energy Storage and Conversion

Donghai Wang, Pacific Northwest National Laboratory, Battelle/PNNL, MS:K3-59, Richland, WA 99354

My primary research interests consist of using advance materials synthesis techniques in the development of new nanostructured oxide for energy storage and conversion application such as battery, capacitors, photovoltaic and catalysis. The focus of my graduate research at Tulane University was synthesis, characterization and applications of macroscopic hierarchical metal or semiconductor nanowire networks. I demonstrated a generalized and efficient approach to synthesize macroscopic hierarchical nanowire networks with well-controlled diameter and mesoscale arrangement, which will be of great interest for sensor, photovoltaic, and other applications. Metal nanowires self support to form networks with high electrochemical active surface area, which are further applied in enzymatic glucose sensing. Semiconductor CdSe nanowire networks show tunable optical properties dependent on nanowire diameter and have been demonstrated as a good electron acceptor in CdSe nanowire network/ polymer photovoltaic devices. This research also describes self-assembly behavior of composite mesostructures under physical confined environment. Currently I am working on novel nanostructured metal oxide for catalysis application as a post-doctoral researcher at Pacific Northwest National Laboratory. I have demonstrated an efficient generalized strategy combing self-assembly technique and nanocrystal synthesis towards mesoporous crystalline metal oxide composed of well-defined metal oxide nanocrystals. The novel materials show extraordinary performance with high activity and stability as gold nanocatalyst supports. Novel nanostructured metal oxide with well-controlled morphology has been studied as electrode in Li ion battery showsing excellent capacity retention and high rate capability on cycling. My future research will focus in the following areas: 1) Develop methods in large scale solution synthesis of one-dimensional oxide nanostructures. 2) Novel assembly technique for the nanomaterials in device fabrication of battery, capacitors, photovoltaics. 3) Synthesis of nanocrystals with well-defined shape and its application in model catalysis involving electrocatalysis, hydrogen generation and Au nanoparticle catalysis.

6bn Measuring and Modeling Fundamental Parameters from Gas Phase Electrophoresis

Leonard F. Pease III, Process Measurements Division, National Institute of Standards and Technology, 100 Bureau Drive, Gaithersburg, MD 20899

Here we present a rapid and quantitative means to measure particle/cluster size distributions, from which we determine fundamental properties of molecular/particle interactions via analytical modeling. In electrospray-differential mobility analysis (ES-DMA) the balance between electrical and drag forces controls the trajectory of particles or clusters that were charged during the electrospray process. Stepping an electrode centered within an annular analysis chamber through a series of voltages guides particles or clusters with a specific size, which enter at the outside of the annulus, into a collection slit in the electrode. Collected particles then pass into a condensation particle counter where they are enumerated via light scattering to provide a size distribution. Our contributions include extending the range of samples examined by this method and providing analytical models to determine fundamental parameters. Highlights include determining the surface coverage and Kuhn length of DNA coated nanoparticles, deriving the surface charge and packing structure of gold colloids from aggregation kinetics, measuring the concentration of viral particles necessary to quantify the viral load for the biotechnology community, and performing biochemical assays. The measurements and modeling allow us to address both fundamental science questions and important applied challenges.

6bo Biomedical and Energy Applications of Lipids

Liangfang Zhang[1], Steve Granick[2], and Robert Langer[1]. (1) Department of Chemical Engineering, Massachusetts Institute of Technology, 77 Massachusetts Ave., E25-342, Cambridge, MA 02139, (2) Departments of Materials Science and Engineering, Chemical Engineering, Chemistry, and Physics, University of Illinois at Urbana-Champaign, 1304 W. Green St., Urbana, 61801

The main goal of this research program is to develop lipid-based biomaterials for healthcare and other medical applications and to understand the fundamental sciences underlying this problem. In addition, I am also interested in exploring energy-related applications of lipids. By combining the application prospects of lipid systems with the fundamental sciences behind them, I am expecting to exploit their synergies.

The significance in the biomedical field is that lipid assemblies, liposomes in particular, have been widely used as drug delivery carriers because of their ability to encapsulate drugs with high loading efficiency, to prolong drug circulation half-life, and to release drugs at a sustained rate or in an environmentally responsive manner. In addition to 11 liposomal drugs that have been approved for clinical use, numerous other liposome-based therapeutics are currently under various stages of clinical and preclinical development. I have many ideas to extend capabilities in this direction. The significance in the burgeoning field of bioenergy is that biomass such as soybean, palm, and algae and living organisms provides potential sources of lipids that could yield up to 9 kcal-gram-1 of energy. By converting lipids into fatty acid monoalkyl esters, I have ideas how to produce renewable biodiesel fuels from this route.

These research plans differ significantly from those of my postdoctoral and Ph.D. research advisors. Robert Langer at MIT has taught me much about biomedical research, but without the accompanying fundamental studies that I plan to include in my own research. Steve Granick at Illinois has taught me much about fundamental studies, but he does not carry them on to the next step of engineering applications, as I plan to do in my own research.

My current postdoctoral research at MIT has already led to 6 patent disclosures as well as 5 published or submitted papers. It focuses on designing lipid-related platform bionanotechnologies for developing nanocarriers for drug delivery in a targeted manner, which includes atherosclerosis therapy, cancer therapy and vaccine delivery. My Ph.D. research at Illinois led to 10 published and 3 submitted papers. At that time I extensively studied lipids, lipid membranes and their interactions with polymers, proteins, and nanoparticles from the perspective of fundamental sciences.

6bp Micro/nano/molecular Engineering of Soft Materials for Drug/gene Delivery and Biosensing

Jingjiao Guan, Ohio State University, 140 west 19th. Ave., Columbus, OH 43210 and L. James Lee, Chemical and Biomolecular Engineering, The Ohio State University, 125A Koffolt, 140 West 19th, Columbus, OH 43210.

My past research focuses on the development of novel polymer micro/nanofabrication and DNA manipulation techniques for biomedical applications. Specifically, I have developed a series of polymer microfabrication methods and fabricated various drug delivery devices with structures and functions that are difficult or impossible for conventional technology. I have also developed a novel strategy for stretching and patterning DNA chains into well-defined arrays with potential applications ranging from gene chip to nanowire/nanochannel-based biosensing. In addition, I have been working on micro/nanofluidic production of highly engineered multi-functional nanodevices for the delivery of DNA/RNA therapeutics. As an assistant professor, I will seek to establish an interdisciplinary research program that specializes in multi-scale engineering of soft materials for biomedical applications. In particular, my future research will consist of two thrust areas: 1) developing novel biosensing technologies such as nanofluidic devices for DNA sensing/separation and DNA microarrays based on stretched DNA, and 2) developing multi-functional nanodevices for the efficient and safe delivery of therapeutic DNA/RNA in anticancer therapy.

References:

1. J. Guan, L. J. Lee, Generating Highly Ordered DNA Nanostrand Array, Proceedings of National Academy of Science, U. S. A. (2005) 102: 18321.

2. J. Guan, B. Yu, L. J. Lee, Forming Highly Ordered Arrays of Functionalized Polymer Nanowires by Dewetting on Micropillars, Advanced Materials (2007) 19: 1212.

3. J. Guan, H. He, D. Hansford, L. J. Lee, Fabrication of Particulate Reservoir-Containing, Capsule-Like, and Self-Foldable Polymer Microstructures for Drug Delivery, Small (2007) 3: 412.

4. J. Guan, N. Ferrell, B. Yu, D. J. Hansford, L. J. Lee, Simultaneous Generation of Hybrid Arrays of Micro/Nanoparticles and Nanowires by Dewetting on Micropillars, Soft Matter (2007) published online, DOI: 10.1039/b709910j.

5. H. He, J. Guan, L. J. Lee, An Oral Delivery Device Based on Self-Folding Hydrogels, Journal of Controlled Release (2006) 110: 339.

6. J. Guan, N. Ferrell, L. J. Lee, D. Hansford, Fabrication of Polymeric Microparticles for Drug Delivery by Soft Lithography, Biomaterials (2006) 27: 4034.

7. J. Guan, H. He, L. J. Lee, D. Hansford, Self-Folding of Hydrogel Three Dimensional Microstructures, Journal of Physical Chemistry B (2005) 109: 23134.

8. J. Guan, A. Chakrapani, D. Hansford, Polymer Microparticles Fabricated by Soft Lithography, Chemistry of Materials (2005) 17: 6227.

6bq High Surface Area Ocvd Deposited Pyrrole-Co-Thiophene-3-Acetic Acid Conducting Copolymer Films for Resistance-Based Sensing Applications

Sreeram Vaddiraju[1], Kris Senecal[2], and Karen K. Gleason[1]. (1) Department of Chemical Engineering, Massachusetts Institute of Technology, 25 Ames Street, Bldg. 66, Rm. 419, Cambridge, MA 02139, (2) Macromolecular Science Team, US Army Research, Development and Engineering Command, Natick, MA 01760

Sensing of biological analytes using conducting polymers not only requires the presence of a functionality on the surface, but also the availability of high surface. Towards this goal, uniform, highly conductive coatings of –COOH functionalized conducting polymer were obtained on spun-coated polyacrylonitrile fiber membranes using the oxidative chemical vapor deposition (oCVD) approach. This functionalization was achieved by co-polymerizing pyrrole with thiophene-3-acetic acid, using iron chloride ($FeCl_3$) as the oxidizing agent. Scanning electron microscopy of the obtained coatings showed that the conducting polymer coating is highly conformal to the surface of the fibers. Further, the presence of –COOH groups on the surface was confirmed by both FTIR and XPS analysis. Conductivities of 2-10 S/cm were observed in the obtained co-polymer films. The amount of –COOH functionality present on the surface was optimized by varying the ratio of thiophene-3-acetic acid to pyrrole in the feed mixture. In addition to spun-coated fiber membranes, conformality of the oCVD deposited conducting polymer coatings in trenches and porous membranes is also studied and will be discussed.

6br What Can Fundamental Research on Metal and Metal Oxide Surfaces Contribute to Solutions of Global Energy and Environmental Problems?

Ling Zhou, School of Engineering and Applied Sciences, Harvard University, 29 Oxford Street, Cambridge, MA 02138 and Robert J. Madix, Division of Engineering and Applied Sciences, Harvard University, Pierce Hall, Cambridge, MA 02138.

"Don't Forget Long-Term Fundamental Research in Energy" is highlighted in a recent Science perspective by Prof. Whitesides and Prof. Crabtree. When energy and environmental concerns have been at the forefront of global attention, fundamental research in these areas is extremely important, since understanding phenomena relevant to energy and environment can lead to new technologies with high energy efficiency and low environmental impact.

Many of the chemical processes in the production, transformation, and storage of energy, as well as for environmental protection, involve catalytic metals and metal oxides, e.g. the operation of fuel cells, the refining of crude oil to fuels, the gasification and liquefaction of coal, the conversion of solar energy, the storage of hydrogen, the oxidation of hydrocarbon and alcohol fuels, the cleanup of exhaust gas from internal combustion engines. In the development of these technologies, surface science studies on metals and metal oxides have played and will continue to play important roles. Such research seeks to understand chemical reactivity through studies of the interactions of atoms, molecules, and ions with surfaces, the making and breaking of chemical bonds on surfaces, and the energy transfer processes within and between molecules on surfaces. Ultimately, fundamental research on metal and metal oxide surfaces will lead to the development of such advances as efficient combustion systems with reduced emissions of pollutants, improved catalysts for clean and efficient production of fuels and chemicals, new multi-functional materials in energy devices, environmental remediation, and so on.

In my research career I have experience from surface chemistry to chemical engineering. I have focused on the partial oxidation of hydrocarbons on metal and metal oxide catalysts, such as n-butane selective oxidation to maleic anhydride on VPO catalysts, methanol oxidation on Cu model catalysts, and olefin epoxidation on Ag and Au model catalysts. I have also studied the structure of cerium oxide, a well-known functional material applied in numerous technologies relevant to energy and environment. Moreover, I am well versed in a wide range of related techniques from micro-reactor systems to ultra-high vacuum systems with which to realize the proposed research. My solid background in chemistry and physics equips me to pursue my research interests. As a faculty candidate I would also be very enthusiastic about designing and teaching courses on modern areas of chemistry and chemical engineering. My goal is to contribute to the education of future chemical engineering through research, teaching, and service to the community.

6bs Self-Assembly of Magnetorheological Fluids Confined in Microfluidic Devices

Ramin Haghgooie, MIT, 77 Massachusetts Ave. 66-053, Cambridge, MA 02139

Confinement of two-dimensional (2D) colloidal systems in the lateral directions can cause important changes in the properties of the system. Likewise, the relaxation of strictly 2D colloidal systems to quasi-2D confinement can induce drastic changes in structure. Of particular interest is the structure of self-assembled magnetorheological (MR) fluids confined in the thin-slit geometry. This system is of interest from both a fundamental science standpoint as a model for the effects of confinement on self-assembly as well as for practical applications indicated by the increasing exploitation of self-assembly for microfluidic applications. Self-assembled MR fluids have been used as structural components in microfluidic systems offering a fast, inexpensive alternative to traditional photolithographic techniques. The characteristic length-scales in microfluidic devices have continued to shrink down to the colloidal size, making studies of self-assembly under extreme confinement a timely undertaking. It has become essential to study these systems in order to enable the design of meaningful applications using such technology.

My doctoral thesis focused upon the self-assembly of MR fluids in confined geometries. In order to study this problem, I developed a versatile Brownian dynamics (BD) code capable of simulating MR colloids of any shape interacting in any general geometry. I used the BD code to study the self assembly of spherical colloids confined in two-dimensional (2D) channels and discovered a number of interesting phenomena that were subsequently published (1,2). In addition to the simulation work, I performed an extensive set of experiments on the self-assembly of MR colloids in 2D microfluidic channels and showed the first experimental observation of re-entrant melting as a function of confining geometry (3). In addition to the 2D studies, I investigated the factors controlling self-assembly in the thin-slit geometry. I elucidated the important physical phenomena affecting the self-assembly of dilute MR fluids in this common microfluidic geometry. I showed how the system transitions from 2D to 3D behavior as the confinement is relaxed from a monolayer to a channel of finite thickness (4,5).

During the course of my Ph.D. I also completed a Masters of Chemical Engineering Practice in which I gained invaluable industrial experience. Since graduating, I took a position as a Lecturer of Chemical and Biological Engineering at Tufts University and spent the 2006-2007 academic year teaching graduate level Transport Phenomena, graduate level Thermodynamics, two undergraduate laboratory courses, and a graduate level course in microfluidic technology. This experience has enabled me to sharpen my teaching skills and given me a great head-start in my development as an academic researcher and teacher. In February of 2007, I was awarded the MIT/MGH postdoctoral fellowship in translational research (2 awards out of more than 50 applicants) and I am currently a postdoctoral fellow in the BioMEMS Resource Center at Massachusetts General Hospital.

(1) Haghgooie, R. and Doyle, P. S.; "Structural analysis of a dipole system in two-dimensional

channels", *Phys. Rev. E*, **70**, 061408, (2004).

(2) Haghgooie, R. and Doyle, P. S.; "Structure and dynamics of repulsive magnetorheological colloids in two-dimensional channels", *Phys. Rev. E*, **72**, 011405, (2005).

(3) Haghgooie, R., Li, C., Doyle, P. S.; "Experimental study of structure and dynamics in a monolayer of paramagnetic colloids confined by parallel hard walls", *Langmuir*, **22**, 3601, (2006).

(4) Haghgooie, R. and Doyle, P. S.; "MR fluid structure in quasi-2D", *Europhys. Lett.*, **77**, 18002, (2007).

(5) Haghgooie, R. and Doyle, P. S.; "Transition from 2D to 3D behavior in the self-assembly of magnetorheological fluids confined in thin-slits", *Phys. Rev. E*, (in press).

Web Page: web.mit.edu/rhaghgoo/www

6bt Computational and Experimental Studies of Protein-Self Assembly with Applications in Nanotechnology and Medicine

Troy Cellmer, Laboratory of Chemical Physics, National Institutes of Health, 9000 Rockville Pike, Building 5, Room 106, Bethesda, MD 20892

My research will be focused on protein self-assembly with applications in biomedicine and nanotechnology. On a fundamental level, it will involve uncovering mechanisms of self-assembly, as well as elucidating the relationship between peptide sequence and the structure of the resulting assemblies. This information will then be exploited to develop new therapies for assembly-related diseases, as well as fabricate novel nanomaterials.

As a graduate student my primary focus was computational studies of multi-protein systems. As a post-doc, I have utilized fluorescence spectroscopy to study ultra-fast protein folding. My research proposal relies on the convergence of both approaches, experimental and computational, to improve our basic understanding of assembly-related processes, and develop new technologies.

6bu Probing the Molecular Mechanism of ATP Bioenergy Conversion: A Multifaceted Approach Combining Computational Modeling with Single Molecule Analysis

Jung-Chi Liao, Bioengineering, Stanford University, 345 Sheridan Ave #105, Palo Alto, CA 94306

Adenosine triphosphate (ATP) is a universal fuel for living systems. Many enzymes in key metabolic pathways hydrolyze or synthesize ATP. Therefore, it is important to understand the molecular mechanism of ATP reaction and its energy conversion. Furthermore, this understanding facilitates the optimization of metabolic engineering applications such as improving the efficiency of biofuel productions. I have adopted multifaceted approaches including molecular dynamics (MD), homology modeling, biochemistry, molecular biology, and single molecule experiments to decipher the molecular mechanism of ATP energy conversion.

To first characterize the free energy landscape of ATP, we have calculated its conformational equilibria in solution using MD augmented with umbrella sampling methods [1]. Comparing the free energy function with several crystallographic structures of ATP analogs, we were able to calculate the energy stored in the proteins when ATP binds.

We then applied a variety of computational techniques to study the bioenergy conversion for several different ATP hydrolysis/synthesis proteins, including myosin [2,3], helicases [4,5], ATP synthase [6], and transcription terminator [7]. Our results provided new insights and mechanistic views of these anabolic and catabolic processes, and furthermore suggested testable hypotheses for experimental studies. Several computational packages for biochemical and biophysical experiments were developed in these works to facilitate future studies of other enzymes [2,3,5].

Recently, we have been probing the reverse directionality of myosin VI [8], which is the only myosin moving toward the (-) end of actin filaments. We designed several constructs with an alpha-actinin artificial lever arm fused at different locations beyond the converter along the wild type lever arm (i.e., the calmodulin binding region). Importantly, computational analysis using homology modeling and molecular dynamics was implemented to eliminate structures with potential steric hindrance [9]. The artificial constructs moved in a predicted manner and toward the (-) end of the actin filament. Thus, we have integrated computational modeling with single molecule experiments to show that the normal calmodulin-bound lever arm is not an integral structural or mechanistic component of the reverse stroke. Furthermore, we were able to change the moving direction of a molecular motor by engineering as few as 18 amino acids. This work demonstrated the ability to engineer bioenergy conversion of an enzyme using a multifaceted approach.

1. **Liao, J.-C.**, Sun, S., Chandler, D., Oster, G. (2004) The conformational states of Mg-ATP in water. European Biophysical Journal, 33, 29-37.

2. **Liao, J.-C.**, Spudich, J.A., Parker, D., Delp, S.L.(2007) Extending the absorbing boundary method to fit dwell-time distributions of molecular motors with complex kinetic pathways. Proceedings of the National Academy of Sciences, USA, 104(9), 3171-3176.

3. Tang, S., **Liao, J.-C.***, Dunn, A., Altman, R.B., Spudich, J.A., Schmidt, J. (2007) Predicting allosteric communication in myosin via a pathway of conserved residues. Journal of Molecular Biology, doi: 10.1016/j.jmb.2007.08.059. *as a co-first author

4. Zheng, W., **Liao, J.-C.***, Doniach, S., Brooks, B.R. (2007) Toward the mechanism of dynamical couplings in Hepatitis C virus NS3 helicase using elastic network model. Proteins: Structure, Function, and Bioinformatics, 67, 886-896. *as a co-first author

5. **Liao, J.-C.***, Jeong, Y., Kim, D.-E., Patel, S. S., Oster, G. (2005) Mechanochemistry of T7 DNA helicase. Journal of Molecular Biology, 350(3), 452-575. *as a co-first author

6. Xing, J., **Liao, J.-C.**, Oster, G. (2005) Making ATP. Proceedings of the National Academy of Sciences, USA, 102(46), 16539-126546.

7. Adelman, J.L., Jeong, Y.-J., **Liao, J.-C.**, Patel, G., Kim, D.-E., Oster, G., Patel, S. S. (2006) Mechanochemistry of transcription termination factor Rho. Molecular Cell, 22, 611-621.

8. **Liao, J.-C.**, Bryant Z., Elting, M.W., Delp, S.L., Spudich, J.A., (2007) Dissecting the reverse power stroke of myosin VI. (in preparation)

9. **Liao, J.-C.**, Bryant Z., Delp, S.L., Spudich, J.A., (2007) Computer-aided engineering of molecular motors to move toward opposite directions. Biomedical Engineering Society 2007 Annual Meeting, Los Angeles, California.

Web Page: www.stanford.edu/~liaojc/

6bv Novel Nanomaterials Development in Membrane Electrode Assembly for Proton Exchange Membrane Fuel Cells

Zhongwei Chen, University of California, Riverside, A242 Bourns Hall, Chemical Engineering, Riverside, CA 92507 and Yushan Yan, Chemical and Environmental Engineering, University of California, Riverside, Bourns Hall A 242, Riverside, CA 92521.

An energy economy based on hydrogen could resolve growing concerns about America's energy supply, national security, air pollution and greenhouse gas emissions, as hydrogen as a fuel produces near-zero emissions and can be produced by domestically available resources. There is a strong incentive for developing commercially available hydrogen powered fuel cells to power automobiles, homes and businesses with near-zero pollution or greenhouse gasses.

Fuel cells are clean, compact and modular energy generation devices that have the potential to revolutionalize the production of electricity and thermal energy. Depending on the type of electrolyte used, there are various kinds of fuel cells, one of which, the proton exchange membrane fuel cell (PEMFC) is attracting enormous attention due to its great potential in transportation, residential and portable applications. The "heart" of the PEMFC is "membrane electrode assembly" (MEA) which is made of a thin proton exchange membrane coated on both sides with a catalyst layer and carbon fiber paper. The MEA enables the essential reaction to happen between hydrogen and oxygen that produces electricity. While many breakthroughs have been made on PEMFC over the past few years, both technical and economic barriers for PEMFC commercialization still exist. The two major technical barriers specifically associated with the PEMFC are: (i) high fuel permeability and low proton conductivity of proton exchange membrane at high temperature (>100 oC); the state-of-the-art electrolyte material for PEMFC, perfluorosulfonic acid (PFSA) such as Nafion, requires humidification to conduct protons. Membranes made with PFSA require close to 100% relative humidity to achieve proton conductivity high enough for fuel cell applications (~0.1 S/cm), dictating humidification of the fuel and oxidant streams. This requirement and the thermal stability of PFSA limit the operating temperature of the fuel cell to <100ºC (typically 60–80ºC), while adding to the complexity, size, weight, and cost of the fuel cell system. Proton exchange membrane with reduced need for external humidification and the ability to operate at temperatures above 100°C would simplify the fuel cell systems and advance their development for automotive and stationary applications. (ii) insufficient durability and catalytic activity of carbon supported Pt catalysts. The Pt nanoparticles are prone to aggregation under the operating conditions in the long-run, which results in Pt surface area loss and hence performance loss over a period of time. In addition the catalyst support i.e. carbon black is susceptible to corrosion, especially in transportation applications under the actual driving cycle conditions. As the carbon is corroded away, Pt nanoparticles will be lost from the carbon support and aggregate into larger particles.

The objective of my Ph. D research is to develop novel proton exchange membrane and catalyst for the MEA of PEMFC, which is divided into two main research areas:

I. To develop Nafion/zeolite nanocomposites proton exchange membrane with low fuel permeability and high proton conductivity for operation up to 120°C or higher with low humidity.

Zeolite materials happen to be primarily composed of silica, and depending on the zeolite structure type, it can have very high surface area due to its microporosity. We expect the Nafion/zeolite nanocomposites proton exchange membrane should have such properties: (a) Higher working temperature: the porous structure of zeolite should retain water molecules well above its boiling point; (b) Higher proton conductivity: sulfonated zeolite nanocrystals are much better proton conductivity than other inorganic materials such as silica; (c) Lower fuel permeability: the nanochannels of zeolite nanocrystals can allow protons to pass through while blocking the passage of fuels such as methanol and hydrogen. First, zeolite nanocrystals has been synthesize by our novel microemulsion and microwave method, then Nafion/zeolite nanocomposite membranes are fabricated by casting method and in-situ method successfully and they can be used to improve performance of methanol feed PEMFC. Third, Nafion/Zeolite nanocomposite membranes will be demonstrated to work well in PEMFC at higher temperature (120-140 oC).

II. To develop novel nanotube-based electrocatalysts with high durability and catalytic activity.

(i) Develop and demonstrate functionalized CNTs and conductive polymer nanotubes as excellent candidates for catalyst support for PEMFC. Increasing the utilization of Pt in electrode structure and improving durability by developing novel nanotube-based catalyst support such as: carbon nanotubes (CNTs) and conductive polymer nanotubes. CNTs have been used to replace traditional carbon powders as the catalyst support in PEMFCs because of their unique structural, mechanical, and electrical properties. It was found that CNTs can improve mass transport and Pt utilization due to their elongated morphology, and they are more corrosion resistant than carbon black. Single walled, double walled and multi-walled CNTs supported Pt (2-3 nm) and Pt-alloy electrocatlysts (Pt/SWNT, Pt/DWNT and Pt/MWNT) have been prepared successfully. Double walled CNT has shown the best catalytic activity and durability among them. The optimization of MEA by them is still on the way. At the same time, conductive polymer nanowires and nanotubes have also been developed and used as Pt and Pt-alloy electrocatalysts catalyst, which have shown improved catalytic activity for the PEMFC.

(ii) Develop and demonstrate Platinum nanotubes (PtNTs) and Pt-alloy-NTs as high durable and catalytic activity catalysts for PEMFC. PtNTs and Pt-alloy-NTs have been synthesized successfully and directly used as a supportless catalyst with the goal of greatly reducing the use of Pt and improving the catalytic activity and durability. PtNTs have unique tube morphology with dimensions at multiple length scales that combine a number of remarkable properties unavailable from any other single catalyst system, which will enhance the catalytic activity and durability.

All these benefits of using the Nafion/zeolite nanocomposite membrane, the novel CNTs supported Pt catalyst and PtNTs and Pt-alloy-NTs can eventually lead to the high proton conductivity, lower fuel permeability, enhanced catalytic activity, increased Pt utilization, and higher durability of MEA, thus effectively improving the performance and durability and reducing the cost of MEA which are still the major obstacles for commercialization of the PEMFC technology.

References

1. Chen, Zhongwei; Waje, Mahesh M; Li, Wenzhen; Yan,Yushan. Supportless Pt and PtPd Nanotubes as Electrocatalysts for Oxygen Reduction Reactions. Angewandte Chemie (2007), (VIP), 46(21), ASAP.

2. Chen, Zhongwei; Xu Lianbin; Li, Wenzhen; Waje, Mahesh M; Yan,Yushan. Polyaniline Nanofibers Supported Platinum Nanoelectrocatalysts. Nanotechnology (2006), 17(20), 5254-5259.

3. Chen, Zhongwei; Holmberg, Brett A; Li, Wenzhen; Munoz, Ronnie; Yan, Yushan. Nafion/Zeolite Nanocomposite Membrane by in Situ Crystallization for a Direct Methanol Fuel Cell. Chemistry of Materials (2006), 18(24), 5669-5675.

4. Chen, Zhongwei; Li, Shuang; Yan, Yushan. Synthesis of Template-Free Zeolite Nanocrystals by Reverse Microemulsion-Microwave Method. Chemistry of Materials (2005), 17(9), 2262-2266.

5. Chen, Zhongwei; Holmberg, Brett A.; Yan, Yushan. Nanocomposite membranes based on UV-curable crosslinked polytetrafluoroethylene and acid functionalized zeolite nanocrystals for fuel cell applications. PMSE Preprints (2005), 92 68.

6. Wang, Xiqing; Liu, Rui; Waje, Mahesh M.; Chen, Zhongwei; Yan, Yushan; Bozhilov, Krassimir N.; Feng, Pingyun. Sulfonated Ordered Mesoporous Carbon as a Stable and Highly Active Protonic Acid Catalyst. Chemistry of Materials (2007), 19(10), ASAP

7. Waje, Mahesh M; Li Wenzhen; Chen Zhongwei; Yushan Yan, 2006. Effect of Catalyst Support Morphology on the PEMFC Performance, ECS Transactions-Cancun, Volume 3.

8. Li Wenzhen; Wang Xi; Chen Zhongwei; Waje, Mahesh M; Yushan Yan. Pt-Ru/DWNTs as a high-performance anode catalyst for direct methanol fuel cells. Journal of Physical Chemistry:B (2006), 110 (31), 15353-15358

9. Wang, Xin; Li, Wenzhen; Chen, Zhongwei; Waje, Mahesh M; Yan, Yushan. The Durability Investigation of Carbon Nanotube as Catalyst Support for Proton Exchange Membrane Fuel Cell Electrode. Journal of Power Sources (2006), 158(1), 154-159.

10. Li, Wenzhen; Wang, Xin; Chen, Zhongwei; Waje, Mahesh M; Yan, Yushan. Carbon Nanotube Film by Filtration as Cathode Catalyst Support for Proton-Exchange Membrane Fuel Cell. Langmuir (2005), 21(21), 9386-9389.

6bw Nanomaterials for Energy Conversion

Michael P. Tate, School of Chemical Engineering, Purdue University, 480 Stadium Mall Dr, West Lafayette, IN 47907 and Hugh W. Hillhouse, School of Chemical Engineering and Energy Center at Discovery Park, Purdue University, 480 Stadium Mall Drive, West Lafayette, IN 47907.

Developing cost effective energy conversion devices is a key challenge of the 21st century. In particular, advances in nanoscale science and engineering may help photovoltaic and thermoelectric power generation devices provide a cheap and abundant supply of electricity and enable our energy economy to achieve carbon-neutrality. The poster will summarize my research on self-assembled nanostructured films to solve some of the key fundamental challenges necessary to develop high efficiency, lost-cost nanostructured thermoelectric and photovoltaic energy conversion devices. In addition, I will present plans on my future research.

Thermoelectric devices allow one to recover electricity from waste heat, but current thermoelectric conversion efficiencies are below the level at which wide spread use becomes practical. However, nanowire arrays with diameters less than the thermal de Broglie wavelength are predicted to boost the efficiency of thermoelectric devices. A promising approach to produce nanowire arrays of such a small diameter (less than 10 nm) is electrochemical deposition within highly ordered nanoporous films formed by evaporation induced self-assembly of surfactants with metal oxide oligomers. However, key elements of our understanding of processing variables that control the formation of film topology remains elusive, in part due to the lack of characterization methods and analytical tools. Here, I will summarize my results on the development of grazing incidence small angle x-ray scattering (GISAXS) tools to quantitatively characterize highly-ordered nanostructured films,[1, 2] and show how we have used these tools to reveal a new understanding of how to control structure and topology of nanoporous films.[3-8] In particular we have developed a simple reproducible method to self-assemble highly ordered and oriented double-gyroid structure nanoporous films with facile mass transport.[4] This key advance enables these nanoporous films to be used to template 4 nm diameter nanowire arrays with high fill fraction. Finally, I will present results on the electrochemical deposition of bismuth telluride nanowire arrays within the nanoporous films and the development for use as nanostructured thermoelectric devices.[9]

REFERENCES

1. Tate, M. P.,* V. N. Urade, J. D. Kowalski, T. C. Wei, B. D. Hamilton, B. W. Eggiman and H. W. Hillhouse,* Simulation and Interpretation of 2D Diffraction Patterns from Self-Assembled Nanostructured Films at Arbitrary Angles of Incidence: from Grazing Incidence (above the critical angle) to Transmission Perpendicular to the Substrate, Journal of Physical Chemistry B, 110, pp. 9882-9892 (2006).

2. Tate, M. P.* and H. W. Hillhouse,* General Method for Simulation of 2D GISAXS Intensities for

Any Nanostructured Film Based on a Discrete Fourier Transform: Implementation of the Distorted Wave Born Approximation, Journal of Physical Chemistry C, Articles ASAP, (2007).

3. Eggiman, B. W.,* M. P. Tate* and H. W. Hillhouse,* The Rhombohedral Structure of Highly Ordered and Oriented Self-Assembled Nanoporous Silica Thin Films, Chemistry of Materials, 18, pp. 723-730 (2006).

4. Urade, V. N., T.-C. Wei, M. P. Tate, J. D. Kowalski and H. W. Hillhouse, Nanofabrication of Double Gyroid Thin Films, Chemistry of Materials, 19, pp. 768-777 (2007).

5. Tate, M. P.,* B. W. Eggiman, J. D. Kowalski and H. W. Hillhouse,* Order and Orientation Control of Mesoporous Silica Films on Conducting Gold Substrates Formed by Dip-Coating and Self-Assembly: A Grazing Angle of Incidence Small-Angle X-ray Scattering and Field Emission Scanning Electron Microscopy Study, Langmuir, 21, pp. 10112-10118 (2005).

6. Tanaka, S., M. P. Tate, N. Nishiyama, K. Ueyama and H. W. Hillhouse, Structure of Mesoporous Silica Thin Films Prepared by Contacting PEO106-PPO70-PEO106 Films with Vaporized TEOS, Chemistry of Materials, 18, pp. 5461-5466 (2006).

7. Tate, M. P.,* C. M. Muzzillo and H. W. Hillhouse,* Mass Transport Rate Control During Spin Coating of Self-Assembled Films, Submitted, (2007).

8. Urade, V. N., L. Bollmann, J. D. Kowalski, M. P. Tate and H. W. Hillhouse,* Controlling Interfaction Curvature in Nanoporous Silica Films formed by Evaporation Induced Self-Assembly from Nonionic Surfactants: II Effect of Processing Parameters on Film Structure, Langmuir, 23, pp. 4268-4278 (2007).

9. Tate, M. P.* and H. W. Hillhouse,* 4 nm Nanowire Networks of Bismuth Telluride, Submitted, (2007).

* Indicates major and corresponding authors

6bx Sustainability, Green Engineering and Industrial Ecology

Arunprakash T. Karunanithi, U.S. Environmental Protection Agency, 26 W. Martin Luther King Drive, Cincinnati, OH 45268

Sustainability can be described as finding a suitable regime that supports the social and economic development of humans without major consequences to the environment (land, atmosphere and ocean) and the ecological services on which people depend.

By applying green engineering principles in process and product design one could greatly reduce the human impact on the environment. During the past decade the field of industrial ecology has evolved and it takes a holistic view on industrial sustainabilty.From the process engineering perspective integration of green engineering and industrial ecology principles in the process/product design framework is critcal and extremely challanging. The first part of the poster presents a framework for this integration.

The second part of the poster deals with sustainable environmental management.Here we look at sustainability from a macroscopic systems perspective.We study the comlex relationships,feedbacks and linkages between environmental, economic, ecological, social and climate systems through information theory. Work related climate regime changes etc will be presented

6bz Simulation Studies of Phase Behavior and Crystal Structures of Colloidal Suspensions

Antti-Pekka Hynninen, Dept. of Chemical Engineering and PRISM, Princeton University, Engineering Quadrangle, Princeton, NJ 08544

I will discuss a number of successful examples of computer simulation studies that help to understand the behavior of experimental colloidal suspensions. As a first example, I will show how simulations can help identify and predict novel crystal structures observed in experimental suspensions of oppositely charged colloids. The structures include analogs of known binary atomic structures, doped fullerene C_{60} structures, but also novel structures that do not have an atomic of molecular analog. Second, I use simulations to better understand the phase diagram of charged colloids in an external magnetic- or electric field, where the colloidal pair potential is dominated by the dipole-dipole interaction. As a third example, I will show that simulations on binary hard-spheres lead to a proposal of exciting experiments that could produce photonic crystals with the wavelength in the visible region. Finally, I will discuss how methods frequently used in theoretical colloid science could be used to study systems that are of interest in biology, such as viral capsid self-assembly and protein-protein interactions.

6ca Large-Scale Synthesis of Salt and Metal Nanoparticles by Flame Synthesis and Application of Magnetic Nanobeads in Separation Technology

Robert N. Grass, Institute for Chemical and Bioengineering, ETH Zurich, Wolfgang-Pauli-Strasse 10, Zurich, Switzerland

To date aerosol flame synthesis is the most important manufacturing process for ceramic nanoparticles including silica, titania, ceria and zinc oxide with a total production volume of several million metric tons per year. The process advantages include low cost, scalability and purity of the formed product. It has been however limited to a very narrow range of oxidic product compounds. This contribution presents investigated strategies to overcome these limitations. I was able to successfully show, that a modified flame process (patented by ETH) can not only be used to synthesize salt nanoparticles (Grass et al. Chem. Commun. 2005) but also non-noble metal nanoparticles such as bismuth (Grass et al. J. Nanoparticle Res. 2006) and cobalt (Grass et al. J. Mater. Chem. 2006).

This new synthesis tool allowed the fast and low-cost preparation of a large range of highly interesting materials with novel electronic, optical and mechanical properties. This very market oriented research led to the development of improved sensors (Athanassiou et al. Nanotechnology 2006; Luechinger et al. Langmuir 2007), more stable bulk nanocrystalline metals (Grass et al. Nanotechnology 2007), metal matrix nanocomposites with beneficial mechanical properties (Grass et al. J. Mater. Chem. 2007) and magnetic nanobeads which can be used for magnetic separation applications.

The separation of compounds from mixtures is one of the major problems in chemical engineering. The complexity and cost of separation largely increases if the compounds which are to be separated have a similar chemical structure, are sensitive to increased temperatures and are initially present at low concentrations. To overcome these difficulties for the separation of proteins and antibodies, biochemists are currently using magnetic microbeads. The technology utilizes magnetic polymer beads which carry a surface functionality. The beads react with the target compound which can then be retrieved from the mixture together with the bead by magnetic forces. The high cost of the beads and the rather poor quality in respect of magnetic properties and low binding capacities very much limited the application of such materials to specialized research laboratories and ultra high-value biotechnological applications.

Besides low cost an ideal material for the use in magnetic separation has to fulfill a range of constraints including: small size (nm scale), good magnetic properties (ideally metallic Co, Fe, Ni), stability over a wide temperature and pH range and enabling the covalent and stable bonding of ligands. All these demands can be met by applying covalently functionalized carbon coated cobalt nanoparticles (Grass et al. Angew. Chem. Int. Ed. In press) which were prepared by flame synthesis and functionalized by the use of diazonium chemistry yielding chloro-, nitro-, and amino terminated nanobeads.

References:

[1] R. N. Grass, T. F. Albrecht, F. Krumeich, W. J. Stark, J. Mater. Chem. 2007, 17, 1485.

[2] R. N. Grass, E. K. Athanassiou, W. J. Stark, Angewandte Chemie 2007, accepted.

[3] R. N. Grass, M. Dietiker, R. Spolenak, W. J. Stark, Nanotechnology 2007, 18, 035703.

[4] R. N. Grass, W. J. Stark, Chem. Commun. 2005, 1767.

[5] R.N. Grass, W. J. Stark, J. Nanopart. Res. 2006, 8, 729.

[6] R. N. Grass, W. J. Stark, J. Mater. Chem. 2006, 16, 1825.

[7] N. A. Luechinger, S. Loher, E. K. Athanassiou, R. N. Grass, W. J. Stark, Langmuir 2007, 23, 3473.

[8] E. K. Athanassiou, R. N. Grass, W. J. Stark, Nanotechnology 2006, 17, 1668.

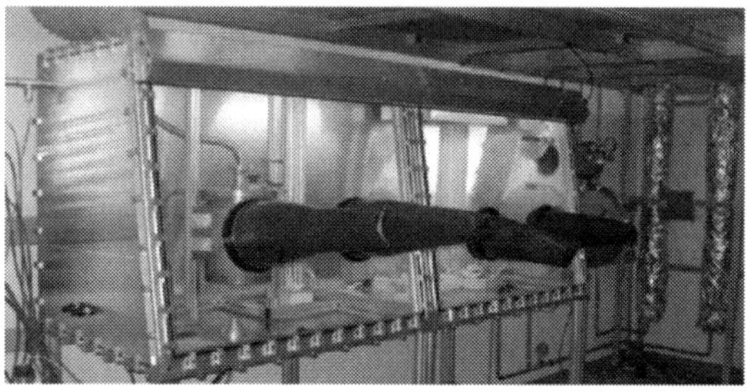

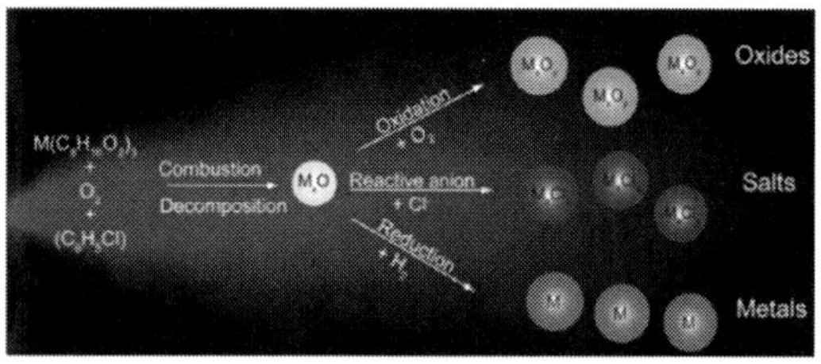

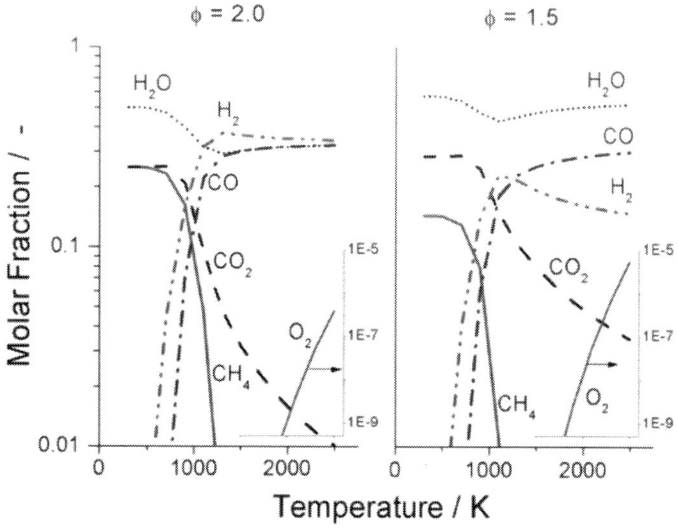

6cb Combining Colloidal Chemistry and Microfluidics

Rhutesh K. Shah, School of Engineering and Applied Sciences, Harvard University, 40 Oxford Street, Cambridge, MA 02138

Poly(N-isopropylacrylamide) or PNIPAM microgels are polymeric particles consisting of cross-linked three dimensional networks. They swell and shrink significantly by absorbing and expelling large quantities of water and this property makes them extremely valuable for a number of applications including macromolecular delivery, sensors, chemical separations, etc. This poster will describe the synthesis and characterization of novel structures formed by the directed assembly of colloidal PNIPAM microgels in droplets. Two specific structures designed: (i) microgel vesicles, i.e., colloidosomes made out of microgels and (ii) gelospheres, i.e., cross-linked spherical particles formed by aggregation of microgels in droplets. We employ microfluidic techniques to disperse aqueous suspensions of microgels in oil. The use of these techniques allows the formation of monodisperse emulsions and, thus, monodisperse self-assembled structures. Such structures can be of immense use for sequestering macromolecules and releasing them in a controlled way in numerous applications.

6cc Properties of Surfaces and Films from Viscous Liquids to Elastic Solids

Hongbo Zeng[1], Jacob N. Israelachvili[1], Matthew Tirrell[2], and L. Gary Leal[3]. (1) Chemical Engineering, University of California, 3357 Engineering II, Santa Barbara, CA 93106, (2) Chemical Engineering, University of California, Santa Barbara, Engineering 2, Room 3357, Santa Barbara, CA 93106-5080, (3) Chemical Engineering, University of California, Santa Barbara, Engineering 2, Rm 3357, Santa Barbara, CA 93106

Surface deformations during the coalescence of two polymer melt films were studied by use of a surface forces apparatus (SFA). A new type of well-ordered periodic transient surface fingering pattern was observed during the coalescence, which eventually disappeared, leaving smooth polymer-air interfaces. The life-times of the transient well-ordered patterns depend on the viscosities and film thickness of the polymers. These observations are contrast to the conventional understanding that liquid-liquid coalescence usually occurs with the deforming surfaces remaining smoothly curved at all stages, with no abrupt changes or esoteric shape transitions. In further, the transitions and dynamic transients associated with the adhesion and/or coalescence of surfaces ranging from viscous liquids to elastic solids were studied in great details. This talk will address the properties of surfaces and films from liquid-like to solid-like state and types of instabilities observed during their approach, adhesive contact, coalescence and detachment, and discuss common situations where similar effects likely occur.

6cd Sustainable Alternatives for Chemical Processing

Jason P. Hallett, Department of Chemistry, Imperial College, South Kensington Campus, London, United Kingdom

My research focus is on reducing the environmental impact of chemical processing through the development of sustainable chemical processes that are both economically competitive and environmentally responsible. This will involve developing techniques involving novel solvents, such as ionic liquids, supercritical fluids, and aqueous systems that combine reaction efficiency and novel separations techniques to improve catalytic efficiency and facilitate product purification. The combination of research in the areas of thermodynamics, tunable solvents, green chemistry and homogenous organometallic and enzymatic catalyst recovery presents unique opportunities to integrate these concepts into the development of sustainable processes. The successful development of these processes will have major impact on the synthesis of pharmaceutical compounds and the production of commodity chemicals and biofuels from renewable resources. The use of new sustainable technologies to achieve these goals is a major part of the future of the chemical industry.

6ce Cardiac Tissue Engineering Using Embryonic Stem Cell Derived Cardiomyocytes and Novel Biomaterials

Elizabeth A. Lipke, Department of Biomedical Engineering, Johns Hopkins University, Traylor Rm #703, 720 Rutland Ave., Baltimore, MD 21205

My future research program will combine my unique expertise in chemical engineering fundamentals, biomaterials, stem cell biology, and cardiac electrophysiology. The long-term goals are (1) determine the effects of material properties, controlled drug delivery, and mechanical and electrical stimuli on stem cells and use this knowledge to design systems for large-scale cell differentiation, (2) improve the electrophysiological function of tissue engineered cardiac grafts by influencing tissue structure and cellular interconnectivity and (3) create materials that will enhance re-endothelialization of the diseased or damaged vascular wall through specific interactions with endothelial and endothelial progenitor cells. I intend to explore the relationship between scaffold material properties and cellular function, including gene expression, differentiation, proliferation, and migration. Potential funding sources for these efforts include the NIH, NSF, and American Heart Association.

This research plan is the culmination of my doctoral work in chemical engineering at Rice University and my undergraduate and post-doctoral work in biomedical engineering at Johns Hopkins University. During my graduate research, I showed that nitric oxide-generating hydrogels influenced key components of the restenosis cascade both in vitro and in vivo. Localized drug delivery was achieved at the site of vascular injury using these poly(ethylene glycol) copolymers, which can be rapidly photopolymerized in situ. These materials can be tailored to provide cell-type specific adhesion ligands, degrade in response to cellular enzymes, and present immobilized growth factors. My postdoctoral research focuses on the differentiation of cardiomyocytes from mouse and human embryonic stem cells and the electrophysiological characterization of these cells. Embryonic stem cell-derived cardiomyocytes (ESC-CMs) have the potential to supply large numbers of cells for cardiac regeneration. Before ESC-CMs can be used for cardiac repair, however, more must be known about their electrophysiology and their ability to functionally incorporate into native cardiac tissue. Our studies show for the first time that ESC-CMs can electrically integrate with neonatal rat ventricular myocytes and by themselves can form an electrophysiologically functional tissue substrate. I currently have my students investigating the influence of mechanical stretch and electrical stimulation on mouse ESC differentiation into cardiomyocytes. We hypothesize that adding these stimuli, and thereby more closely mimicking the in vivo environment, will cause the cells to reach a more fully differentiated state and potentially increase the percentage yield of cardiomyocytes. I also am studying the potential of nanopatterned PEG hydrogels to direct cardiac alignment and thereby influence the cardiac electrophysiology of the resulting tissue. By creating anisotropic structures, our engineered tissue better mimics the native myocardium.

Integrating teaching and research comes naturally to me, and my educational background has prepared me to teach a broad range of courses. As a lecturing professor at Rice University in the Spring of 2005, I used my background to illustrate to my students how the skills they were learning in the tissue culture laboratory course could be applied to do sophisticated research. Currently, I directly supervise two undergraduate student researchers and one graduate student researcher. During my daily interactions with my students, I teach them critical thinking skills through directed questioning and convey key concepts to them through "mini lectures" and by compelling them to read the literature. Previous students that I have mentored have gone on to successful graduate and medical school careers and one is currently pursuing a faculty position.

6cf Self-Assembly of Functional Rod-Coil Block Copolymers

Bradley D. Olsen, Chemical Engineering, University of California, Berkeley, 201 D Gilman, Berkeley, CA 94720-1462 and Rachel A. Segalman, University of California at Berkeley, Dept of Chemical Engineering, 201 D Gilman, Berkeley, CA 94720-1462.

While block copolymer self-assembly offers an elegant route for the self-assembly of structures on the 10-100 nm length scale, many functional polymers, such as helical proteins and conjugated semiconducting polymers, do not follow these classical self-assembly behaviors. Nonetheless, new applications in organic photovoltaics and biotechnology rely on controlled nanoscale morphologies incorporating biological and semiconducting polymers. Due to extended conjugation or hydrogen-bonded helical structure, such functional polymers have rodlike shapes rather than the classical Gaussian coil chain shape; incorporation of these rodlike polymers into block copolymers results in an interplay between liquid crystalline interactions and microphase separation. To understand the thermodynamics of functional rod-coil block copolymers, we have synthesized a model system with accessible phase transitions that we have used establish a universal bulk phase diagram and investigate self-assembly in thin films.

Phase space in rod-coil block copoymers is four dimensional; the dimensionless variables characterizing the system include the traditional repulsive interaction between the two blocks and the coil volume fraction as well as the strength of the liquid crystalline interaction and the geometrical ratio of the rod and coil block sizes. The aligning liquid crystalline interactions favor microphases with little curvature, resulting in a stable lamellar phase at low temperatures in most block copolymers. At high coil fraction and high geometric asymmetry between the rod and coil, hexagonal packing of rectangular rod aggregates is observed. At higher temperatures, the rod and coil blocks become miscible and nematic and isotropic phases are observed. Independent experiments measure the strengths of liquid crystalline interactions between the rods and repulsions between the rod and coil, allowing the phase diagram to be converted to a universal set of coordinates applicable to all rod-coil diblock copolymers.

Many applications of rod-coil block copolymers, such as organic photovoltaics, require both the creation of nanoscale structure and on the details pattern formation in thin films. The thin film state introduces additional complexities due to surface energy of the interfaces and geometric confinement normal to the substrate. Selective segregation of the coil block to the supported film interface orients lamellae preferentially parallel to the substrate in films, and perpendicularly oriented lamellae form defects between parallel grains. The lamellae have high bending moduli due to the liquid crystalline interactions between rods. The high bending moduli and in-plane liquid crystalline packing of the rod blocks lead to unusual grain shapes, non-traditional defect structures, and the potential for new handles in controlling long range order.

Publications:

14. "Crystalline Structure in Thin Films of DEH-PPV Homopolymer and PPV-b-PI Rod-Coil Block Copolymers." **B.D. Olsen**, D. Alcazar, V. Krikorian, M.F. Toney, E.L. Thomas, and R.A. Segalman. Manuscript in preparation.

13. "Square Grains in Asymmetric Rod-Coil Block Copolymers." **B.D. Olsen** and R.A. Segalman. Manuscript in preparation.

12. "Hierarchical Structure Control in Block Copolymers with Magnetic Fields." Y. Tao, H. Zohar, **B.D. Olsen**, and R.A. Segalman. Submitted.

11. "Non-Lamellar Phases in Asymmetric Rod-Coil Block Copolymers at Increased Segregation Strengths. **B.D. Olsen** and R.A. Segalman. Submitted.

10. "Domain Size Control by Self-Assembly of Rod-Coil Block Copolymers and Homopolymers Blends." Y. Tao, **B.D. Olsen**, V. Ganesan, and R.A. Segalman. *Macromolecules* **2007**, *40*, 3320-3327.

9. "Thin Film Structure of Symmetric Rod-Coil Block Copolymers." **B.D. Olsen**, X. Li, J. Wang, R.A. Segalman. *Macromolecules* **2007**, *40*, 3287-3295.

8. "Phase Transitions in Asymmetric Rod-Coil Block Copolymers." **B.D. Olsen** and R.A. Segalman. *Macromolecules* **2006**, *39,* 7078-7083.

7. "Higher Order Liquid Crystalline Structure in Low-Polydispersity DEH-PPV." **B.D. Olsen**, S.-Y. Jang, J.M. Lüning, and R.A. Segalman. *Macromolecules* **2006**, *39*, 4469-4479.

6. "Polymeric Nanocoatings by Hot-Wire Chemical Vapor Deposition (HWCVD)." K.K.S. Lau, Y. Mao, H.G. Pryce Lewis, S.K. Murthy, **B.D. Olsen**, L.S. Loo, and K.K. Gleason. *Thin Solid Films* **2006**, *501*, 211-215.

5. "Structure and Thermodynamics of Weakly Segregated Rod-Coil Block Copolymers." **B.D. Olsen** and R.A. Segalman. *Macromolecules* **2005**, *38*, 10127-10137.

4. "Peptide Attachment to Vapor Deposited Polymeric Thin Films." S.K. Murthy, **B.D. Olsen**, and K.K. Gleason. *Langmuir* **2004**, *20*, 4774-4776.

3. "Effect of Filament Temperature on the Chemical Vapor Deposition of Fluorocarbon-Organosilicon Copolymers." S.K. Murthy, **B.D. Olsen**, and K.K. Gleason. *J. App. Poly. Sci.* **2004,** *91*, 2176-2185.

2. "Making Thin Polymeric Materials, Including Fabrics, Microbicidal and Also Water-Repellent." J. Lin, S.K. Murthy, **B.D. Olsen**, K.K. Gleason, A.M. Klibanov. *Biotechnology Letters* **2003**, *25*, 1661-1665.

1. "Initiation of Cyclic Vinylmethylsiloxane Polymerization in a Hot-Filament Chemical Vapor Deposition Process." S.K. Murthy, **B.D. Olsen**, and K.K. Gleason. *Langmuir* **2002**, *18*, 6424-6428. Structure of Symmetric Rod-Coil Block Copolymers." **B.D. Olsen**, X. Li, J. Wang, R.A. Segalman. *Macromolecules* **2007**, *40*, 3287-3295.

8. "Phase Transitions in Asymmetric Rod-Coil Block Copolymers." **B.D. Olsen** and R.A. Segalman. *Macromolecules* **2006**, *39*, 7078-7083.

7. "Higher Order Liquid Crystalline Structure in Low-Polydispersity DEH-PPV." **B.D. Olsen**, S.-Y. Jang, J.M. Lüning, and R.A. Segalman. *Macromolecules* **2006**, *39*, 4469-4479.

6. "Polymeric Nanocoatings by Hot-Wire Chemical Vapor Deposition (HWCVD)." K.K.S. Lau, Y. Mao, H.G. Pryce Lewis, S.K. Murthy, **B.D. Olsen**, L.S. Loo, and K.K. Gleason. *Thin Solid Films* **2006**, *501*, 211-215.

5. "Structure and Thermodynamics of Weakly Segregated Rod-Coil Block Copolymers." **B.D. Olsen** and R.A. Segalman. *Macromolecules* **2005**, *38*, 10127-10137.

4. "Peptide Attachment to Vapor Deposited Polymeric Thin Films." S.K. Murthy, **B.D. Olsen**, and K.K. Gleason. *Langmuir* **2004**, *20*, 4774-4776.

3. "Effect of Filament Temperature on the Chemical Vapor Deposition of Fluorocarbon-Organosilicon Copolymers." S.K. Murthy, **B.D. Olsen**, and K.K. Gleason. *J. App. Poly. Sci.* **2004,** *91*, 2176-2185.

2. "Making Thin Polymeric Materials, Including Fabrics, Microbicidal and Also Water-Repellent." J. Lin, S.K. Murthy, **B.D. Olsen**, K.K. Gleason, A.M. Klibanov. *Biotechnology Letters* **2003**, *25*, 1661-1665.

1. "Initiation of Cyclic Vinylmethylsiloxane Polymerization in a Hot-Filament Chemical Vapor Deposition Process." S.K. Murthy, **B.D. Olsen**, and K.K. Gleason. *Langmuir* **2002**, *18*, 6424-6428.

6cg Multiscale Modeling of Viscoelastic Flow and Complex Fluids in Micro/nanofluidics

Xin Hu, NSEC center for Affordable Nanoengineering of Polymer Biomedical Devices (CANPBD), The Ohio State University, Room 125A, 140 West 19th Ave., Columbus, OH 43210 and Ly James Lee, Chemical and Biomolecular Engineering, The Ohio State University, 125 Koffolt Labs, 140 W. 19th Ave., Columbus, OH 43210.

Conventional flow simulation based on constitutive equations (CE) cannot predict different viscoelastic flow behaviors because it does not establish the relationship between microscopic structure of polymer chain and macroscopic properties of viscoelastic fluids. A mesocopic concurrent multiscale approach named CONNFFESSIT ('Calculation of Non-Newtonian Flow: Finite Elements and Stochastic Simulation Technique') has been applied to solve this problem by combining the Brownian dynamics simulation (BDS) and the finite element method (FEM). Currently the simplest dumbbell model is used to mimic single linear polymer chain and has successfully achieved stable simulation results at a higher Weissenberg number than the conventional methods. More advanced models using bead-spring and bead-rod chains are under investigation in order to simulate viscoelastic flow in complex geometries with much higher Weissenberg number, which is commonly observed in polymer processing.

The coarse-grained Brownian dynamics simulations using the bead-spring and bead-rod chain models have also been used to simulate single polymer dynamics in micro/nanofluidics. Hydrodynamic and electrokinetic micro/nanofluidic flows are first solved by the macroscopic finite element method and the calculated flow and electric fields are imported as an input for Brownian dynamics simulation. Thus this is a hierarchical multiscale simulation. Previous work is concentrated on DNA dynamics in micro/nanofluidic flows and the quantitative and qualitative agreements have been obtained between the simulation and experimental results. Currently the simulation efforts are focus on the hydrodynamic interactions (HI) between DNA chain and the confined geometries using a concurrent multiscale simulation technique named SRD ('Stochastic Rotation Dynamics').

References:

1. X. Hu, Z. Ding, and L. J. Lee, "Simulation of 2D Transient Viscoelastic Flow Using the CONNFFESSIT Approach", Journal of Non-Newtonian Fluid Mechanics 127, 107-122 (2005).

2. X. Hu, C. Liu, G. Xu, and L. J. Lee, "Viscoelastic Flow in Micro-Injection Molding", Proceedings of Society of Plastic Engineers Annual Technical Conference, Paper #: 0450 (2007).

3. Y.-J. Juang, S. Wang, X. Hu, and L. J. Lee, "Dynamics of Single Polymers in a Stagnation Flow Induced by Electrokinetics", Physical Review Letters 93, 268105 (2004).

4. X. Hu, S. Wang, Y.-J. Juang, and L. J. Lee, "A Five-Cross Microfluidic Network Design Free of

Coupling between Electrophoretic Motion and Electroosmotic Flow", Applied Physics Letters 89, 084101 (2006).

5. Y.-J. Juang, X. Hu, S. Wang, L. J. Lee, C. Lu, and J. Guan, "Electrokinetic Interactions in Microscale Cross-Slot Flow", Applied Physics Letters 87, 244105 (2005).

6. S. Wang, X. Hu, and L. J. Lee, "Dynamic Assembly by Electrokinetic Microfluidics", Journal of the American Chemical Society 129, 254-255 (2007).

7. H. C. Jung, W. Lu, S. Wang, L. J. Lee, and X. Hu, "Etching of Pyrex Glass Substrates by Inductively Coupled Plasma Reactive Ion Etching for Micro/Nanofluidic Devices", Journal of Vacuum Science & Technology B 24(6), 3162-3164 (2006).

8. X. Hu, S. Wang, and L. J. Lee, "Single Molecular Dynamics in a Tapering Contraction-Expansion Microchannel with Electrophoresis", (to be submitted).

9. S. Wang, X. Hu, and L. J. Lee, "Electrokinetic Transport of Rigid Particles and DNA Molecules through Polymeric Nanonozzle Array", (to be submitted).

10. Y. Yang, M. C. Cheng, X. Hu, D. Liu, R. Goyette, L. J. Lee, and M. Ferrari, "Low Pressure Carbon Dioxide Enhanced Polymer Self-Diffusion Below the Glass Transition Temperature", Macromolecules 40(4), 1108-1111 (2007).

6ch A Systems Biology Approach to Protein Translation

Hermioni Zouridis, Chemical and Biological Engineering, Northwestern University, 2145 Sheridan Rd., Evanston, IL 60208 and Vassily Hatzimanikatis, Laboratory of Computational Systems Biotechnology, EPFL, CH-1015, Lausanne, Switzerland.

Translation, or protein synthesis, is a process that is central to cellular function and well conserved among all living organisms. Understanding the translation mechanism is important in many areas of medicine and biotechnology. Recent advances in genomics, microarray technology, and proteomics make possible the mapping between mRNA and protein expression levels in cells and have led to a wealth of information on protein synthesis at the systems, or genome wide, level. Mathematical modeling and computational studies are vital for integrating this vast information in order to characterize protein synthesis networks and predict how translational machinery responds under genetic or environmental perturbations.

Translation is essentially a polymerization process facilitated by the ribosome on an mRNA template consisting of initiation, elongation, and termination phases. Initiation occurs with binding of the ribosome to the ribosomal binding site near the 5' end of the mRNA. During elongation the ribosome facilitates assembly of the polypeptide chain with one amino acid added at each codon along the length of the mRNA. Amino acids are delivered to the ribosome by tRNAs that serve as adapter molecules between the amino acid and the codon occupying the ribosomal A site. Termination involves release of the completed peptide from the ribosome near the 3' end of the mRNA. Several ribosomes can simultaneously translate the same mRNA, forming a structure called a polysome.

We developed a gene sequence specific mechanistic model for the translation machinery which accounts for all the elementary steps of the translation mechanism. Specifically, our model includes all the elementary steps involved in the elongation cycle at every codon along the length of the mRNA. We performed a sensitivity analysis in order to determine the effects of kinetic parameters and concentrations of the translational components on protein synthesis rate. Utilizing our mathematical framework and sensitivity analysis, we investigated the translation kinetic properties of a single mRNA species in E. coli. We propose that translation rate at a given polysome size depends on the complex interplay between ribosomal occupancy of elongation phase intermediate states and ribosome distributions with respect to codon position along the length of the mRNA, and this interplay leads to polysome self-organization that drives translation rate to maximum levels.

We expanded our model to account for the non-specific binding of tRNAs to the ribosomal A site, and we find that the competitive, non-specific binding of the tRNAs is the rate limiting step in the elongation cycle for every codon. By introducing our model in terms of the Michaelis – Menten kinetic framework, we determine that these results are due to the tRNAs that do not recognize the ribosomal A site codon acting as competitive inhibitors to the tRNAs that do recognize the ribosomal A site codon. We also expanded our sensitivity analysis to determine the contribution of elongation cycle kinetic parameters of each codon on the overall translation rate. Our sensitivity analysis predicts that different configurations of codons along the length of the mRNA control translation rate at different polysome sizes. We also observe that the relative position of codons along the mRNA determines the optimal protein synthesis rate and the rate limiting effect of the individual codons.

We applied our expanded mechanistic framework and sensitivity analysis to the translation of every protein in E. coli in order to determine the translational efficiency of each gene. Because of the important role codons play in translation kinetics, codon usage patterns have been correlated with gene characteristics such as expression level, intragenic position, and length. However, codons have varying elongation kinetics due to different tRNA availabilities, codon – anticodon compatibilities, and the multiple elementary steps and translational components involved in the elongation cycle at every codon. Hence, applying our mechanistic framework to E. coli genes aids in understanding the complex, nonlinear interplay between codon usage and protein synthesis properties, and characterizing how these properties relate to patterns such as gene expression levels and function in cells. Our results have implications in design of rational protein production systems, wherein quantitative knowledge of responses of protein expression to genetic or environmental perturbations can be used to optimize a cellular system towards the production of a protein of interest.

6ci Modulating Catalytic Properties at the Gas-Solid Interface

Raj Ganesh Pala, Chemistry and Biochemistry, University of California, Santa Barbara, Department of Chemistry and Biochemistry 9510, University of California, Santa Barbara, CA 93106-9510

My present research interests are in the area of heterogeneous catalysis with an emphasis on CH bond activation in oxidative dehydrogenation reactions, partial oxidation and total oxidation reactions (1, 2). In the past, I have modeled:

1) Early stages of epitaxial thin film growth (3)

2) Ion selectivity observed in selectivity filters present in biological membrane proteins (4)

3) Effects of elastic strain on adsorptive and catalytic properties (5)

4) Influence of oxide supported metal clusters on reactant diffusion and kinetics of product formation (6).

In all the above research efforts, I have used density functional calculations and force field calculations to support the models and mechanisms proposed. In the near future, I will be pursuing my interests in heterogeneous catalysis and will venture into the area of modeling electrochemical systems.

My post doctoral work with Prof. Horia Metiu at the University of California at Santa Barbara involves the use density functional calculations to investigate how the catalytic properties of oxide surfaces can be modulated by replacing some of the host cations with other metal dopants. Many oxidation reactions operate via the Mars and Van Krevelen mechanism in which the oxygen atom from the oxide surface is supplied to the reactant that is getting oxidized, leaving behind a surface oxygen vacancy. Hence, the energy needed to create an oxygen vacancy from the oxide surface can be used to calibrate the propensity of the surface towards the Mars and Van Krevelen mechanism (1).

We have performed calculations on doped ZnO, MgO, CaO, BaO and CdO to identify the trends in the energetics of surface oxygen vacancy formation. We show that the doped oxide can operate in two qualitatively different methods with respect to the manner in which the oxygen can be supplied to the reactant that is oxidized. In one scenario, the substituted dopant may be "activated" and this leads to the generation of activated adsorbed oxygen molecules and also to the generation of peroxo complexes. In the second scenario, the dopant activates the oxide surface by facilitating the creation of a surface oxygen vacancy. The reactive oxygen species generated in the two different scenarios also leads to differences in reactivity of the oxide in oxidation reactions. Based on these predictions, the different forms of active oxygen species have been characterized by spectroscopic probes and by using various reaction chemistries (7).

We also elucidate how the oxide surface can be doped to facilitate CH bond activation which is important in oxidative dehydrogenation reactions. When many reaction channels are possible in an undoped oxide surface, doping also makes it possible to increase the selectivity by facilitating a particular reaction channel while suppressing others. Our calculations also support the experiments performed on methanol chemistry on doped ZnO (7).

My doctoral work was with Prof. Feng Liu at the University of Utah, and involved research mainly in three areas namely, early stages of epitaxial thin film growth, using externally applied strain to modulate adsorption and catalysis on a metal surface and how reactant diffusion and product evolution kinetics are influenced by oxide supported metal clusters.

In the area of epitaxial growth, we proposed a new concept called "critical epi-nucleation" to distinguish nucleation on surfaces with and without reconstruction (3). On a reconstructed surface, the critical classical nucleus is stable against dissociation, but may not yet break the underlying surface reconstruction. Consequently, there must exist a "critical epi-nucleus" that is not only stable but also has established the epi-configuration by removing the reconstruction of the underlying substrate. We illustrated this concept by first-principle calculations of homonucleation on reconstructed Si(001) surface where the critical epi-nucleus consists of six adatoms.

In the area of modifying adsorptive and catalytic properties using elastic strain, we demonstrated a model for determining the adsorptive and catalytic properties of strained metal surfaces based on linear elastic theory. We employed first-principle calculations of CO adsorption on Au and K surfaces and CO dissociation on Ru surface (4). The model involved a single calculation of the adsorption-induced surface stress on the unstrained metal surface, which determines quantitatively how adsorption energy changes with external strain. The model is generally applicable to both transition and non-transition metal surfaces, as well as to different adsorption sites on the same surface. Extending the model to both the reactant and the transition state of surface reactions allows determination of the effect of strain on surface reactivity.

In the area of catalysis using oxide supported metal clusters, experiments on the CO oxidation reaction using seven-atom Au clusters deposited on TiO_2surface correlated CO_2 formation with oxygen associated with Au clusters. We performed first principle calculations using a seven-atom Au cluster supported on a reduced TiO_2 surface to explore potential candidates for the form of reactive oxygen (5). These calculations suggested a thermodynamically favorable path for O_2 diffusion along the surface Ti row, resulting in its dissociated state to be bound to Au cluster and TiO_2 surface. Subsequently, CO can approach along the same path and react with the dissociated O_2 to form CO_2. The origin of the slow kinetic evolution of products observed in experiments was also investigated and was attributed to the strong binding of CO_2 to the Au cluster and the surface simultaneously.

1) Raj Pala and Horia Metiu "Modification of the oxidative power of ZnO surface by substituting some

surface Zn atoms with other metals" (In press, J. Phys. Chem. C)

2) Raj Pala and Horia Metiu (In preparation)

3) Raj Pala and Feng Liu, "Critical epi-nucleation on reconstructed surfaces and a model study of Si (001) homoepitaxy", Phys. Rev. Lett. 95, 136106 (2005).

4) Raj Pala, B.Chanda, S.K. Gupta, M.K.Mathew and J.Chandrasekhar, "Modeling of ion permeation in calcium and sodium channel selectivity filter", Proteins: structure, function and genetics, 38, 384 (2000).

5) Raj Pala and Feng Liu "Determining the adsorptive and catalytic properties of strained metal surface using adsorption-induced stress", J. Chem. Phys., 120, 7720 (2004).

6) Raj Pala and Feng Liu "Nature of reactive O_2 and slow CO_2 evolution kinetics in CO oxidation by TiO_2supported Au cluster", J. Chem. Phys., 125, 16, 144714(2006)

7) M. Sushchikh, R. Pala, J.N. Park, T. Wei, H. Metiu, E.W.McFarland (in preparation)

6cj Towards Solid State Silicon Nano and Microwire Photovoltaic Devices

Michael A. Filler[1], Brendan M. Kayes[1], Morgan C. Putnam[2], Michael D. Kelzenberg[3], and Harry A. Atwater[1]. (1) Applied Physics, California Institute of Technology, 1200 E. California Blvd, MC 128-95, Pasadena, CA 91125, (2) Chemical Engineering, California Institute of Technology, 1200 E. California Blvd, MC 128-95, Pasadena, CA 91125, (3) Electrical Engineering, California Institute of Technology, 1200 E. California Blvd, MC 128-95, Pasadena, CA 91125

Traditional solar cell architectures require expensive, high-quality crystalline materials due to their long carrier diffusion lengths. However, nano and microwire arrays have the potential to enable low-cost, high-efficiency solar cells via efficient radial minority carrier collection in materials with low minority carrier diffusion lengths. This can be achieved by creating an optically thick array of wires aligned normal to a substrate where each wire has a radial, solid state pn junction, enabling light absorption along the entire wire length but minority carrier collection radially along the wire minimum dimension [1].

Silicon nano and microwires for photovoltaic applications have been grown by a vapor-liquid-solid (VLS) chemical vapor deposition (CVD) process at atmospheric pressure on Si(111) substrates with $SiCl_4/H_2$ and Au, Cu, and Ni catalysts. Transmission electron microscopy (TEM) indicates that wires are single crystalline and grow along the <111> direction. While Au is traditionally employed as a catalyst for VLS wire growth, our studies show that Cu and Ni can also catalyze the growth of vertically aligned wires. In particular, Ni is not a deep-level trap in silicon and is significantly cheaper than Au, an important consideration for large-area applications such as photovoltaics.

Vertically aligned, densely packed, large-area arrays of silicon nano and microwires are fabricated by either self-assembling or templating metal catalysts on Si(111) substrates. Arrays with wire diameters between 50 and 250 nm are created via drop casting and self-assembly of Au colloid. Highly ordered microwire arrays are fabricated by photolithographic patterning and etching of a thermally grown oxide, evaporation of catalyst metal, and resist lift-off. The oxide serves as a surface diffusion barrier for metal catalyst atoms, preventing droplet ripening and maintaining pattern fidelity during the growth process. Arrays with nearly 100% vertically aligned wires, exceeding 75 μm in length can be achieved over areas of at least 1 cm^2. Achievement of large-area arrays is an important step toward realizing a functioning device.

Final device fabrication requires properly doping wires to obtain radial pn junctions, passivating surfaces to reduce recombination rates, conformally depositing an optically transparent material for structural support, and deposition of front and back electrical contacts. Formation of radial pn junctions can be achieved through a combination of *in situ* doping during wire growth and *ex situ* diffusion doping. The optical and electrical properties of arrays and individual wires passivated by thermal oxidation and organic functionalization are investigated with photoluminescence (PL) and near-field scanning optical microscopy (NSOM). Conformal filling of regions between these high-aspect ratio wires can be accomplished with spin-on-glasses, CVD deposited films, as well as polymeric materials.

[1] B. M. Kayes, H. A. Atwater, and N. S. Lewis, J. Appl. Phys., 97, 114302 (2005).

6ck Metabolic Engineering of the Terpenoid and Indole Pathways in *Catharanthus Roseus* Hairy Roots

Christie A.M. Peebles, Rice University, Dept of Bioengineering, MS-142, PO Box 1892, Houston, TX 77251-1892, Susan I. Gibson, University of Minnesota, 322 Cargill Building for Microbial and Plant Genomics, 1500 Gortner Ave., St. Paul, MN 55108, Jacqueline V. Shanks, Chemical and Biological Engineering, Iowa State University, 2114 Sweeney Hall, Ames, IA 50011, and Ka-Yiu San, Department of Bioengineering, Rice University, MS-142, PO Box 1892, Houston, TX 77005.

Plants produce a wide variety of secondary metabolites many of which are economically important in the areas of pharmaceuticals, nutrition, flavors and fragrances, fine chemicals, cosmetics, etc. In most cases, production of these compounds are limited to plants because synthetic routes of producing these structurally complex molecules are often more expensive than producing them natively within the plant even though the yields within the plant are small. Thus it is desirable to use the techniques of metabolic engineering to increase the yield of specific metabolites within the plant.

We have chosen to study the terpenoid indole alkaloid (TIA) pathway in *Catharanthus roseus* hairy roots as a model system because it produces two pharmaceutically important anticancer drugs vinblastine and vincristine which are used in the treatment of cancers such as lymphoma and leukemia. Over the course of my research, our focus has been to increase the flux down the terpenoid pathway and to combine those efforts with our previous success of increasing the flux down the indole pathway to increase the overall availability of the precursors to the TIAs. We have used the tools of precursor feeding studies, elicitor studies, and expression profiling of target genes by Q RT PCR to better understand the bottlenecks within the indole and terpenoid pathways. Using this information, we then constructed transgenic *C. roseus* hairy root lines by using molecular biology tools to overexpress one or more strategic genes in the terpenoid and/or indole pathway under the control of an inducible promoter system. In addition, we determined the effects of these genetic manipulations on TIA production by focusing on the levels of 6 TIA compounds, where the authentic standards were available. We have begun to further characterize other unidentified metabolites from these transgenic hairy root lines using LCMS and tandem LSMS/MS data.

6cl Effects of Confinement and Interfaces on Structural Relaxation of Thin Polymer Films above, below, and at the Glass Transition

Rodney D. Priestley, Chemical and Biological Engineering, Northwestern University, 2145 Sheridan Rd E136, Evanston, IL 60208

Over the past decade, there has been considerable interest in the effect that nanoscale confinement has on the properties of glass forming materials. The field has been driven because of both scientific and technological reasons. Examining the properties of glass formers confined to the nanoscale may provide vital information about the underlying physics of the material. With the emergence of nanotechnology, some glass formers, in particular, polymeric glass formers, will be used at increasingly smaller length scales. An understanding of how polymeric properties are impacted by nanoconfinement is essential to their full utilization in nanotechnology applications. The focus of my research has been investigating how confinement and interfaces impact the alpha relaxation dynamics, the glass transition temperature (Tg), and physical aging of polymeric glasses using fluorescence and dielectric spectroscopy.

The study of the effects of confinement on Tg of polymers began more than a dozen years ago. Since that time, it has been observed that Tg can undergo major changes relative to bulk Tg with confinement. In freely suspended films and films supported on a substrate that lack attractive substrate interactions, Tg decreases with nanoscale confinement. However, for films supported on a substrate that possess attractive interactions, Tg increases with nanoscale confinement. For suspended films and films supported on a substrate that lack attractive substrate interactions, the effect is associated with how the free surface locally reduces Tg, as evidenced by a fluorescence/multilayer method. Using the previously developed fluorescence/multilayer method, we have extensively investigated why Tg increases in a system (poly(methyl methacrylate)/silica) that possesses attractive interactions. The effect is associated with how the attractive substrate interactions increase Tg at the substrate interface. The attractive substrate interactions not only perturb Tg at the substrate but also do so in way that overcomes the opposing effects of the free surface. In addition, we have investigated related systems such as poly(ethyl methacrylate), poly(propyl methacrylate), and poly(isobutyl methacrylate) supported on silica.

Although the glass transition is associated with the cooperative dynamics (alpha relaxation dynamics) of many repeat units, how it is impacted by nanoconfinement has not been intensively investigated. Using dielectric spectroscopy few studies have reported a reduction in the peak temperature of the alpha process (T*f*Ñ) and a broadening of the alpha relaxation dynamics in polystyrene with confinement at temperatures above Tg. Such studies are extremely cumbersome since polystyrene possesses a weak dipole. To address this issue we have attached molecular dipoles to polystyrene to enhance its dielectric activity. In agreement with previous studies we observe a reduction T*f*Ñ and a broadening of the alpha relaxation dynamics with nanoscale confinement. Since the glass transition is related to the alpha relaxation dynamics, it must be that these dynamics at the interfaces are different from the bulk. However, because of experimental limitations no direct measurement of the complete alpha relaxation time distribution at the interfaces has been conducted. We have developed a novel multilayer/dielectric spectroscopy technique to allow for such measurements. For the first time, a reduction in the peak temperature of the alpha process and a broadening of the alpha-relaxation process are observed at the polystyrene interface. The results suggest that the impact of nanoconfinement on the alpha process originates from interfacial effects just as the impact of nanoconfinement on Tg originates from interfacial effects.

When cooled below Tg polymers are in a non-equilibrium state of higher specific volume and specific energy. As a result, if annealed below Tg, polymers will undergo a spontaneous structural relaxation process toward their equilibrium state. This relaxation process, often referred to as physical aging, can result in a time dependence of end-use properties of critical technological importance such as increased modulus, increased brittleness and reduced permeability. To date, conventional techniques for monitoring aging of bulk polymers are incapable of doing so for nanoconfined polymers. We have developed novel fluorescence methods to monitor aging in thin polymer films. More importantly, our technique allows for the monitoring of aging at specific locations in films near interfaces. For systems that lack attractive interactions with the substrate, aging is unaffected by confinement. However, for systems that possess attractive interactions with the substrate, aging is strongly suppressed. A novel use of the fluorescence/multilayer approach allows for monitoring aging for the first time at the free surface and substrate polymer interface. At both interfaces aging is reduced with a nearly complete suppression of aging observed at the substrate interface when aged deep in the glassy state. Our work indicates that nanoconfinement and interfacial effects strongly alter aging and that the development of polymer glasses that do not physically age may be possible. Using dielectric spectroscopy the molecular origins for the suppression of physical aging in nanoconfined systems have been investigated and will be presented as well.

6cm Engineering Extremophilic Chaperones for Biocatalysis and Nanobiotechnology

Timothy Whitehead, UC Berkeley, 201 Gilman Hall #1462, Berkeley, CA 94720

Extremophilic chaperones are promising because of their restorative functions within the cell, their ability to facilitate protein folding under harsh conditions, and their overall minimal protein folding systems (1). An often-overlooked aspect of chaperones is their exquisite quaternary structures, with different proteins forming ring structures (chaperonins), clamps (HSP70, prefoldin), spheres (sHSP), and filaments (chaperonins, γ prefoldin). In this presentation I describe use of chaperones for engineering of stress pathways in mesophilic hosts and exploitation of chaperones to build tunable nanostructures.

We have investigated the effects of heterologous expression of hyperthermophilic molecular chaperones on the survival phenotypes of *E. coli* TUNER strain. We first demonstrate that individual heterologous expression of three different molecular chaperones (prefoldin, small heat shock protein, or thermosome) can increase the temperature tolerance of *E. coli*. We demonstrate that chaperone function of these proteins as responsible for the survival phenotype in *E. coli*. We extend this approach to test for increased survivability against the biofuels ethanol and butanol. We conclude by testing additivity effects of alternative protein-stabilizing solutes (e.g., trehalose) with these hyperthermophilic chaperones.

Protein self-assembly is a particularly attractive option for bottom-up creation of tunable smart materials and hybrid inorganic/organic functional devices. Proteins can potentially serve as scaffolds for a host of nanostructured materials by self-assembling them at precise spatial distances and directing their placement onto solid supports at defined locations. We wish to find a general approach to build protein assemblages of arbitrary geometry over multiple length scales. To achieve this goal, the following benchmarks must be overcome: (A) a suitable protein template must be identified. (B) Control of the length of the protein template must be demonstrated. (C) Genetic modification of the template must be established. (D) Absolute control over a single dimension (e.g., length) must be engineered. These initial four steps directly lead to creation of more complex shapes and geometries. This presentation elucidates the progress we have made in these areas.

We have recently discovered a new type of filamentous protein (the γ prefoldin, or γ-PFD) from the hyperthermophile *Methanocaldococcus jannaschii* (2, 3). The filaments are polydisperse in length (microns) and monodisperse in width (8.5 nm) and height (4 nm). They are also remarkably stable, maintaining their structure up to at least 100ºC and in 7 M guanidinium chloride. With the overall aim of utilizing γ-PFD filaments as structural components in 3-D nano-scale architectures, we describe here the kinetic scheme behind γ-PFD filament formation and demonstrate control over filament length by rational design of an extension resistance mutant protein (so-called TERMs, or thermophilic extension resistant mutants) (4). We then demonstrate self-assembly of this filament with functionalized ends and a different functionalized interior. The next phase of this project involves using powerful genetic selection methodology to create periodic 3-D architectures of arbitrary geometry, and progress toward this end will be discussed.

1. Laksanalamai, P., Whitehead, T. A. & Robb, F. T. (2004) Nat Rev Microbiol 2, 315-24. 2. Boonyaratanakornkit, B. B., Simpson, A. J., Whitehead, T. A., Fraser, C. M., El-Sayed, N. M. & Clark, D. S. (2005) Environ Microbiol 7, 789-97. 3. Whitehead, T. A., Boonyaratanakornkit, B. B., Hollrigl, V. & Clark, D. S. (2007) Protein Science 16, 626-634. 4. Whitehead, T.A, Meadows, A.L., Clark, D.S., submitted.

6cn Computational Chemistry for Better Catalysis

N. Aaron Deskins, Chemical and Materials Sciences Division, Pacific Northwest National Laboratory, 902 Battelle Boulevard, Richland, WA 99352

My research interests consist of using modern computational chemistry methods, particularly density functional theory, to assist in the development of new materials for energy production that minimize harmful effects to the environment. Understanding the catalytic reactions that take place over several oxide metals has been the motivation for my research, and I plan to use this knowledge to build on future research. I would like to further study the photoreactivity of TiO_2, and especially the reactivity of TiO_2 nanostructures.

The focus of my graduate research at Purdue University was the modeling of ceria, CeO_2, as a catalyst and support in the water-gas shift (WGS) reaction. This reaction is important because it removes CO, which is a poison to proton exchange membrane (PEM) fuel cells, while producing H_2. I used density functional theory to simulate the doping of ceria with transition metals[1], and examined on the effect of these dopants on the reaction mechanisms. As part of this research I also studied the adsorption and reaction of several pertinent surface species, such as CO, H_2O, formates, and carbonates. My results showed that doping of ceria significantly changes the electronic structure of the surface, which enhances the WGS reactivity. Recent work has involved understanding the interactions of methanol and formaldehyde with ceria[2][3]. I also had experience modeling Pt and surface reconstruction removal by O_2 adsorption[4].

My current position as a post-doctoral researcher at the Pacific Northwest National Laboratory involves studying TiO_2 for photocatalytic purposes. TiO_2, titania, is used for pollutant molecule degradation, and also has the possibility for solar energy production. Charged species (electrons and holes) are created upon light adsorption, and can react with surface adsorbates. I have developed an atomic-level model for transport of these charge carriers[5][6], as well as modeled TiO_2 interfaces[7] that are crucial for charge transfer between nanoparticles. In collaboration with experiment I am also examining the adsorption and reactivity of carboxylic acids on TiO_2. Carboxylic acids are reasonably photo-active, so represent a prototypical reactant. Other projects involve the adsorption and migration of transition metal clusters over γ-alumina, Al_2O_3, an important industrial catalyst support.

[1] N. A. Deskins, A. Phatak, F. Ribeiro, K. T. Thomson, *Transition Metal Doping of Ceria and Subsequent Reduction from First-Principles Simulation*, Submitted to Journal of Physical Chemistry C.

[2] D. Mei, N. A. Deskins, M. Dupuis, Q. Ge, *Methanol Adsorption on the Clean $CeO_2(111)$ Surface : A*

Density Functional Theory Study, Journal of Physical Chemistry C, 111, 10514-10522 (2007).

[3] D. Mei, N. A. Deskins, M. Dupuis, *A Density Functional Theory Study of Formaldehyde Adsorption on Ceria*, Accepted in Surface Science.

[4] N. A. Deskins, J. Lauterbach and K. T. Thomson, *Lifting the Pt{100} surface reconstruction through oxygen adsorption: A density functional theory analysis*, Journal of Chemical Physics, 122, 184709 (2005).

[5] N. A. Deskins, M. Dupuis, *Electron Transport via Polaron Hopping in Bulk TiO_2: Density Functional Theory Characterization*, Physical Review B, 75, 195212 (2007).

[6] N.A. Deskins, M. Dupuis, *Intrinsic Hole Migration in TiO_2*, In preparation

[7] N. A. Deskins, S. Kerisit, K. M. Rosso, M. Dupuis, *Molecular Dynamics Characterization of Rutile-Anatase Interfaces* [Cover Article], Journal of Physical Chemistry C, 111, 9290-9298 (2007).

6co Modeling Biological Ligand Design with Pharmaceutical Accuracy

Michael R. Shirts, Department of Chemistry, Columbia University, Havemeyer Hall MC 3131, New York, NY 10027

Recent developments in computational processing power [1] and in theoretical methods [2] have made it possible to compute absolute ligand binding free energies to a precision of under 1 kcal/mol. This level of precision makes it possible to separate errors in sampling from inadequacies of molecular mechanics force fields, and begins to bring the design of drugs with high affinities from an art to a process of nanoscale engineering.

I compare experimental free energies of binding of FKBP-12 with a variety of ligands with extensive explicit solvent absolute free energies via molecular dynamics, with converged calculations being within 1 kcal/mol of experiment, along with a number of theoretical tools that make these precise and accurate calculations possible. I also present advances in continuum solvent Monte Carlo methods for calculating ligand binding affinity on this and other systems.

[1] M. Shirts and V. S. Pande, Science, 290:1903 (2000)

[2] M. R. Shirts and V. S. Pande, J. Chem. Phys. 122:134508 (2005)

Web Page: www.columbia.edu/~mrs2144/MRSCV.html

6cp Directed Differentiation and Tissue Engineering of Keratinocytes Derived from Human Embryonic Stem Cells

Christian M. Metallo, Lin Ji, Juan J. de Pablo, and Sean P. Palecek. Chemical and Biological Engineering, University of Wisconsin-Madison, Madison, WI 53706

Engineered tissue constructs with regenerative capacity require the incorporation of relatively large quantities of functional, lineage committed progenitors. Given their enhanced proliferative capacity and pluripotency, human embryonic stem cells (hESCs) are an attractive source of cells for tissue engineering. However, highly efficient processes must be developed to direct hESCs through developmental pathways to the appropriate lineages in order to exploit the technological potential of these cells. Formulation of such processes requires integration of engineering strategies with developmental biology concepts to effectively produce cells and tissues for scientific, diagnostic, and therapeutic use. Here we employ quantitative analysis of differentiated hESC populations to identify key signaling factors involved in ectodermal lineage specification; our primary focus is the generation of epithelial precursors that can be used to engineer regenerative constructs for the skin or related tissues. Application of retinoic acid (RA) to undifferentiated hESC populations, in conjunction with bone morphogenetic protein (BMP) signaling, efficiently mediates epithelial rather than neural differentiation. This differentiation process can be effectively used to generate essentially pure populations of hESC-derived keratinocytes under defined, feeder-free conditions. When coupled with the self-renewal capacity of hESCs, this method can produce relatively large quantities of functional, non-transformed keratinocytes. Furthermore, hESC-derived keratinocytes exhibit the capacity to terminally differentiate and form coherent epithelial sheets when presented with the appropriate cues. This differentiation process may therefore be used to produce epithelial tissues for scientific or clinical use.

6cq Engineering the Ribosome for the Development of New Technologies

Lydia M. Contreras and Matthew P. DeLisa. Cornell University, Chemical and Biomolecular Engineering, Ithaca, NY 14853

Engineering the Ribosome for the Development of New Technologies Recent data has linked ribosomes to a number of important cellular processes that extend beyond protein biosynthesis; some of these include protein folding, secretion regulation and targeting, and translation regulation. The realization that ribosomes dynamically connect polypeptide synthesis to other major cellular processes has led to the development of several powerful biotechnological applications such as a method known as in vitro ribosome display for isolating improved functional protein variants from large combinatorial libraries. Additionally, several key properties of ribosomes have led to the possibility of extending its use for protein purification and solubility enhancement. While the development of these technologies has been facilitated by the vast amount of structural and dynamic data that has been collected on the ribosome in recent years, a great deal about this complex machinery remains unknown. In particular, lack of knowledge about how the interactions that take place during protein translocation inside the ribosome may impact the downstream processing of proteins has hindered the development of new technologies and the extension of already existing ones. Our research has focused on exploring how the ribosome can be further used to improve the soluble expression of complex proteins in bacteria by: (1) understanding how the passage of proteins from the ribosome to the cytosol can affect its folding properties, (2) developing in vivo ribosome-mediated technologies for applications in protein engineering based on the interactive properties of the ribosome, and (3) engineering the ribosome for an efficient protein purification tool. Our progress so far indicates that ribosomes are remarkably "engineerable" and hold great promise for the development of numerous diverse technologies.

6cs Automating Development of Genome-Scale Metabolic Networks: *Clostridium Acetobutylicum*, a Known Singularity and Biofuel Production

Ryan S. Senger, Chemical and Biological Engineering, Northwestern University, 2145 Sheridan Road, Tech E250, Evanston, IL 60208-3120 and Eleftherios T. Papoutsakis, Dept. of Chemical Engineering, Delaware Biotechnology Institute, 15 Innovation Way, Newark, DE 19711.

Solventogenic non-pathogenic clostridia are of major importance for the production of biofuels through the acetone, butanol and ethanol (ABE) fermentation. However, much is still not understood by the scientific community regarding the molecular events that lead to a cascade of clostridial sigma factor-modulated regulatory programs. These programs generally include acidogenesis during vegetative growth followed by solventogenesis and multiple stages of cellular differentiation (sporulation). To investigate metabolic flux re-routing in response to clostridial regulatory programs, and to develop targets for further metabolic engineering, a genome-scale metabolic network was constructed for *Clostridium acetobutylicum* ATCC 824 using novel semi-automated technology. Multiple biochemical reaction, enzymatic and transporter databases were interfaced with genomic information and a technique of reverse metabolic network engineering to create a specialized metabolic network for this organism. For example, in spite of an "incomplete" TCA cycle, *C. acetobutylicum* ATCC 824 can biosynthesize all *L*-amino acids due to heavy incorporation of the urea cycle and a specialized function of the acetylornithine transaminase (ArgD, e.c. 2.6.1.11, CAC2388). In further development of the network, genetic algorithms were imposed with a known membrane protein selectivity criterion to address a singularity of the metabolic network concerning acetate and butyrate uptake in the production of acetone.

6ct Enabling Microscopic Simulators to Perform System-Level Analysis of Viscoelastic Flows

Zubair Anwar and Robert C. Armstrong. Department of Chemical Engineering, Massachusetts Institute of Technology, 77 Massachusetts Avenue, 66-258, Cambridge, MA 02139

State-of-the-art simulations for viscoelastic flow are currently limited in their ability to perform accurate analysis of stability and flow transitions in viscoelastic fluids of industrial importance. This is primarily due to the need to couple directly the macroscopic conservation equations with a microscopic kinetic theory model in order to accurately describe such flows. While this "hybrid" approach makes it possible to study coarse-grained models of polymeric liquids, there is no simple path either to obtaining long-term dynamics for a set of flow parameters or to providing accurate analyses of stability and flow transitions due to the unavailability of closed equations to which numerical techniques may be applied.

To address this problem, we utilize the coarse time-stepper based approach to stability and bifurcation analysis, which has shown potential to enable models at a 'fine' (microscopic/stochastic) level of description to be used to perform modeling tasks at a 'coarse' (macroscopic/systems) level [1]. The motivation for such an analysis is a framework for obtaining macroscopic time information from appropriately initialized calls to a microscopic simulator for only short times. The process of casting either a Fokker-Planck equation or an equivalent stochastic differential equation from polymer kinetic theory as a time-stepper allows for obtaining stationary states and performing stability/bifurcation analysis of the unavailable closed form macroscopic system. We present several examples that focus on application of this method at equilibrium and for homogeneous and complex flows.

Equilibrium Bifurcation Diagram of the Doi Model with the Onsager Excluded Volume Potential [2]

We present the equilibrium Doi model with the Onsager excluded volume potential as a system of dynamical equations for the basis functions chosen for discretization in orientation space and without any closure approximations. While a time-stepper based description only allows access to the stable steady states of the problem, with an iterative, matrix-free Newton solver and an equation for pseudo-arc length continuation, we were able to obtain the bifurcation diagram for the system at equilibrium, capturing both the stable and unstable states of the isotropic and nematic branches. We also show that most of the eigenvalues of the linearized system lie in a tight cluster about zero, with only a few eigenvalues leaving this cluster near a fold bifurcation.

Dilute Solution of Bead-spring Chains in Homogeneous Shear and Uniaxial Elongational Flows

The stochastic differential equation for a free draining bead spring chain with 5 FENE springs undergoing steady homogeneous shear or uniaxial elongation flow was used to construct a time-stepper for the stress contribution of each chain segment to total fluid stress. For a given set of chain segment stresses, the time-stepper evolved a representative ensemble of chain configurations over a fixed time horizon, before returning the evolved ensemble averaged chain segment stresses. This time-stepper was used in conjunction with an iterative, matrix-free Newton solver to obtain stationary states for the unclosed model that are in excellent agreement with the steady states computed from a dynamic simulation. The eigenvalues of the system at steady state for a range of Weissenberg numbers were shown to lie in a tight cluster, which allowed the use of an inexact Newton method to obtain stationary states from a stochastic model.

Pressure-Driven Flow of Non-Interacting Rigid Dumbbells in a Planar Channel [2]

In this work, a micro-macro simulation of a non-homogeneous flow was recast as a black-box integrator that took a given distribution of dumbbell orientations for the flow domain and returned an evolved distribution after integration over some specified time horizon. The integrator also computed consistent flow fields at each intermediate time-step. Having formulated the black box code, we were then able to make calls to it from a Newton-GMRES solver for the purpose of obtaining the steady state of the underlying nonlinear system, given an initial guess of isotropy and without any modification to the micro-macro simulation algorithm.

[1] I. G. Kevrekidis, C. W. Gear, and G. Hummer. Equation-free: The computer-aided analysis of comptex multiscale systems. AICHE Journal, 50(7):1346–1355, 2004.

[2] Z. Anwar, and Robert C. Armstrong. Using Newton-GMRES for viscoelastic flow time-steppers. J. Non-Newtonian Fluid Mech., Submitted.

6cu New Materials for Reduced Cost, High Performance, Micro Direct Methanol Fuel Cells

William Mustain, Chemical and Biomolecular Engineering, Georgia Institute of Technology, 311 Ferst Dr., Atlanta, GA 30332

Direct methanol fuel cells (DMFCs) have the potential to provide clean, efficient, sustainable power for a broad spectrum of applications. Heat engines are highly competitive in high power-density applications where device size and fuel infrastructure is not a primary concern. However, at smaller sizes, electrochemical devices become more competitive because they are simpler than heat engines, requiring no moving parts, while providing higher energy densities since they scale as function of their surface area, not volume. Thus, the natural advantages of electrochemical devices can be most easily used in portable and low-power devices where room temperature operation, limited support infrastructure, and simple designs can be exploited. Also, liquid-based fuel cells provide higher energy density compared to Li-ion batteries (approximately 10 X), lower cost, and are environmentally friendly. There are several factors that should be addressed before commercialization of these devices is seriously considered, including: low open circuit voltage and performance of passive cells, cost, fuel loss (mostly through crossover), fuel delivery and integration. Therefore, it is critical to develop new materials that alleviate these technical concerns and to target power levels and applications where fuel cells are truly competitive with established technologies and market insertion is facile. The presented work has been done with the intention of creating new fuel cells and designs which address some of the fundamental problems with state-of-the-art devices.

Thin-film catalyst/glass composite electrodes for methanol oxidation and oxygen reduction reactions

Carbon-supported platinum and platinum-ruthenium alloy nanoparticles were incorporated into a porous silicon dioxide glass matrix. The glass was prepared by the sol-gel hydrolysis reaction between TEOS and water in the presence of methanol. During the hydrolysis reaction, the nanoparticles are introduced to the system, which is then deposited on the cell substrate and imposed to several curing steps where the solvent is evaporated and the reaction completed. The resulting structure was porous, homogeneous and the catalyst was well dispersed. Finally, the catalyst islands were grown by electroless deposition until the propagation threshold was met and the resulting layer was used as both catalyst and current collector. The electrochemical performance of the resulting films for both methanol oxidation and oxygen reduction was studied by ex-situ cyclic voltammetry in sulfuric acid electrolyte, polarization measurements in a micro DMFC and ac impedance in a two-compartment glass cell. Experimental results indicate that the electrochemical performance of the DMFC is improved compared to cells prepared with Nafion®-based catalyst layers, where a 50 mV increase in the open circuit potential and higher currents are realized for fuel cells prepared with the composite catalyst.

High activity, methanol tolerant Co-Pd alloy electrocatalysts for the oxygen reduction reaction

Cobalt-palladium (CoPdx) bimetallic electrocatalysts of various compositions were prepared and their physical and electrochemical properties elucidated. The electrocatalytic activity electrodeposited polycrystalline CoPdx electrodes for the oxygen reduction reaction was compared using cyclic voltammograms obtained in oxygen saturated, 0.5 M H2SO4 at 25 oC. CoPd3 was identified as having the highest activity, showing peak currents comparable to a polycrystalline Pt electrode under identical experimental conditions. The rotating disk electrode technique was used to collect kinetic data for the oxygen reduction on CoPd3 and the activity of the binary electrocatalyst was evaluated. It was found that the ORR on CoPd3 has a modest activation energy, 52 kJ/mol, and a Tafel slope of approximately 60 mV/decade, indicating that the rate determining step is a chemical stop following the first electron transfer step and likely includes the breaking of the O-O bond. Also, the electrochemical activity of carbon supported cobalt-palladium alloy electrocatalysts of various compositions was investigated for the oxygen reduction reaction in a 5 cm2 single cell polymer electrolyte membrane fuel cell. Polarization experiments have been conducted at various temperatures between 30 and 60 oC and it appears that the catalyst with a nominal cobalt-palladium atomic ratio of 3:1, CoPd3, exhibits the best performance of all studied catalysts, exhibiting a catalytic activity comparable to a commercial Pt catalyst. The CoPd3 catalyst also exhibits excellent chemical stability, with the current decreasing by only 10 % at 0.8 V over 25 hours. The CoPd3 catalyst also exhibits superior tolerance to methanol crossover poisoning than Pt.

Novel passive fuel delivery and ion-transport mechanism for DMFCs

In this work, a new type of passive direct methanol fuel cell was developed to allow the fuel cell to operate with limited support infrastructure and no moving parts. In a departure from previous work, this passive DMFC utilizes a stationary fuel phase and allows the anode to "float" in the fuel reservoir. During operation, ions are generated at the floating anode electrode (PtRu), conducted through the solution phase to the liquid solution saturated wicking material (cotton, cellulose, etc.). Also, a small amount of an earth metal salt (i.e. sodium bisulfate) was added to the aqueous liquid fuel (10-12 M methanol) in order to increase the ionic conductivity of the solution phase. The wicking material moves the ions to the solid proton exchange membrane where they are transported to the cathode and consumed. The current invention addresses one of the most significant issues associated with conventional passive DMFCs: consistent and predictable fuel delivery to the electrode surface. In providing an ion pathway, instead of a fuel delivery device, one intrinsic flaw of wicking a binary fuel mixture is addressed. All fibrous or porous wicking materials transport liquids selectively, dictated by the surface tension of the liquid, resulting in inconsistent concentration gradients and unpredictable performance for long-term operation. Both polarization and chronoamperometric results indicate that the device shows excellent performance stability and its operation is not affected by cell orientation.

Research Interests

Nafion®-free micro DMFCs; Anion-exchange membranes; Novel oxygen reduction and methanol oxidation catalysts in acid and alkaline media; Kinetic evaluation of carbonate formation and consumption in anionic fuel cells; Predictive modeling for electrochemical systems with a focus on the oxygen reduction reaction in acid media; Reaction pathway and kinetic decomposition for methanol oxidation.

References

1. W.E. Mustain, S. Prakash, H. Kim and P. Kohl, "Fully Passive High Energy Density Hybrid Power Module", Invention disclosure filed May 2007.

2. S. Prakash, W.E. Mustain and P. Kohl, "Ionic Conductivity of Phosphorous-doped SiO2 Ultra Thin Films", to be submitted May 2007.

3. W.E. Mustain, S. Prakash, H. Kim and P. Kohl, "Thin-film Pt-Glass and PtRu-Glass Composite Catalyst Layer for Fuel Cells", Invention disclosure filed May 2007.

4. W. Mustain and P. Kohl, "Characterization of Thin-Film Electrodes on Proton-Conducting Glass Membranes for Micro DMFC Applications", 211th Meeting of the Electrochemical Society, May 2007.

5. W.E. Mustain, K. Kepler and J. Prakash, "Investigations of carbon-supported CoPd3 catalysts as oxygen cathodes in PEM fuel cells", Electrochem. Comm., 8 (2006) 406.

6. W.E. Mustain, K. Kepler and J. Prakash, "CoPdx oxygen reduction electrocatalysts for polymer electrolyte membrane and direct methanol fuel cells", Electrochim. Acta, 52 (2007) 2102.

7. W.E Mustain and J. Prakash, "Kinetics and Mechanism for the Oxygen Reduction Reaction on Polycrystalline Cobalt-Palladium Electrocatalysts in Acid Media", J. Power Sources, In Press.

8. W. Mustain and J. Prakash, "Cobalt-Palladium Electrocatalysts for Oxygen Reduction in Acid Media", 211th Meeting of the Electrochemical Society, May 2007.

9. W.E. Mustain and P. Kohl, "Floating Anode Ion Wicking Direct Methanol Fuel Cell", Invention disclosure filed May 2007.

6cv Insights on Biological and Material Properties of Nanoscale Systems from Multiscale Modeling, Simulation, and Experiment

Elaine R. Chan, Semiconductor Electronics Division, Electronics and Electrical Engineering Laboratory, NIST, 100 Bureau Drive, Gaithersburg, MD 20899-8120

My research is focused on understanding the physical mechanisms that govern the properties of nanoscale materials and biological systems. Multiscale modeling and simulation, together with experiment, provide a powerful framework for identifying the origin of these properties and determining how they govern the system meso- and macro-scopic behavior. I discuss here the self-assembly of hybrid organic/inorganic polymer-tethered nanoparticles and the structure-function relationship of bacterial pore-forming toxin proteins. Specific topics of future interest that expand on these research experiences will also be discussed.

Self-assembly in hybrid organic/inorganic nanostructured materials

Self-assembly is a promising route for potentially designing new nanoparticle-based materials and devices with novel properties. The motivation for this concept is based on the notion that biomimetic principles can be utilized to precisely organize the particles into ordered structures. However, knowledge of the processes that occur during self-assembly is required to fabricate these materials and devices. Because the significant length and time scales of these processes are difficult to gain access to experimentally, a multiscale modeling and simulation approach is useful for investigating materials assembly. This approach can also be used to verify the reliability of different computational approaches and to calculate properties that can be verified experimentally.

Here I present computer simulation results for the self-assembly of nanoscale cubic polyhedral oligomeric silsesquioxane (POSS) molecules functionalized with polymeric tethers. The development of coarse-grained models and interaction potentials, on the basis of insights from quantum mechanical calculations and explicit atom simulations, for studying polymer-tethered POSS assemblies will be discussed. The mesoscale particle simulations predict self-assembly of the molecules into bulk structures similar to those found in coil-coil block copolymer and surfactant systems. The influence of various thermophysical and material parameters on self-assembly, including temperature, concentration, nanoparticle shape, polymer tether composition, and the number of tether substituents, is examined. This work is a collaborative effort with Sharon Glotzer and John Kieffer (University of Michigan), Matthew Neurock (University of Virginia), and Clare McCabe and Peter Cummings (Vanderbilt University).

Structure-function properties of bacterial pore-forming toxin proteins

Recent bioterrorism acts emphasize the need to gain a fundamental understanding of bacterial pathogenicity. Although advances have been made toward elucidating the structure of several anthrax toxins, the molecular mechanism(s) of action are not completely understood. It is currently believed that late stage anthrax infection is caused by the transport of toxic proteins secreted by the bacterium *Bacillus anthracis* through an ion channel formed by a third protein that is secreted by the organism. This hypothesis can be tested, in part, by precise measurements and theoretical predictions of the ionic current flowing through single anthrax channels. Because the crystallographic structure of the nanopore is not yet available, knowledge of the pore's thermophysical and electrophysiological properties is important for validating and improving existing models of channel structure. This information will aid the development of more effective therapeutic agents against anthrax, and may also provide the aid the sensitive and selective detection of anthtrax toxins using single nanopores.

To test the structural model for the anthrax nanopore, the single channel current-voltage relationship is measured in the presence of monovalent electrolyte solutions and under a wide range of electrolyte concentrations and pH values. Advanced statistical signal processing algorithms are applied to the stochastic fluctuations in the single-channel ionic current time series to identify the number, value and lifetimes of all the conductance states. Insights from these experimental and theoretical measurements on pore structural behavior will be discussed. This work is being performed in collaboration with John Kasianowicz and Vincent Stanford (NIST), Tam Nguyen (SAIC-NCI), and Maria Kurnikova (Carnegie Mellon University).

6cw Development of a Biomems Pulsatile Hormone Delivery System

Eric E. Nuxoll, Department of Pharmaceutics, University of Minnesota, 9-177 Weaver Densford Hall, 308 Harvard St. SE, Minneapolis, MN 55455

People requiring rhythmic delivery of drugs, notably Gonadotropin Releasing Hormone (GnRH), currently rely on catheter-based subcutaneous or intravenous delivery systems with a host of inconveniences and dangers. A hydrogel-based delivery system has been demonstrated which may replace the current therapy with a simple, implantable device.

This system consists of a reservoir covered by a hydrogel membrane which undergoes a phase transition (and consequent permeability change) at particular pH values. The reservoir contains the therapeutic agent along with the enzymes glucose oxidase and catalase. In its swollen state, the membrane is quite permeable, allowing external glucose to diffuse into the reservoir where it is enzymatically converted to gluconic acid. The subsequent drop in pH causes the hydrogel to collapse and dramatically decrease its permeability. Glucose can no longer enter the reservoir (and therapeutic agent can no longer leave) until the chemical signal relaxes (e.g. faster-diffusing protons exit the system) and the membrane returns to its permeable state. The cycle then repeats.

This system has been demonstrated and refined on the lab-bench scale (~80 mL total volume) under a variety of conditions and parameter values, but many challenges remain before a viable implantation is achieved. Among these challenges are the miniaturization and biocompatibilization of the bench-scale system. To miniaturize the device, fabrication techniques adapted from the semiconductor industry have been used to develop new reservoirs and membrane mounts. The main challenge in biocompatibilization is protecting the stimuli-sensitive hydrogel from immunoproteins without hindering the passive transport of glucose or hormone. A composite nanoporous membrane is being developed to be used as a protective covering for this purpose.

6cx Molecular Engineering of Surfaces for Biomaterials and Biosensors

Matthew Bernards and Shaoyi Jiang. Chemical Engineering, University of Washington, Benson Hall, Box 351750, Seattle, WA 98195

My research efforts have been focused on two emerging areas of bionanotechnology centered on biological interfaces. The first of these areas is protein-biomaterial interactions where I have completed a series of fundamental studies in an effort to develop mechanisms for controlling protein orientation and conformation for guiding the design of biomaterials. The second area that I have focused on is the development of super-low fouling surface coatings for a number of applications. These projects have allowed me to collaborate with a number of other researchers in a highly interdisciplinary atmosphere.

Protein-Biomaterial Interactions

It has been suggested that it is possible to influence the cellular response to implanted biomaterials, especially at early interaction times, by controlling the orientation, conformation, and composition of the initial layer of proteins that adsorb to the material upon implantation. One portion of my research has focused on developing three mechanisms for controlling protein orientation and conformation for applications in biomaterials and biosensors. These three mechanisms include specific protein-protein interaction driven orientation, specific protein-substrate interaction driven orientation, and charge-driven protein orientation.

The first mechanism, specific protein-protein interaction driven orientation, was probed by comparing the cellular adhesion properties of two bone proteins, bone osteopontin (OPN) and bone sialoprotein (BSP) when they were specifically bound to collagen type I or randomly adsorbed to control substrates. During natural bone formation cells first lay down a collagenous matrix composed of primarily type 1 collagen. After this collagen network is formed, non-collagenous proteins bind to the collagen matrix and then the matrix is mineralized to form mature bone. Both OPN and BSP have been implicated in this process and are known to have specific binding interactions with collagen. All of the substrates were prepared with identical amounts of adsorbed OPN or BSP as determined by radiolabeled adsorption isotherms. This allows for a direct comparison of the cell binding ability of these proteins on all of the surfaces examined. In this study it was determined that OPN has a favorable orientation imparted to it for cell binding when it is specifically bound to collagen I. At the same time, it was found that the cellular adhesion properties of BSP are a direct result of the conformational flexibility of the protein. When directly compared to each other, it was determined that OPN is more important than BSP for cellular adhesion to collagen.

The second mechanism for controlling protein orientation is that of specific protein-substrate interaction driven orientation. In this study OPN and BSP were specifically bound to hydroxyapatite (HAP), a mineral that mimics the main component of bone. The results of this work indicate that BSP has a more favorable orientation or conformation than OPN for cellular adhesion to HAP. However, this positive effect on cell binding is eliminated when the surface roughness of the underlying HAP substrate becomes too great.

The final mechanism is charge-driven protein orientation. This mechanism has been used to control the orientation of antibodies for biosensor applications and osteopontin for cell adhesion by others. I utilized this mechanism to compare the roles that both orientation and conformation have on the cellular adhesion properties of adsorbed vitronectin (VN), a naturally labile protein. The orientation of VN was controlled with charged self-assembled monolayers and the conformation was varied with changes in pH. In this study it was found that conformation plays a greater role than orientation in the cell binding properties of VN. Additionally, the maximum cell binding was obtained under conditions with an acidic pH as typical of that found in a wound healing environment.

Through these fundamental studies I have been able to elucidate an understanding of when orientation or conformation will dominate the bioactivity of a protein. There is a trade-off between the conformational flexibility of the protein structure in the area surrounding the bioactive site of interest and the location of the bioactive site relative to the site of orientation modulation. At the same time, it has been shown that it is possible to control the orientation and conformation of adsorbed proteins using these three mechanisms for biomaterials applications.

Super-low Fouling Surface Coatings

The control of non-specific protein adsorption is useful throughout a number of industries including biomedical materials, biosensors, and ship coating applications. Recent advances in this field have suggested that surface coatings developed from a mixture of charged components that maintain an overall neutral surface charge, greatly reduce the non-specific protein adsorption. My recent research efforts have focused on using atom transfer radical polymerization (ATRP) to graft zwitterionic and mixed charge monomers to surfaces to test their non-fouling properties.

The main focus of this research has been to develop and test a novel class of polymeric brushes formed by ATRP from mixed solutions of monomers with various charged groups. The polymer brushes were characterized by atomic force microscopy (AFM) and electron spectroscopy for chemical analysis (ESCA) and the protein adsorption to these surfaces was measured using a surface plasmon resonance biosensor. In this work it was determined that in order to achieve super-low fouling surfaces, the monomers must be well mixed on the surface in equal ratios. Additionally, the reaction conditions must be closely monitored in order to achieve this desired surface ratio. Further work is on-going to determine the importance of the polymer brush thickness on its non-fouling properties.

Research Directions

One portion of my proposed plan of research is centered on the development of biomimetic materials for tissue engineering applications. In order to develop tissue engineering scaffolds or biomaterial coatings that closely resemble native tissue, an understanding of the role that specific proteins play in specific tissues must be developed. One way to develop this understanding is to break down native tissues into their components and characterize the interactions of specific proteins with each component at the molecular level. This detailed functional information can be combined with recombinant protein technology and nanoscale processing techniques to develop 3-dimensional biomimetic implants with molecular level control. It will result in scaffolds which have superior properties to traditional scaffolds because they closely mimic the native tissue they are replacing.

In my future research, I would also like to work on the development of an immobilized, bioactive cell-ligand substrate for use in an SPR sensor. The goal of this work would be to immobilize cancer cell ligands to a sensor platform in a way that allows them to maintain their native biological properties. This substrate would rapidly advance the development of specific ligand targeting drugs by providing an array-based diagnostics sensor for the analysis of novel ligand-targeting drug delivery vehicles in a rapid and quantitative method. Healthy, normal cell ligands would also be included in the array to analyze the targeting specificity of the drug delivery vehicles to cancer cells versus normal cells.

6cy Micro/nano Electrochemical Systems for Sensor and Energy Applications

Ilwhan Oh, Department of Chemical and Biomolecular Engineering, University of Illinois at Urbana-Champaign, 294 RAL, MC-712, 600 S. Mathews Ave., Urbana, IL 61801

Following two topics summarize the works accomplished in my Ph.D and post-doctoral research.

(1) Surface electrochemistry of oxygen reduction and formic acid oxidation

Several underpotential deposits (UPD) show electrocatalytic activity for oxygen reduction. Bi, Tl, and Pb UPD on Au(111) were studied with surface science tools in order to verify structure-function relation. Proposed mechanism involves a bimetallic interaction between the adsorbed oxygen molecule and the catalyst surface.

(2) Microfabricated electrochemical sensor for toxic vapor

Electrochemical sensor for toxic vapors, such as organophosphate and hydrogen cyanide, is developed. In order to build a fast and sensitive sensor, critical aspects include fabrication of high-area gas-liquid interface, thin-electrolyte cell, and the sensing electrodes with low detection limit.

For the future research, I propose to work on the following problems:

(1) Electrochemical sensor for peroxide explosives

Electrochemistry can play a unique and significant role in microfabricated systems. Compared to optical transducer, fabrication and measurement of electrochemical transducer is much simpler and cheaper. Also, copper electrodeposition for high-speed Cu/low-k interconnect is a representative example of electrochemistry contribution to microfabrication. Future research opportunities and challenges include electrochemical fabrication and application of nanowires, fabrication of MEMS-compatible reference electrode, and incorporation of enzymes or other biomaterials into electrochemical sensors and energy sources.

(2) Electrochemistry of graphene materials

Beside fabrication techniques and device design, one of the most important pillars in the microchemical systems is materials that constitute the microchemical systems. Recently, a very interesting carbon material, named graphene, is reported. The graphene can be grown as an ultrathin graphite layer grown on a single-crystal silicon carbide by thermal decomposition. This novel material is compatible with the conventional microfabrication techniques and has the electronic property similar to that of the carbon nanotube. While initial research activities were mainly on solid-state electronic properties of the epitaxial graphene, chemical and electrochemical aspects of the epitaxial graphene are rarely studied. On the other hand, it is well known that metallic or semiconductor nanowires can be fabricated onto a graphite surface by an edge-decoration electrodeposition. The edge decoration, combined with lithography, can be used to fabricate metallic and semiconducting nanostructures of custom shape and size.

6cz Catalytic Routes for the Production of Fuels and Chemicals

Carsten Sievers[1], Pradeep K. Agrawal[1], Christopher W. Jones[1], and Johannes A. Lercher[2]. (1) School of Chemical and Biomolecular Engineering, Georgia Institute of Technology, 311 Ferst Drive, N.W., Atlanta, GA 30332-0100, (2) Department of Chemistry, Technische Universität München, Lichtenbergstrasse 4, Garching, 85747, Germany

During my doctoral studies with Prof. Johannes A. Lercher at the Technische Universität München, Germany, I was working on fundamental aspects of different heterogeneously catalyzed reactions. An essential part of this work was the physico-chemical characterization of the catalysts with a focus on IR and solid state NMR spectroscopy.

Polarization and activation of alkanes on zeolite LaX. La^{3+} exchanged faujasite type zeolites are of great interest for catalytic applications, such as cracking and alkylation. When alkanes were adsorbed on zeolite LaX (faujasite with a low Si/Al ratio), strong polarization of the sorbates was observed [1]. This polarization was particularly strong for alkanes, which are branched in the 2 position. It was shown by ^{13}C CP NMR spectroscopy that a partial positive charge is induced on secondary and tertiary carbon atoms, whereas a partial negative charge was found on tertiary protons. The polarization is strong enough to readily activate octane isomers with multiple branchings at ambient temperature [2]. The activation occurs via hydride abstraction forming a surface carboxy group/carbenium ion. Once reactive species are present on the surface, typical carbenium ion reactions (cracking, alkylation, isomerization, hydride transfer) are observed.

Isobutane/2-butene alkylation over lanthanum exchanged faujasite type zeolites. Premature deactivation is the key problem for the industrial implementation of solid acid catalysts for industrial isobutane/2-butene alkylation. Deactivation was investigated by performing the alkylation reaction for different times on stream [3]. After reaction, the catalysts were characterized by various physicochemical techniques. It was shown that the deactivation is caused by a combination of site poisoning and pore mouth plugging. Moreover, LaX (Si/Al = 1.1) and LaY (Si/Al = 2.4) were compared as catalysts for isobutane/2-butene alkylation [4]. The catalyst lifetime and alkylate yield were superior for LaX. The differences were assigned to the higher concentration of strong Brønsted acid sites in LaX. This is in contrast to the commonly accepted trend that the strength of Brønsted acid sites in zeolites increases with increasing Si/Al ratio. The reasons for this trend are the presence of residual sodium in LaY and a stronger distortion of the zeolite framework by polyvalent lanthanum cations in LaX.

Tailoring catalytic activity and selectivity in supported ionic liquids. Immobilizing metal complexes in thin films of ionic liquids on silica supports is a novel approach for combining the selectivity of heterogeneous catalysts with the possibility of facile separation of heterogeneous systems. The interaction of the immobilized metal complexes with the supported ionic liquids was investigated by ^{1}H MAS NMR spectroscopy. It was shown that tight solvent cages are formed around the complexes [5]. The selectivity in these cages can be very different from reactions in classic solvents. The enatiomeric excess for the hydrogenation of acetophenone to (*S*)-1-phenylethanol exceeded the ee found in methanol by far [6]. Depending on the reaction conditions high selectivities were found for the Markownikoff and *anti*-Markownikoff product for the hydroamination of styrene with aniline [7].

I recently started my work as a postdoctoral fellow at the Georgia Institute of Technology under the supervision of Prof. Christopher W. Jones and Prof. Pradeep K. Agrawal. My research is focused on catalytic pathways for the production of fuels and hydrogen from woody biomass.

Production of fuels and hydrogen from lignocellulosic biomass. Utilization of renewable renewable feedstocks is a research field of rapidly growing interest. Although a few industrial processes (e.g. ethanol from corn) have been implemented, it is clear that economic viability and sustainability of these technologies require further research. Several innovative concepts have been demonstrated in the last years but many catalytic studies have only been performed with model compounds. Our research is focused on the development of an integrated route to fuels and hydrogen from woody biomass. The first step in the chemical processing of biomass feedstock is depolymerization of the lignin and holocellulose fractions of the wood by hydrolysis. The dissolved fractions are analyzed by HPLC allowing the quantification of primary hydrolysis and degradation products. Analysis of the residues by ^{13}C CP MAS NMR provides information on the concentrations of lignin and carbohydrates as well as on chemical modifications, which occur during the depolymerization process. With the information on dissolution and degradation yields, process conditions can be optimized to provide a higher selectivity to sugars or other intermediates, which will be used as feedstock for catalytic upgrading. Moreover, novel reaction media are investigated for reforming of biomass and model compounds. First experiments indicate that reforming can be achieved under surprisingly mild conditions.

Processing of biological feedstocks will be a key part in my future research. In particular, I am planning to focus on the interaction of key compounds in biomass reforming with heterogeneous catalysts. These fundamental studies will be complemented by work on specific application for the productions of chemicals from biomass. Additionally, I am interested in enantioselective heterogeneous catalysis.

References

[1] C. Sievers, A. Onda, R. Olindo, and J.A. Lercher, J. Phys. Chem. C 111 (2007) 5454-5464.

[2] C. Sievers, A. Onda, A. Guzman, K.S. Otillinger, R. Olindo, and J.A. Lercher, J. Phys. Chem. C 111 (2007) 210-218.

[3] C. Sievers, I. Zuazo, A. Guzman, R. Olindo, H. Syska, and J.A. Lercher, J. Catal. 246 (2007) 315-324.

[4] C. Sievers, J.S. Liebert, M.M. Stratmann, R. Olindo, and J.A. Lercher, Appl. Catal. A-Gen. submitted for publication.

[5] C. Sievers, O. Jimenez, T.E. Müller, S. Steuernagel, and J.A. Lercher, J. Am. Chem. Soc. 128 (2006) 13990-13991.

[6] K.L. Fow, S. Jaenicke, T.E. Müller, and C. Sievers, J. Mol. Catal. A-Chem. (2007) accepted for publication.

[7] O. Jimenez, T.E. Müller, C. Sievers, A. Spirkl, and J.A. Lercher, Chem. Commun. (2006) 2974-2976.

6da Solvent Design for Pharmaceutical Process

Charles Acquah, Chemical Engineering, University of Connecticut, Storrs, CT 06269

Control of crystal morphology in industrial crystallization processes is very important since this affects the ease of separation, washing, drying, packaging, handling and storage of crystals. In the case of pharmaceutical products, crystal morphology affects drug dissolution, bioavailability and tableting characteristics. Factors such as cooling rate, agitation speed, agitation time, degree of supersaturation, seeding and presence of impurities affect the crystal morphology. In addition to these, the type of solvent chosen for a crystallization process has a strong influence on morphology. Investigating the full impact of solvents on morphology of crystals will help us gain insights useful for solvent selection and solvent design strategies for crystallization processes. Our studies involve three aspects, namely: experiments, molecular dynamics and modeling. So far a comprehensive experimental study of crystal shape distribution of four carboxylic acids re-crystallized from six different solvents have been undertaken. Solvents were chosen to cover a wide range in polarity ranging from non-polar to highly polar. Crystals were characterized using Optical Microscopy and Scanning Electron Microscopy techniques. Powder X-Ray Diffraction patterns obtained from crystals was used to rule out polymorph formation. Experimental results reveal that low-viscous solvents with low hydrogen-bonding solubility parameter give higher aspect ratio (ratio between two major dimensions) crystals while polar solvents with comparatively higher hydrogen bonding ability yield lower aspect ratio crystals. However, for high-viscous polar solvents, high aspect ratio crystals were observed. The next phase of the research employs molecular dynamics to explain observed morphological trends. Since crystal-solvent interactions are mainly controlled by interfacial tension between the two phases, the goal here is to use molecular dynamics simulations to extract key interfacial thermodynamic quantities such as the surface free energy which play key role in determining crystal morphology. The final phase of the research is aimed at developing a semi-empirical model for predicting crystal shape of carboxylic acids grown from solution.

6db Copper Electrodeposition Onto Resistive Ruthenium Substrates in the Presence of Solution Additives

Jennifer Younker, Chemical and Biomolecular Engineering, Universisty of Illinois at Urbana-Champaign, 600 S. Mathews Ave., Urbana, IL 61801

There is much interest in direct seedless electrodeposition of copper onto thin resistive barrier layers, such as ruthenium. Removing the need for a copper seed layer is a critical step toward reaching the goal of submicron feature sizes, where copper seed layers have led to the development of defects and voids. In order for deposition onto a barrier layer to be feasible, the barrier layer must prevent copper diffusion into the dielectric, act as an adhesion layer for the copper, and be conductive. Ruthenium meets these requirements due to its low resistivity and capacity for preventing copper diffusion.

In addition to direct deposition onto barrier layers, the influence of electrodeposition solution additives on copper nucleation and growth morphology on resistive electrodes is not well understood. As device dimensions decrease, the role of solution additives on copper metallization, specifically nucleation and nucleation cluster density, will become increasingly important.

Copper electrodeposition experiments on resistive ruthenium electrodes were performed for understanding the chemistry of the additives and their effect on copper nucleation. Results of these experiments provide information about the dependence relationship of nucleation density on the nucleation mechanism, nuclei growth, plating solution additive concentrations, and operating conditions. The solution additive chemistry information gained from these first experiments will be utilized in computational tools to predict the growth of thin copper films on novel barrier layers. The results obtained from the computational tools will assist in the design of nucleation and overgrowth experiments leading to thin film formation. The impact of this research will be the ability to "tune" the additive chemistry to achieve high nucleation density and rapid lateral growth and coalescence, as well as provide valuable experimental information to validate simulation tools developed by research group members.

I have a strong history of classroom teaching in Chemical Engineering, General Engineering, and high school chemistry. I have a genuine passion for teaching and am highly committed to education. My teaching experience combined with my research experience in microelectronics, electrochemistry, and materials makes me feel confident in my ability to teach a variety of Chemical Engineering courses.

6dc Nanocomposites of Polyurethane Elastomers

J. J. Huang[1], S. M. Liff[1], G. H. McKinley[1], and M.C. Boyce[2]. (1) Mechanical Engineering and Institute for Soldier Nanotechnologies, Massachusetts Institute of Technology, Building NE 47-511, 77 Massachusetts Avenue, Cambridge, MA 02139, (2) Mechanical Engineeering, Massachusetts Institute of Technology, Building NE 47-511, 77 Massachusetts Avenue, Cambridge, MA 02139

Elastomeric materials with high stiffness, strength and ductility are desirable for impact loading like blast and ballistic events. We have fabricated nanocomposites of a series of polyurethane elastomers with the same types of hard and soft segment but different concentration, and different chain extenders. A solvent exchange method was used in preparation of nanocomposites. The formed nanocomposites were extensively characterized by transmission electron microscopy, X-ray diffraction, differential scanning calorimetry, dynamic mechanical analysis, tensile testing, and split Hopkinson pressure bar techniques. Our results show that good exfoliation of the clay was achieved, and that the well-exfoliated nanocomposites have dramatically improved moduli (in some cases, the increase in moduli is about 16 fold). The stress-strain behavior at both low and high rates of the nanocomposites shows non-linear behavior, and the nanocomposites show higher stress levels than the neat polyurethane materials. At high strain rates, the polyurethane nanocomposites seem to be insensitive to rates.

6dd Multiscale Approaches towards Nanomaterials Design: Microporous Thin Films and Nanoparticle Assemblies

Mark A. Snyder, Department of Chemical Engineering and Materials Science, University of Minnesota, 421 Washington Ave. SE, Minneapolis, MN 55455

Microporous thin films and ordered nanoparticle assemblies and coatings compose a rich subset of current nanotechnology initiatives with exciting applications spanning high-resolution molecular separations and chemical sensing, fuel cell membranes, nanoscale materials templating, colloidal lithography, protective coatings, and even biological applications. Realization of these materials demands rational, hierarchical device design in which structure-properties relations are developed to bridge molecular details with those at the nanoscopic, microscopic, and device levels. As such, recent nanotechnology roadmaps have espoused a synergistic theme, recognizing the need for intimate combination of theory, modeling, synthesis, and characterization for developing structure-properties relations directed at device design and structure and defect control.

Through my doctoral (**Dept. of Chemical Engineering, University of Delaware with Professor Dion Vlachos**) and postdoctoral (**Dept. of Chemical Engineering and Materials Science, University of Minnesota with Professor Michael Tsapatsis**) work, we have focused on several key nanomaterials issues including: 1) benign, facile synthesis of oxide nanoparticles as building units for nanomaterial devices, 2) fabrication of mono- and multilayer ordered nanoparticle coatings, 3) assembly of ordered nanoparticle-crystals, 4) characterization of microporous thin film polycrystallinity and defects, and 5) development of multscale computational frameworks to derive structure-properties relations for predicting device performance.

I. Multiscale approaches for elucidating structure-properties relations of diffusion through polycrystalline, microporous thin films (Ph.D. work with Prof. Dion Vlachos)

The ordered crystalline microstructure, nanometer-sized pores, cation tunability, and established techniques for thin film fabrication make microporous materials (e.g., zeolites) attractive for applications in high-resolution separations, membrane reactors, and chemical sensors.[1] The strongly coupled scales associated with molecular transport through *polycrystalline* membranes and lack of comprehensive *structure-properties relations* challenge the current design paradigm. To this end, we have developed multiscale modeling approaches (i.e., both continuum[2] and stochastic mesoscopic theories[3,4]) for bridging disparate scales associated with molecular diffusion through zeolite membranes. Models of permeation through both prototype and realistic membranes have provided insight into the role of crystal terminations, strong adsorbate-adsorbate interactions, and underlying diffusion mechanisms in permeation.[5,6]

Motivated by device sensitivity to crystal terminations, we have developed new protocols for advancing laser scanning confocal microscopy as a *quantitative, non-destructive* technique for imaging polycrystalline features of zeolite membranes, with interesting implications for even more *general materials characterization*. Efforts to push the quantitative limits of this technique have included correlative studies,[7] comprehensive statistical image analysis,[8] spatially-resolved pseudo-porosimetry studies using different sized dye molecules, and fabrication and application of calibration standards with nanometer scale features.[9] This combination of modeling and non-destructive characterization has enabled development of structure properties relations linking polycrystallinity with membrane performance specifically for NaX zeolite films.[8]

II. Benign nanoparticle synthesis: Tuning particle properties, controlling ordered assembly, realizing multi- and monolayer nanoparticle films, and encapsulating living cells (Postdoctoral work with Prof. Michael Tsapatsis)

While my doctoral work revolved around films consisting of intergrown single crystals of a microporous material (zeolite), my postdoctoral work has explored the *synthesis, characterization*, and *assembly* of monodisperse oxide (i.e., silica) nanoparticles into porous continuous films. Realization of *nanoparticle-crystals* and controlled assembly of large-area nanoparticle films has implications spanning technologies of coatings and colloidal lithography to chemical sensing and biology. While silica nanoparticles are attractive for such applications, challenges to device fabrication in each of these areas derive, at least in part, from the lack until recently of a simple means for synthesizing stable and monodisperse nanoparticles, and the limited capabilities for rationally tuning particle size, porosity, order, and monolayer continuity. Complicating efforts especially in the biological arena are the harsh conditions (e. g., high pH) associated with the current paradigm for silica nanoparticle formation (i.e., hydrolysis of tetraethylorthosilicate in highly basic solutions in the presence of tetrapropylammonium cations). We have addressed a number of these key issues involving benign synthesis of oxide nanoparticles (i.e., silica[10]) and the ability to tailor their size and ordered assembly.[11] In particular, we have explored biomimetic approaches as a facile route for synthesizing stable silica nanoparticles,[10] and have identified numerous handles (e.g., pH, silica content, hydrolysis and hydrothermal ageing temperature) for tuning particle diameter from less than 5 nm to more than 20 nm. In addition to particle size, we have identified conditions for controlling long-range (10 μm) ordered assembly of these nanoparticles, and for realizing continuous mono- and multilayer coatings over square centimeter areas.[11]

Finally, we have harnessed the benign nature of this novel silica nanoparticle synthesis for applications in the biological arena. Through further neutralization of the lysine-silica sols by addition of compatible peptide oligomers (e.g., di-lysine), we are able to synthesize clear silica gels with tunable porosity and morphology, and have employed them for facile living cell encapsulation. The sustained viability of the encapsulated cells serves as proof-of-concept for their use as implantable "cellular reactors" capable of producing a metabolite of interest while being protected from immunological attack.

References

1. **Snyder, M. A.** ; Tsapatsis, M., *Angewandte Chemie* **2007**, In press.

2. **Snyder, M. A.** ; Vlachos, D. G.; Katsoulakis, M. A., *Chemical Engineering Science* **2003**, 58, 895.

3. **Snyder, M. A.** ; Vlachos, D. G., *Molecular Simulation* **2004**, 30, 561.

4. **Snyder, M. A.** ; Chatterjee, A.; Vlachos, D. G., *Computers & Chemical Engineering* **2005**, 29, 701.

5. **Snyder, M. A.** ; Vlachos, D. G., *Journal of Chemical Physics* **2005**, 123.

6. **Snyder, M. A.** ; Vlachos, D. G., *Journal of Chemical Physics* **2005**, 123.

7. **Snyder, M. A.** ; Lai, Z.; Tsapatsis, M.; Vlachos, D. G., *Microporous and Mesoporous Materials* **2004**, 76, 29.

8. **Snyder, M. A.** ; Vlachos, D. G.; Nikolakis, V., *Journal of Membrane Science* **2007**, 290, 1.

9. **Snyder, M. A.** ; Vlachos, D. G., *Microporous and Mesoporous Materials* **2007**, 102, 101.

10. Davis, T. M.; **Snyder, M. A.** ; Krohn, J. E.; Tsapatsis, M., *Chemistry of Materials* **2006**, 18, 5814.

11. **Snyder, M. A.**; Lee, J. A.; Davis, T. M.; Scriven, L. E.; Tsapatsis, M., **2007**, *Submitted.*

6de Engineering Novel Surfaces to Control Cell Adhesion and for Drug Delivery Applications

Srivatsan Kidambi, Center for Engineering in Medicine, Massachusetts General Hospital, Shriners Hospital, Harvard Medical School, 51 Blossom Street, Boston, MA 02114

Over the past decades, the development of new methods for fabricating thin films that provide precise control of the three-dimensional topography and cell adhesion has generated lots of interest. These films could lead to significant advances in the fields of tissue engineering, drug delivery and biosensors which have become increasingly germane areas of research in the field of chemical engineering. The ionic layer-by-layer (LbL) assembly technique called "Polyelectrolyte Multilayers (PEMs)", introduced by Decher in 1991, has emerged as a versatile and inexpensive method of constructing polymeric thin films, with nanometer-scale control of ionized species. PEMs have long been utilized in such applications as sensors, eletrochromics, and nanomechanical thin films but recently they have also been shown to be excellent candidates for biomaterial applications. In this project, we engineered these highly customizable PEM thin films to engineer in vitro cellular microenvironments to control cell adhesion and for drug delivery applications. PEM films were engineered to control the adhesion of primary hepatocytes and primary neurons without the aid of adhesive proteins/ligands. We capitalized upon the differential cell attachment and spreading of primary hepatocytes and neurons on poly (diallyldimethylammoniumchloride) (PDAC) and sulfonated polystyrene (SPS) surfaces to make patterned co-cultures of primary hepatocytes/fibroblasts and primary neurons /astrocytes on the PEM surfaces. In addition, we developed self-assembled monolayer (SAM) patterns of m-d-poly(ethylene glycol) (m-dPEG) acid molecules onto PEMs. The created m-dPEG acid monolayer patterns on PEMs acted as resistive templates, and thus prevented further deposits of consecutive poly(anion)/poly(cation) pairs of charged particles and resulted in the formation of three-dimensional (3-D) patterned PEM films or selective particle depositions atop the original multilayer thin films. These new patterned and structured surfaces have potential applications in microelectronic devices and electro-optical and biochemical sensors. The PEG patterns developed are tunable at certain salt conditions and be removed from the PEM surface without affecting the PEM layers underneath the patterns. These removable surfaces provide an alternative method to form patterns of multiple particles, proteins and cells. This new approach provides an environmentally friendly and biocompatible route to designing versatile salt tunable surfaces. Finally, we illustrate the use of PEM films to engineer aptamer and siRNA based drug delivery systems.

6df Time-Series Transcriptomic Analysis of a Systematically Perturbed Arabidopsis Thaliana Liquid Culture System: A Systems Biology Perspective

Bhaskar Dutta, University of Maryland, 2113 Chemical and Nuclear Engineering Building, College Park, MD 20742-2111 and Maria Klapa, Institute of Chemical Engineering and High-Temperature Chemical Processes, Foundation for Research and Technology-Hellas (FORTH), Stadiou St., Platani, Patras, Greece.

Revealing the gene regulation network has been one of the main objectives of biological research. Studying such a complex, multi-scale and multi-parametric problem requires educated fingerprinting of cellular physiology at different molecular levels under systematically designed perturbations. Conventional biology lacked the means for holistic analysis of biological systems. In the post-genomic era, advances in robotics and biology lead to the development of high-throughput molecular fingerprinting technologies. Transcriptional profiling analysis using DNA microarrays has been the most widely used among them.

My Ph.D. thesis concerns the dynamic, transcriptional profiling analysis of a systematically perturbed plant system. Specifically, *Arabidopsis thaliana* liquid cultures were subjected to three different stresses, i.e. elevated CO_2 stress, salt (NaCl) stress and sugar (trehalose) applied individually, while the latter two stresses were also applied in combination with the CO_2 stress. The transcriptional profiling of these conditions involved carrying out 320 microarray hybridizations, generating thus a vast amount of transcriptomic data for *Arabidopsis thaliana* liquid culture system. To upgrade the dynamic information content in the data, I developed a statistical analysis strategy that enables at each time point of a time-series the identification of genes whose expression changes in statistically significant amount due to the applied stress. Additional algorithms allow for further exploration of the dynamic significance analysis results to extract biologically relevant conclusions. All algorithms have been incorporated in a software suite called MiTimeS, written in C++. MiTimeS can be applied accordingly to analyze time-series data from any other high-throughput molecular fingerprint.

The experimental design combined with existing multivariate statistical analysis techniques and MiTimeS revealed a wealth of biologically relevant dynamic information that had been unobserved before. Due to the high-throughput nature of the analysis, the study enabled the simultaneous identification and correlation of parallel-occurring phenomena induced by the applied stress. Stress responses comparisons indicated that transcriptional response of the biological system to combined stresses is usually not the cumulative effect of individual responses. In addition to the significance of the study for the analysis of the particular plant system, the experimental and analytical strategies used provide a systems biology methodological framework for any biological system, in general.

6dg Analysis of Transcription Networks

Mark P. Brynildsen and James C. Liao. Chemical and Biomolecular Engineering, University of California, Los Angeles, 420 Westwood Plaza, 5531 Boelter Hall, Los Angeles, CA 90095

Organisms use a variety of cellular networks to coordinate responses from internal and external stimuli. With the advent of high-throughput technologies, such as DNA microarray and ChIP-chip binding assays, it is now possible to study these networks at the system level. In particular, DNA microarray is a popular technique that has generated a multitude of gene expression data. Gene expression is modulated by transcription factors (TF) through a protein-DNA interaction network that dictates which TF controls which gene promoter. This network, commonly referred to as the transcription network, integrates signals from upstream pathways into a cohesive cellular response. Understanding the transcription network and signals transduced by it (transcription factor activities, TFA) is key to advancements in metabolic engineering, including biofuel production, and drug discovery and design. Fundamental questions our research has addressed include, how are transcription networks structured, what capabilities do these structures entail, how do transcription networks respond to environmental stress, and how do we identify the direct targets of stimuli from gene expression data and other available sources. To this end, we have characterized the ability of transcription networks to generate gene expression (1), developed a method to cope with uncertainty in transcription networks caused by network dynamics (eg. due to environment and stress) and noise (2), designed a method to identify transcription networks and TFAs solely from gene expression data (3), developed techniques to infer functional roles for TFs (4, 5), and are currently attempting to identify direct targets of butanol in Escherichia coli through the use of DNA microarray experiments and bioinformatics techniques we have developed, in an effort to increase the tolerance of E. coli to this biofuel.

1. Brynildsen MP, Tran LM, Liao JC. (2006) Versatility and connectivity efficiency of bipartite transcription networks. Biophys J., Oct 15;91(8):2749-59. (Epub 2006 Jun 30)

2. Brynildsen MP, Tran LM, Liao JC. (2006) A Gibbs sampler for the identification of gene expression and network connectivity consistency. Bioinformatics, Dec 15;22(24):3040-6 (Epub 2006 Oct 23)

3. Brynildsen MP, Wu TY, Jang SS, Liao JC. (2007) Biological network mapping and source signal deduction. Bioinformatics, (Epub May 15)

4. Yang YL, Suen J, Brynildsen MP, Galbraith S, Liao JC. (2005) Inferring yeast cell cycle regulators and interactions using transcription factor activities. BMC Genomics, Jun 10;6(1):90.

5. Tran LM, Brynildsen MP, Kao KC, Suen JK, Liao JC. (2005) gNCA: a framework for determining transcription factor activity based on transcriptome: identifiability and numerical implementation. Metab Eng., Mar;7(2):128-41.

6dh Targeted Intracellular Delivery of Drugs and Genes

Yah-el Har-el, Russell H. Morgan Dept. of Radiology and Radiological Science, Johns Hopkins University, 1150 Orleans St., Cancer Research Building 2, Room 4M62, Baltimore, MD 21231

With the advent of genomic biomarker signatures of disease and discoveries of new genotype–phenotype associations, the concept that therapy can be customized to the individual is emerging. My future research interests lie in the targeting of cancer cells for diagnosis and treatment using nanoscale delivery systems. Between my graduate and post-doctoral work, I have gained experience in both polymeric and liposomal drug and gene delivery systems as well as targeted and non-targeted delivery for both therapeutic and diagnostic purposes. I am interested in bringing these together to continue and also to expand upon the research I have begun in using DNA-based engineered constructs to deliver anticancer agents to the cell nucleus.

My PhD thesis, with Justin Hanes, in the Chemical & Biomolecular Engineering Department at Johns Hopkins, focused on the synthesis and characterization of improved non-viral gene delivery systems; specifically high-density nanocomplexes composed of numerous copies of plasmid DNA and the cationic polymer polyethylenimine (PEI). This included the encapsulation of DNA and PEI/DNA nanocomplexes within polymer microparticles. As an additional project, I also studied the effect of delivering combined chemo- and immunotherapy via polymer systems to malignant glioblastoma. In order to complete these studies, I became proficient in making a variety of polymer-based drug delivery systems and characterizing them using sizing techniques from the nano- to the micro-scale (dynamic and static light scattering, transmission and scanning electron microscopy (TEM and SEM), and the Coulter Principle). These studies were primarily *in vitro* experiments, consisting of cell culture where FACS and protein expression assays were used. While obtaining my PhD, I was awarded an Abel Wolman Fellowship, awarded on a school-wide basis to outstanding beginning PhD students in engineering, and I passed the Chemical Engineering Qualifying Exam with Honors. I also spent 4 semesters as a teaching assistant for a course entitled, “Engineering Aspects of Controlled Drug Delivery”.

As a post-doctoral fellow, I am working with George Sgouros, in the Radionuclide Therapy and Dosimetry Research Lab at the Johns Hopkins School of Medicine. I began my post-doc using alpha particle emitting radionuclides to target and treat ovarian cancer metastases with antibody and immunoliposomal delivery systems, and have recently begun targeting prostate cancer using aptamers. Since starting, I have applied for, and received a National Research Service Award (NRSA) from the National Cancer Institute to radiolabel DNA with alpha particles in order to generate a more efficient cell kill due to intracellular perinuclear localization of the alpha-emitter. I am also involved in collaborations in which liposomes encapsulate magnetic resonance (MR) contrast agents as imaging tools, or as models for drug delivery. The liposomes targeting specificity is analyzed using FACS, radioactivity, and/or confocal microscopy *in vitro*, or *in vivo* using small animal imaging (μ SPECT/CT, MR, fluorescence). Both the ovarian cancer work and the contrast agent work have been submitted as abstracts at this conference.

Web Page: www.hopkinsmedicine.org/RTD_lab/lab_members/har_el.html

6di Applications of Flux Analysis to Plant Oilseeds

Doug K. Allen, Department of Plant Biology, Michigan State University, S-302 Plant Biology Building, East Lansing, MI 48824

For most crops the component of agronomic importance is the seed. Plant seeds store oil, protein and carbohydrate in a dense form and offer great potential to produce novel biochemicals and petroleum replacements. Nevertheless, much of the biochemistry of seed metabolism is poorly understood. Elucidating fluxes through the network of carbon metabolism in developing seeds is needed to provide a framework for rational metabolic engineering of plants as "green factories". Soybean (Glycine max) seeds are an attractive target for analysis because they provide the world's largest supply of biological oil for food and industrial applications and are the highest producer of protein of any major crop. The complexity of plant metabolism compared to prokaryotes requires more information for successful flux analysis and a greater investment in accurately mimicking in planta growth conditions. To address these challenges we have increased the number of measurements using a variety of stable- and radio-labeling experiments with subsequent analysis by NMR, GCMS, and scintillation counting. To ensure that the conditions of labeling experiments accurately mimic in planta conditions we have optimized medium composition and light levels. Embryos cultured in defined medium containing sugars and amino acids under 30 μmol m^{-2} s^{-1} green light accumulate storage reserves very close to in planta levels (39% protein and 19% oil in culture compared to 40% and 20% in planta, respectively). In addition, plants maintain an ability to refix carbon released as carbon dioxide. Oil biosynthesis is an expensive process from a carbon standpoint, with one carbon being lost to carbon dioxide for every two that go into oil production. Brassica napus seeds that produce large amounts of biomass as oil have evolved to recapture this carbon, but an understanding of the generality and regulation of this process is lacking. Furthermore, a number of parallel pathways occur in different subcellular compartments of plant cells, resulting in some less identifiable intracellular fluxes. Through the use of flux models these issues can be explored more rigorously.

6dk Sensor-Aided Bioprospecting

Karen M. Polizzi, School of Biosciences, University of Exeter, Geoffrey Pope Building, Stocker Rd, Exeter, United Kingdom

Scientific discoveries are often fueled by technologies that allow visualization of events taking place at the cellular level. One example, is the discovery of fluorescent proteins, such as GFP, which gave researchers the ability to visualize organelles and proteins within cells. The design of genetically-encoded sensors for molecules of interest is the basis for a multi-faceted research program. A sensor for a small molecule of interest allows the rapid screening of compounds that affect cellular metabolism, the idenfication of new metabolic pathways, and facilitates the discovery and evolution of biocatalysts towards the production of molecules that do not have a rapid screening system already available. Following successful sensor design, the projects become modular and can be adapted towards a biomedical, microbiological, or biocatalytic focus with relative ease. The development of these techniques would streamline future discoveries in all areas of biotechnology.

6dl Generation of Renewable Fuels and Chemicals from Lipids Via Supercritical Fluid Processing

Darrell L. Sparks, Chemical Engineering, Mississippi State University, Box 9595, Mississippi State University, MS 39762

Identifying renewable sources for fuels and chemicals is currently a paramount global research initiative. Lipids derived from renewable feedstocks such as vegetable oils, animal fats, and microorganisms are a promising alternative to petroleum. For example, lipids derived from soybean oil are used to produce biodiesel, which is currently the dominant fuel alternative to petroleum diesel. In order to generate fuels and chemicals from lipids, typically the lipids are first extracted from lipid-containing biomass. Traditionally, this unit operation is accomplished by extraction with an organic solvent such as hexane. However, hexane is a volatile, flammable material that can leave behind a toxic residue. Hence, if the extracted lipids are to be further processed into food-grade products, extra processing costs can be incurred to ensure that any remaining hexane is at a safe level. Extraction of lipids with supercritical carbon dioxide is an appealing alternative because carbon dioxide is inexpensive, readily available, non-toxic, and non-flammable. Also, supercritical carbon dioxide is considered a tunable extraction medium because changes in temperature or density of the solvent can greatly alter its solvating effectiveness.

Supercritical carbon dioxide can also be an effective reaction medium. Many lipids are composed of unsaturated fatty acids, and once lipids have been extracted from a biomass matrix, the unsaturated fatty acids can be oxidized in a controlled manner (as opposed to autoxidation) to produce a variety of valuable chemical intermediates. The oxidation products are useful in a spectrum of products such as pharmaceuticals and polymers. By conducting the reaction in a supercritical fluid, several advantages can be realized. For example, the reaction rate and conversion can be improved by increasing pressure. Also, if the solubilities of the reaction products in the supercritical fluid are different, then a supercritical fluid separation can be used to fractionate the products. For oxidation reactions, the use of carbon dioxide as the supercritical fluid is very appealing because it is already completely oxidized; therefore it will not break down during the reaction.

This poster will emphasize the author's graduate work under the guidance of Dr. Rafael Hernandez at Mississippi State University. In the first phase of the author's research, supercritical carbon dioxide was evaluated as an alternative lipid extraction media to hexane. The first lipid source considered was rice bran, a component of rice kernels removed during whitening and polishing. Extraction of rice bran lipids with supercritical carbon dioxide provided similar oil yields as hexane on a gravimetric basis; however, use of supercritical carbon dioxide resulted in a much higher solvent efficiency [1]. The other lipid source was municipal wastewater sewage sludge [2]. Once again, supercritical carbon dioxide was comparable to hexane in terms of oil yield. However, addition of a small amount of methanol to the carbon dioxide increased extraction yield significantly [3]. In the second phase, the oxidation of oleic acid (a common lipid component) was performed in supercritical carbon dioxide. Ozonolysis of oleic acid is the most common oxidation reaction performed commercially to generate azelaic acid and pelargonic acid. Although solubility data for oleic acid are readily available, the solubility of azelaic acid and pelargonic acid had to be determined. Pelargonic acid had a much higher solubility in supercritical carbon dioxide than azelaic acid [4, 5]. In order to fully understand the solubility results, several models were utilized, including the Peng-Robinson equation of state and Chrastil's equation. The effectiveness of these models was evaluated in predicting the solubility behavior of a range of compounds [6].

The author's primary research interests are in the areas of renewable fuels and chemicals, supercritical fluids, and lipids. As part of completing the Ph.D. qualifying exam, the author wrote and successfully defended a NSF-style proposal. As a result, the author has several proposal ideas in the aforementioned areas. Potential funding sources for these areas are the Department of Energy, the Department of Agriculture, and the National Science Foundation.

Though confident in the ability to teach any chemical engineering course at both the undergraduate and graduate levels, the author's teaching interests include mass transfer operations, reaction kinetics, thermodynamics, and elective courses in renewable fuels and lipid chemistry. The author served as a tutor for Mass and Energy Balances for several years and also taught a junior-level undergraduate course entitled Mass Transfer Operations. Hence, the author has experience interacting with students and is excitedly anticipating the opportunity to teach students on a permanent basis.

[1] Sparks, D., R. Hernandez, M. Zappi, D. Blackwell, and T. Fleming, 2006, Extraction of Rice Bran Oil Using Supercritical Carbon Dioxide and Propane, JAOCS 83: 885-891.

[2] Zappi, M., French, T., Hernandez R., Dufreche S., and Sparks, D., "Production of Biodiesel and Other Valuable Chemicals from Wastewater Treatment Plant Sludges," United States Patent Application 20050112735.

[3] Dufreche, S., R. Hernandez, T. French, D. Sparks, M. Zappi, and E. Alley, 2007, Extraction of Lipids from Municipal Wastewater Plant Microorganisms for Production of Biodiesel, JAOCS 84: 181-187.

[4] Sparks, D., L.A. Estévez, N. Meyer, and R. Hernandez, 2007, Solubility of Azelaic Acid in Supercritical Carbon Dioxide, J. Chem. Eng. Data (Under Review).

[5] Sparks, D., L.A. Estévez, K. Barlow, and R. Hernandez, 2007, Solubility of Pelargonic Acid in Supercritical Carbon Dioxide, J. Chem. Eng. Data (In Preparation).

[6] Sparks, D., L.A. Estévez, and R. Hernandez, 2007, Model Evaluation for the Prediction of Solid Solubility in Supercritical Carbon Dioxide, Chemical Engineering Science (In Preparation).

6dm Control of the Interspecies Biofilm Signal Indole in Pathogenic *E. Coli* O157:H7 and *Pseudomonas Aeruginosa* and Proteome and Transcriptome Analysis of Engineered *E. Coli* for Aerobic Mineralization of *Cis*-1,2-Dichloroethylene

Jintae Lee and Thomas K. Wood. Artie McFerrin Department of Chemical Engineering, Texas A&M University, 200 Jack E. Brown Building, MS 3122, College Station, TX 77843-3122

Control of Interspecies Biofilm Signal Indole in Pathogenic *Escherichia coli* O157:H7 and *Pseudomonas aeruginosa* PAO1

Bacteria exist in biofilms in natural, medical, and engineering environments. Due to their high resistance to antibiotics, biofilms cause serious problems to human health such as lung infections, dental disease, and urinary tract infections. Procaryotes and eucaryotes signal not only themselves but also one another; hence, there is competition and interference of cell signals. As a stationary phase signal, indole is secreted in large quantities (500 μM) into rich medium by *Escherichia coli* and has been shown to control several genes (e.g., *astD, tnaB, gabT*), multi-drug exporters, and the pathogenicity island of *E. coli*; however, its impact on biofilm formation has not been well-studied. Also, hydroxyindoles and indigoid compounds can be easily produced from indole in the presence of toluene *o*-monooxygenase (TOM) of *Burkholderia cepacia* G4. Through a series of global transcriptome analyses, confocal microscopy, isogenic mutants, and dual-species biofilms, we found that indole is a non-toxic signal that controls *E. coli* biofilms by repressing motility, inducing the sensor of the quorum sensing signal autoinducer-1 (SdiA), and influencing acid resistance (e.g., *hdeABD, gadABCEX*). We have manipulated biofilm formation in a dual-species biofilm of *E. coli* and *Pseudomonas fluorescens* by controlling indole concentrations (*BMC Microbiol* in revision 2007). Since the addition of indole significantly repressed *ymgB* along with the repression of other acid resistance genes (*hdeABD* and *gadABCEX*), we solved the structure of YmgB to 1.8 Å resolution, which is a biological dimer that has a critical role in acid-resistance in *E. coli* via indole. Also, the exogenous addition of indole, 5-hydroxyindole, and 7-hydroxyindole decreased biofilms of a pathogenic *E. coli* O157:H7 (EHEC) without cell growth inhibition (which suggests their use as potential therapeutics), while the addition of isatin (indole-2,3-dione) increased biofilm formation of EHEC (*Appl Environ Microbiol*on-line 2007). Isatin caused increased motility as a result of flagella synthesis, caused a 20-fold reduction in indole concentrations as a result of repression of *tnaA*, and altered AI-2 transport whereas 7-hydroxyindole inhibited biofilm formation most likely through sulfur metabolism. In the other hand, the addition of indole and 7-hydroxyindole to *Pseudomonas aeruginosa* PAO1 (an opportunistic human pathogen) increase its biofilm formation without cell growth inhibition although this strain does not synthesize these compounds. We also found that indole (1000 μM) induced three porin import genes of *P. aeruginosa, oprH, oprF*, and *oprD*. Indole enhanced sensitivity to the antibiotic imipenem through the amino acid and antibiotic porin protein OprD; therefore, indole increases antibiotic sensitivity by increasing import. 7-hydroxyindole (500 μM) abolished swarming motility and influenced *P. aeruginosa* biofilm formation by altering increasing synthesis of fimbriae and repressing some aspects of quorum sensing.

Additionally, *P. aeruginosa* PAO1 degrades readily indole and 7HI. Since *P. aeruginosa* and *E. coli* are probably present in consortia, this study provides further evidence of the importance of interspecies communication for biofilm formation and the importance of indole.

Proteome and Transcriptome analysis of Engineered *Escherichia coli* for Aerobic Mineralization of *cis* -1,2-Dichloroethylene

Metabolically engineered *Escherichia coli* has previously been used by us to degrade *cis*-1,2-dichloroethylene (*cis*-DCE, *J Biol Chem* 279:46810, 2004; *Environ Microbiol* 6:491, 2004). The strains express the six genes of an evolved toluene *ortho*-monooxygenase from *Burkholderia cepacia* G4 (TOM-Green, which formed a reactive *cis*-DCE epoxide) and either (1) g-glutamylcysteine synthetase (which forms glutathione but is not limited by feedback inhibition) and the glutathione *S*-transferase IsoILR1 from *Rhodococcus*- AD45 (which adds glutathione to the reactive *cis*-DCE epoxide) or (2) with an evolved epoxide hydrolase from *Agrobacterium radiobacter* AD1 (EchA F108L/I219L/C248I which converts the reactive *cis*-DCE epoxide to a diol). Also, we quantified the impact of this metabolic engineering for bioremediation by measuring the changes in the proteome through a shotgun proteomics technique (iTRAQ) by tracking the changes due to the sequential addition of TOM-Green, the glutathione *S*-transferase IsoILR1, and g-glutamylcysteine synthetase and due to adding the evolved EchA vs. the wild-type enzyme to TOM-Green (*J Proteome Res* 5:1388, 2006). We found that the addition of eight genes involving glutathione induced glutathione synthesis and a stress response (induction of *katG*, *ahpF*, and *dps*) as well as repressed fatty acid synthesis, gluconeogenesis, the tricarboxylic acid cycle, and indole synthesis (down-regulation of *tnaA*). Since the effect of adding the evolved epoxide hydrolase was not as definitive, we used DNA microarrays to complement the proteomics study to investigate the impact of the addition of the evolved EchA to TOM-Green; hence, we studied the impact of removing the toxic epoxide from the cell. The whole transcriptome analysis shows that 87 genes are significantly induced or repressed with the addition of the evolved epoxide hydrolase EchA. Among the induced genes, many stress-related genes (*yfiD*, *marA*, *nemA*, *ytfE*, *grxA*, *marR*, *ybbT*, *b3913*, *b3914*, *grxD*, and *ycfR*) were most significantly induced. Heat shock proteins (*ibpA*, *ibpB*, and *htpX*), 50S ribosomal proteins (*rmpABGF*), and several hypothetical genes (*yeaR*, *yogA*, and *yeaQ*) were also induced. Among the repressed genes, genes (*tnaA*, *tnaL*, and *trpL*) involved in the indole biosynthesis were repressed. Both the whole transcriptome analysis and the proteomic approach identified oxygen-activated genes, 50S ribosomal proteins, indole-biosynthesis genes, *soxR*, *ompX*, and *cysK*; hence, the DNA microarray analysis complements the proteomics approach. In order to enhance *cis*-DCE degradation, we investigated *cis*-DCE degradation using an indole-deficient mutant (*tnaA*); our hypothesis is that the initial *cis*-DCE oxidation by TOM-Green will be enhanced by removing the competitive substrate indole. In addition, we studied the impact of the newly identified possible stress-related genes, *yeaR*, *yogA*, *yeaQ*, and *ymgB* on *cis*-DCE degradation. This study shows clearly that cloning eight genes have significantly enhanced the degradation of *cis*-DCE and that both the whole transcriptome analysis and the proteomic approach have helped to understand the impact of the metabolic engineering.

Publications

1. J. Lee, K. Reardon and T.K. Wood, Transcriptome analysis of Engineered *Escherichia coli* for Aerobic Mineralization of *cis*-1,2-Dichloroethylene. in preparation (2007)

2. J. Lee and T.K. Wood, Indole and 7-Hydroxyindole Enhance *Pseudomonas aeruginosa* Biofilm Formation and Antibiotic Sensitivity. to be submitted (2007)

3. J. Lee, R. Page, R. García-Contreras, J. Palermino X. Zhang, O. Doshi, T. K. Wood, and W, Peti, Structure and Function of the *E. coli* Protein YmgB: a Protein Critical for Biofilm Formation and Acid-Resistance. *J. Mol. Biol.* under review (2007)

4. T. Bansal, D. Englert, **J. Lee**, T. K. Wood, and A. Jayaraman, Epinephrine and Norepinephrine Divergently Regulate *Escherichia coli* O157:H7 Chemotaxis, Biofilm Formation, and Gene Expression Compared to Indole. *Infect. Immun.* under review (2007)

5. J. Lee, A. Jayaraman, and T. K. Wood, Indole is an Inter-Species Biofilm Signal Mediated by SdiA. *BMC Microbiol.* accpeted (2007)

6. J. Lee, T. Bansal, A. Jayaraman, W. Bentley, and T. K. Wood, Enterohemorrhagic *Escherichia coli* Biofilms Are Inhibited by 7-Hydroxyindole and Stimulated by Isatin. *Appl. Environ. Microbiol.* on-line (2007)

7. J. Domka, **J. Lee**, T. Bansal, and T. K. Wood, Temporal Gene-Expression in *Escherichia coli* K-12 Biofilms *Environ. Microbiol.* 9: 332-346 (2007).

8. J. Domka, **J. Lee** and T. K. Wood, YliH and YceP Regulate *Escherichia coli* K12 Biofilm Formation By Influencing Cell Signaling. *Appl. Environ. Microbiol.* 72: 2449-2459 (2006).

9. J. Lee, L. Cao, S. Y. Ow, M. E. Barrios-Llerena, W. Chen, T. K. Wood, and P. C. Wright, Proteome Changes after Metabolic Engineering to Enhance Aerobic Mineralization of *cis*-1,2-Dichloroethylene. *J. Proteome Res.* 5: 1388-1397 (2006).

10. L. Cao, **J. Lee**, W. Chen, T. K. Wood, Enantioconvergent Product of (*R*)-1-phenyl-1,2-Ethanediol From Styrene Oxide by Combining the *Solanum tuberosum* and an Evolved *Agrobacterium radiobactor* AD1 Epoxide Hydrolases. *Biotechnol. Bioeng.* 94: 522-529 (2006).

11. T. K. Wood, A. F. G. Barrios, M. Herzberg, **J. Lee**, Motility Influences Biofilm Architecture in *Escherichia coli*. *Appl. Microbiol. Biotechnol.* 72: 361-367 (2006)

12. G. Vardar, Y. Tao, **J. Lee**, and T. K. Wood, Alanine 101 and Alanine 110 of the Alpha Subunit of *Pseudomonas stutzeri* OX1 Toluene-*o*-Xylene Monooxygenase Influence the Regiospecific Oxidation of Aromatics. *Biotechnol. Bioeng.* 92: 652-658 (2005)

13. J. Lee and H. Pedersen, Stable Genetic Transformation of *Eschscholzia californica* Expressing Synthetic Green Fluorescent Proteins. *Biotechnol. Prog.* 17: 247-251, (2001)

14. J. Lee, M. H. Cho and J. Lee, Characterization of an Oxygen-Dependent Inducible Promoter System, the nar promoter, and *Escherichia coli* with an Inactivated *nar* Operon. *Biotechnol. Bioeng.* 52: 572-578 (1996)

15. J. Lee, M. H. Cho, E.-K. Hong, K.-S. Kim and J. Lee, Characterization of the *nar* promoter to Use as an Inducible Promoter. *Biotechnol. Lett.* 18: 129-134, (1996)

6dn Bridging the Scales: From Contact Mechanics to Fluidized Beds

Gustavo G. Joseph, Chemical and Biological Engineering, University of Colorado, Boulder, CO 80309-0424

Flows of dry or wet particulate materials are prevalent in the chemical and pharmaceutical industries, in processes such as slurry transport, spray coating, coagulation, filtration, drying, granulation, agglomeration, and pneumatic transport. Such flows can also be observed in natural phenomena as varied as sedimentation, coastal erosion, debris flows, pollen capture, and planetary ring dynamics. Particulate flows are expected to be prevalent in future space applications, including solids processing on flights, mining of lunar and Martian soils, powdered food and drink handling, etc. Despite their pervasiveness in both industry and nature, particle processes remain poorly understood and yet their understanding is essential in order to tackle the energy challenges of the next few decades, be it with particle-catalyzed fuel cells, clean coal combustors, or coal gasification technologies, to mention a few.

I am presently a Research Associate in Chemical Engineering at the University of Colorado at Boulder, having completed my graduate studies in Mechanical Engineering at the California Institute of Technology in May 2003. My areas of interest include particle technology and multiphase flows in general, thermodynamics, and heat transfer, with a particular interest in rheology, friction, and contact models.

To date, prediction of complex phenomena such as multi-particle drag, species segregation, mixing, entrainment, elutriation, and attrition, to name a few, has essentially been limited to empirical correlation. A physically-based overhaul of the existing predictive models in particle technology is urgently required. My background and experience make me especially well suited for performing research in this area, with respect to both the experimental aspects (how to design, test and analyze systems that identify and quantify the relevant physics in particulate flows) and the multi-phase flow and contact mechanics theory aspects (the creation and validation of models required to treat such systems analytically and in numerical simulations).

I was first exposed to particle technology as an undergraduate researcher at the National University of Mexico. Under the mentorship of Professor Baltasar Mena I modified a version of a DEM code developed at the Lawrence Livermore National Laboratory and used it to simulate various granular flows. My Ph.D. research under the mentorship of Professor Melany L. Hunt at Caltech focused on determining the effect of a viscous fluid on the rebound of a macroscopic particle impacting a flat wall, both in oblique and normal settings. At Colorado, as a Research Associate sponsored by Professor Christine M. Hrenya, I have designed and built a fluidized bed system for studying elutriation and species segregation. The project started as a collaboration through the NSF Grant Opportunities for Academic Liaison with Industry (GOALI) program. A follow-on GOALI project, for which I am a co-principal investigator, was awarded earlier this year.

At both Caltech and Colorado, I have worked closely with undergraduate students. In addition to serving as a teaching assistant, I have directly mentored 10 students in research projects that led to several peer-reviewed publications and conference presentations. This summer, I had the opportunity to teach Fluid Mechanics at the University of Colorado during the June 2007 summer semester, a course that I am teaching again during the present fall semester. In addition, I am presently teaching one of the sections of the First Year Engineering Projects, a nationally-recognized course that gives students the opportunity to put engineering theory into practice early in their undergraduate years through working in teams to design, build and test new products and inventions. Based on my experience in experimental collaborative work and in one-on-one student mentoring, I believe that I have the necessary leadership and teaching skills to direct a diverse group of collaborators and students.

As a faculty member I intend to create a research group that bridges the gap between microscopic (particle-level) physics and predictions at the macroscopic level. The group will grow based on the following initial directions, which I consider crucial for a physical understanding of particulate flows: 1) analysis of the contact interactions in multiphase particulate systems, 2) characterization of the effects of surrounding fluids on the flow of these materials, and 3) development of novel instrumentation necessary for critical validation of theoretical models.

Publication Record

1. Leboreiro J, **Joseph GG**, Hrenya CM (2007) Revisiting the standard drag law for bubbling, gas-fluidized beds, *Powder Technology*, invited paper (in press).

2. Leboreiro J, **Joseph GG**, Hrenya CM, Snider DM, Banerjee SS, Galvin JE (2007) The influence of binary drag laws on simulations of species segregation in gas-fluidized beds, *Powder Technology*, in press.

3. **Joseph GG**, Leboreiro J, Hrenya CM, Stevens AR (2007) Experimental segregation profiles in bubbling gas-fluidized beds, *AIChE Journal*, in press.

4. Muite BK, Hunt ML, **Joseph GG** (2004) The effects of a counter-current interstitial flow on a discharging hourglass, *Physics of Fluids*, 16(9), 3415–3425.

5. **Joseph GG** and Hunt ML (2004) Oblique particle–wall collisions in a liquid, *Journal of Fluid Mechanics*, 510, 71–93.

6. **Joseph GG** and Hunt ML (2002) Wall erosion due to normal particle–wall collisions in a liquid, *Proceedings Fourth World Congress on Particle Technology*, Sydney (New South Wales, Australia).

7. Jaimes J, **Joseph GG**, Geffroy E, Mena B, Herrera–Velarde JR (2002) Pattern formation in oscillatory granular flows, *Revista Mexicana de Física*, 48(6), 534–538.

8. **Joseph GG**, Zenit R, Hunt ML, Rosenwinkel AM (2001) Particle–wall collisions in a viscous fluid, *Journal of Fluid Mechanics*, 433, 329–346.

9. **Joseph GG**, Geffroy E, Mena B, Walton OR, Huilgol RR (2000) Simulation of filling and emptying in a hexagonal-shape solar grain silo, *Particulate Science and Technology* 18(4), 309–327.

10. Zenit R, **Joseph GG**, Hunt ML (1999) The coefficient of restitution for liquid immersed collisions, *Proceedings 3rd ASME / JSME Joint Fluids Engineering Conference*, San Francisco (California, USA).

11. **Joseph G**, Geffroy E, Mena B (1999) Flow of granular materials in a hexagonal silo, *Revista Mexicana de Física*, 45(Suppl. 1), 128–133.

12. **Joseph G**, Mena B, Moreno E, Sansores E, Walton OR (1996) Granular Flow in a Solar Hexagonal Silo, Proceedings XIIth International Congress on Rheology, Aït-Kadi A, Dealy JM, James DF, Williams MC, editors, Québec (Québec, Canada), 795–796.

6do Detection and Analysis of Biomolecules Using Diagnostic Microfluidic Systems

Edgar D. Goluch, Department of Bioengineering, University of Illinois at Urbana-Champaign, 3225 Micro and Nanotechnology Lab, 208 North Wright Street, Urbana, IL 61801

My research focuses on exploratory, multidisciplinary research in microfluidics at the system level.

Development of Diagnostic Systems: It is well known that early detection of cancer onset or recurrence increases the chances of patient survival as well as life expectancy. I have developed a single disposable chip that is capable of carrying out a multi-step process that employs nanoparticles—a biobarcode assay for cancer marker detection. The first application of the system is ultra-high sensitivity detection of prostate specific antigen to monitor recurrence of prostate cancer in patients that have undergone prostatectomy. Target analyte concentrations as low as 500 attomolar (300 total molecules) have been detected in buffer and serum samples. [4, 7] Recently, I have expanded the capabilities of the system to allow for simultaneous detection of multiple cancer markers in a single sample. In its current state, the chip-based detection scheme matches benchtop experiments in sensitivity while taking only a quarter of the time to complete. [3] My current work in this area includes modeling and optimization of the system by measuring the kinetic rates of individual steps in the process and implementing electrical detection of the final output signal.

Analysis of Lipid Bilayer Systems: The cellular membrane is the location of many biochemically and medically important processes. Ion transport, hormone signaling, blood coagulation, and endocytosis are just a few examples of the numerous important cellular processes that occur at the lipid bilayer of the membrane. These important biochemical processes are controlled by membrane receptor proteins, which are the targets of the majority of pharmaceuticals. It is therefore of great interest to develop high-throughput screening assays involving lipid bilayers and membrane-associated proteins. This requires the ability to pattern stable lipid bilayers in a defined manner. Toward this goal, in collaboration with the Sligar Group at the University of Illinois, I have developed a microfluidic method of patterning nanoscale bilayer particles known as Nanodiscs and a multiplexed assay to examine the interaction between Nanodiscs and analytes. [9]

BioMEMS for Cellular Studies: Cell motility is a complex phenomenon that enables numerous functions ranging from searching for foreign organisms to spreading of cancer cells. I have fabricated large-scale micro-contact printing stamps, which the Mrksich group at the University of Chicago uses to investigate the effect that local and global geometric cues have on the movement of cells. [10] I am also currently developing a device that will apply a force to individual cells in an array in order to study the effect on cellular adhesion and structure.

Journal Publications:

1. Ryu, K.S., Wang, X., Shaikh, K.A., Goluch, E.D., Bullen, D., Zou, J., Liu, C., Mirkin, C.A., "Integrated Microfluidic Inking Chip for SPM Nanolithography." Applied Physics Letters, Vol. 85, No. 1, pp. 136-138, July 5, 2004.

2. Goluch, E. D., Shaikh, K.A., Ryu, K.S., Chen, J., Engel, J.M., Liu, C. "A Microfluidic Method for In-Situ Deposition and Precision Patterning of Thin-Film Metal on Curved Surfaces." Applied Physics Letters. Vol. 85, No. 16, pp. 3629-3631, October 18, 2004.

3. Ryu, K.S., Shaikh, K.A., Goluch, E.D., Fan, Z., Liu, C. "Micro Magnetic Stir-Bar Mixer Integrated with Parylene Microfluidic Channels." Lab-on-a-Chip. Vol. 4, No. 6, pp. 608-613, November 29, 2004.

4. Shaikh, K.A., Ryu, K.S., Goluch, E.D., Nam, J.-M., Liu, J., Thaxton, C.S., Chiesl, T.N., Barron, A.E., Lu, Y., Mirkin, C.A., Liu, C. "A Modular Microfluidic Architecture for Integrated Biochemical Analysis," PNAS, vol. 102, no. 28, pp. 9745-9750, July 2005.

5. Ryu, K.S., Shaikh, K.A., Goluch, E.D., Liu, C. "Two-terminal longitudinal hotwire sensor for monitoring the position and speed of advancing liquid fronts in microfluidic channels." Applied Physics Letters. Vol. 88, pp. 104104-104107, March 9, 2006.

6. Chiesl, T.N., Putz, K.W., Babu, M., Mathias, P., Shaikh, K.A., Goluch, E.D., Liu, C., Barron, A.E. "Self-Associating Block Copolymer Networks for Microchip Electrophoresis Provide Enhanced DNA Separation via "Inchworm" Chain Dynamics." Analytical Chemistry. Vol. 78, pp. 4409-4415, July 1, 2006.

7. Goluch, E.D., Nam, J.M., Georganopoulou, D.G., Chiesl, T.N. Shaikh, K.A., Ryu, K.S., Barron, A.E., Mirkin, C.A., Liu, C. "A Bio-Barcode Assay for On-Chip Attomolar-Sensitivity Protein Detection" Lab-on-a-Chip. Vol. 6, pp.1293-1299, September 2006.

8. Li, S., Shaikh, K.A., Szegedi, S.S., Goluch, E.D., Liu, C. "A Micromachined Inking Chip for Scanning Probe Nanolithography Using Local Thermal Vapor Inking Method." Applied Physics Letters. Vol. 89, pp.173125-173128, October 26, 2006.

9. Goluch, E.D., Shaw, A.W., Sligar, S.G., Liu, C. "Parallel Patterning and Analysis of Lipid Bilayer Nanodiscs using a Microfluidic Grid." NanoLetters. (in preparation).

10. Maduram, J., Goluch, E.D., Liu, C., Mrksich, M. "Local and Global Geometric Cues in the Adhesive Microenvironment Modulate Cytoskeletal Architecture and Cell Polarity." PNAS. (in preparation).

Web Page: mntr.micro.uiuc.edu

6dp Engineering Peptides to Build Inorganic Materials

Scott K. Stanley, Polymers Division, National Institute of Standards and Technology, 100 Bureau Drive, Stop 8541, Gaithersburg, MD 20899

My overall research interests lie in chemical self assembly of nanomaterials from the atom up and directing assembly of nanomaterials at various length scales. My PhD (with John G. Ekerdt at UT Austin) focused on understanding surface reactions that lead to nanocrystal formation in chemical vapor deposition and controlling reaction kinetics to spatially direct the self assembly of nanocrystals. My postdoctoral fellowship (NRC-NIST Postdoctoral Fellow with Wen-li Wu and Eric K. Lin) is serving to bridge this PhD expertise to solution based nanocrystal synthesis and assembly of nanomaterials. Currently I am gaining expertise with neutron and x-ray scattering techniques and working with engineered peptides as related to the studies proposed below.

Proposed and Current Studies: Engineering peptides to build inorganic materials

Molecular biology has progressed to the point where previously rare materials (*i.e.*, specialty proteins, peptides, enzymes, etc.) are now readily available and can be prepared through automated synthesis in the laboratory at the gram level. This availability allows one to exploit the efficiency of biological systems to elegantly create materials. Directed evolution and phage display techniques have long been used to identify peptides that interact with inorganic materials such as metals and semiconductors [1, 2]. More recently, peptide sequences have been identified that mediate the formation of nanomaterials from ions in solution through biomineralization [3]. The potential to create complex structures from the atom up using biomolecules is interesting for technological applications and helps provide insights into mechanisms of how biological macromolecules and inorganics interact. Many problems can be addressed through study of engineered peptides that build inorganic materials; specifically I am interested in:

Peptide-ion interactions. Metal ion complexation and coordination with peptides *in vivo* is critical in many biological processes. Biomimetic photosynthetic structures and myriad human health issues such as Alzheimer's disease require an improved understanding of peptide-ion interactions.

Macromolecular chemistry in proteins. Macromolecules are known to control their local chemical environment, allowing access to seemingly unfavorable reaction pathways. Engineered peptides offer model systems to control and understand these effects in larger macromolecular systems.

Fundamentals of biomineralization. The exact mechanism of peptide-mediated nanostructure formation is unclear at present [2]. Conventionally there has been strong interest in biomineralization proteins that reduce ions in solution and build complex, hierarchical inorganic materials such as bone, teeth, and shells. Applying new techniques and capabilities to biomineralization processes can contribute to answering many of the outstanding questions in this area and in newer areas such as implant bio-integration.

Postdoctoral work: Applying x-ray and neutron scattering techniques to nanomaterial synthesis and nanomaterial assembly

My NRC Fellowship at NIST is focused on obtaining expertise working with biomolecules and using neutron and x-ray scattering to study nanomaterial synthesis in solution. Neutron scattering is especially helpful because contrast matching techniques allow study of structure and dynamics of only the polymer or peptides participating in nucleation and growth processes. Currently, I am applying x-ray and neutron scattering techniques to study (1) how image charges affect polyelectrolyte adsorption in systems with widely different relative dielectric properties and (2) how peptides mediate nucleation of nanoparticles in solution. This work is providing the necessary tools and experience to link my PhD expertise in controlling nanomaterial assembly with the proposed studies using engineered peptides.

PhD work: Directed nucleation and growth of nanomaterials in chemical vapor deposition

Surface chemistry governs the self assembly of nanostructures on surfaces in chemical vapor deposition (CVD) processes, such as the assembly of nanocrystals on oxide surfaces for flash memory applications. The initial sub-monolayer of atoms on a surface exhibits unexpected, unique reactive pathways for different oxide surfaces [4-7]. Nanocrystal assembly, which is governed by these surface reactions, is then studied on patterned substrates to direct self assembly of nanocrystals [8, 9]. Directed nucleation and growth of nanocrystals is examined within macroscopic (~100 μm) to nanoscopic (20 nm) features [10, 11]. Below a critical size, nanocrystal formation is sharply affected; however, we found that nanocrystals (singlets, doublets, and triplets) can form even in the smallest 20 nm features through judicious choice of reaction conditions. To extend this level of understanding and control to solution based nanomaterials systems, the new tools and techniques I am learning during my postdoc are required.

Aside from the thesis work described above, I also investigated the growth of Si nanowires by the vapor-liquid-solid (VLS) method [12], investigated surface chemistry of dielectrics with non-linear optical techniques [13], developed a combinatorial approach to screen nanostructure growth conditions [14], and fabricated core-shell nanocrystals for chemical and electrical stability [15].

References

1. Whaley, S.R., et al., Selection of peptides with semiconductor binding specificity for directed nanocrystal assembly. Nature, 2000. 405(6787): p. 665-668.

2. Sarikaya, M., et al., Molecular biomimetics: nanotechnology through biology. Nature Materials, 2003. 2(9): p. 577-585.

3. Naik, R.R., et al., Biomimetic synthesis and patterning of silver nanoparticles. Nature Materials, 2002. 1(3): p. 169-172.

4. Zhu, J.H., W.T. Leach, S.K. Stanl3y, J.G. Ekerdt, X.M. Yan. Growth of high-density Si nanoparticles on Si3N4 and SiO2 thin films by hot-wire chemical vapor deposition. Journal of Applied Physics, 2002. 92(8): p. 4695-4698.

5. Stanl3y, S.K. and J.G. Ekerdt, Surface reactions in chemical vapor deposition of metals and silicon alloys. Surface Science Reports, 2007. Invited Review (In Prep.).

6. Stanl3y, S.K., S.S. Coffee, and J.G. Ekerdt, Interactions of germanium atoms with silica surfaces. Applied Surface Science, 2005. 252(4): p. 878-882.

7. Stanl3y, S.K. and J.G. Ekerdt. Interactions of Ge atoms with hi-k oxide dielectric surfaces. Materials Research Society Symposium Proceedings 879E, Paper Z3.23 (2005).

8. Stanl3y, S.K., et al., Surface reactions and kinetically-driven patterning scheme for selective deposition of Si and Ge nanoparticle arrays on HfO2. Surface Science, 2006. 600(5): p. L54-L57.

9. Stanl3y, S.K., et al., Ge interactions on HfO2 surfaces and kinetically driven patterning of Ge nanocrystals on HfO2. Journal of Vacuum Science & Technology A, 2006. 24(1): p. 78-83.

10. Stanl3y, S.K., S.S. Coffee, and J.G. Ekerdt. Directed self assembly of nanocrystals within macroscopic to nanoscopic features. Materials Research Society Symposium Proceedings 901E, Paper 0901-Ra19-03.1 (2006).

11. Coffee, S.S., S.K. Stanl3y, and J.G. Ekerdt, Directed nucleation of ordered nanoparticle arrays on amorphous surfaces. Journal of Vacuum Science & Technology B, 2006. 24(4): p. 1913-1917.

12. N.N. Kulkarni, J. Bae, C.K. Shih, S.K. Stanl3y, S.S. Coffee, J.G. Ekerdt.., Low-threshold field emission from cesiated silicon nanowires. Applied Physics Letters, 2005. 87(21): p. 3.

13. R. Carriles, J. Kwon, Y.Q. An, S.K. Stanl3y, J.G. Ekerdt, M.C. Downer, J. Price, A.C. Diebold.., Optical characterization of process-dependent charging in hafnium oxide structures. Journal of Vacuum Science & Technology B, 2006. 24(4): p. 2160-2168.

14. Stanl3y, S.K. and J.G. Ekerdt. Combinatorial studies for high density Si and Ge nanoparticle arrays. Materials Research Society Symposium Proceedings 933E, Paper 0933-G05-11 (2006).

15. Stanl3y, S.K. and J.G. Ekerdt. Core-shell Ge nanoparticles on oxide surfaces for enhanced interface stability. Materials Research Society Symposium Proceedings 933E, Paper 0933-G02-07 (2006).

Web Page: polymers.nist.gov/staff/detail.cfm?SID=436

6dq Crowding and Confinement in Fluids and Biological Systems

Jeetain Mittal, Laboratory of Chemical Physics, National Institute of Health, Bethesda, MD 20892

The smooth functioning of a biological system depends sensitively on macromolecular transport, facilitating every process from protein transcription to the enzymes finding their target binding sites. One of the distinctive features of these systems is the presence of high molecular concentrations, commonly termed macromolecular crowding. As a manifestation of this crowding and confinement, the behavior of proteins and other intracellular organisms can deviate sharply from that in homogeneous bulk solution. A viable conceptual starting point to understand these deviations is to study how controlled modifications in the model cellular systems will cause a change in the properties of biological macromolecules. This question of understanding the effect of confinement also has parallels in liquid state theories where one is interested in knowing structure, dynamics and thermodynamics as a function of the physical and chemical characteristics of the confining medium.

Currently, I am a postdoctoral fellow in the Laboratory of Chemical Physics, National Institute of Health, Bethesda. The main objective of our research here is to design theoretical and simulation methods to predict the effects of macromolecular crowding and confinement on the protein stability. We are also developing theoretical methods based on path-integral formalism to understand the dominant folding pathways in the helix-coil transition for proteins. This particular approach has advantages over the conventional molecular simulation techniques in treating much longer time and length scales.

In my PhD dissertation, under the supervision of Dr. Thomas M. Truskett and in collaboration with Dr. Jeffrey R. Errington, we focused on understanding equilibrium and supercooled fluid behavior in diverse types of confining environments, starting from the most basic slit-pore model to more realistic quenched-annealed models for porous media. Some of the important findings from these studies are,

(i) The relationship between excess entropy (with respect to ideal gas state) and self-diffusivity for simple fluids is essentially unaffected by confinement, which allows one to use thermodynamics to "predict" how confinement impacts dynamics [1,2,3]. We also have clarified which definition of average density, based on total volume or particle center accessible volume, is most appropriate for understanding the thermodynamic and kinetic effects of confinement.

(ii) A new equation is proposed for predicting fluid diffusivity in "quenched annealed" models for random porous media [4]. Interestingly, it only requires as input the value of bulk fluid diffusivity and the available space in the system, the latter of which is a well defined thermodynamic quantity and therefore possible to calculate exactly. This work contributes toward resolving a controversy in the field regarding whether "static structure" alone can account for the large differences in dynamics by quenched-annealed systems with indistinguishable pair correlation functions.

(iii) We provided evidence that there is an intimate relationship between excess entropy and the self-diffusivity of supercooled liquids. Given that the reduced transport properties of fluids above their freezing point show a quasi-universal scaling with excess entropy, our simulations suggest that the connection between thermodynamics and dynamics exists across the entire liquid range, *from ideal gas to glass* [5,6].

(iv) The missing link between the structure and mobility of glass-forming liquids in deeply supercooled state is demonstrated to be much simpler (only requires the knowledge of pair correlation function and number density) and broader in context (even valid for systems with anomalous diffusion behaviour such as water models, short-range attractive colloidal system) than previously anticipated [6,7]. It is intimately connected to the two-body excess entropy discussed above.

(v) An "energy landscape based" statistical mechanical theory for nanoscale amorphous films was developed which is based on our hypothesis that the confinement induced shift in the properties of material can be understood in terms of how its energy landscape is changed with respect to the bulk. The theory is able to successfully reproduce several nontrivial experimental trends observed for liquid and glassy films such as shift in bulk thermodynamic phase boundaries, shift in glass transition temperature due to confinement, etc. This landscape based approach is different from current theories in that one can study the thermal, mechanical, and kinetic behavior of a material within the same framework [8].

In my Master's dissertation under the supervision of Prof. Ashutosh Sharma (IIT, Kanpur), we proposed a new mechanism of thin film instability engendered solely by the density variations (for example, due to confinement, layering, defects, and restructuring) that shows the same morphological characteristics as well-known spinodal dewetting [9,10]. This work was aimed at helping a rational design and interpretation of thin film experiments as inverse problem of determining thin film potential from the measurement of instability length scale is shown to be dependent on the uncertainty of density variations.

References

1. **J. Mittal**, J. R. Errington and T. M. Truskett, "Thermodynamics Predicts How Confinement Modifies Hard-Sphere Dynamics", *Phys. Rev. Lett.* **96**, 177804 (2006).

2. **J. Mittal**, J. R. Errington, and T. M. Truskett, "Does confining the equilibrium hard-sphere fluid between hard walls change its average properties?" *J.Chem. Phys.* (submitted)

3. **J. Mittal**, J. R. Errington, and T. M. Truskett, "Relationships between self-diffusivity, packing fraction, and excess entropy in simple bulk and confined fluids", *J. Phys. Chem. B* (submitted).

4. **J. Mittal**, J. R. Errington and T. M. Truskett, "Using Available Volume to Predict Fluid Diffusivity in Random Media", *Phys. Rev. E* **74**, 040102(R) (2006).

5. **J. Mittal**, J. R. Errington and T. M. Truskett, "Relationship between Thermodynamics and

Dynamics of Supercooled Liquids", *J. Chem. Phys.* **125**, 076102 (2006).

6. J. R. Errington, T. M. Truskett, and **J. Mittal**, "Family of entropy based anomalies for a water-like fluid" *J. Chem. Phys.* **125**, 244502 (2006).

7. **J. Mittal**, J. R. Errington, and T. M. Truskett, "Quantitative link between single-particle dynamics and static structure if supercooled liquids" *J. Phys. Chem. B Letters* **110**, 18147 (2006).

8. **J. Mittal**, P. Shah, and T. M. Truskett, "Using energy landscapes to predict the properties of thin films" *J. Phys. Chem. B* **108**, 19769 (2004).

9. A. Sharma and **J. Mittal**, "Instability of Thin Liquid Films by Density Variations: A New Mechanism that Mimics Spinodal Dewetting", *Phys. Rev. Lett.* **89**, 186101 (2002).

10. A. Sharma, **J. Mittal** and R. Verma, "Instability and dewetting of thin films induced by density variations", *Langmuir* **18**, 10213-10220 (2002).

6dr Utilizing Surface Science to Improve the Attachment of a Biopolymer Coating on Titanium, An Implant Metal

Holly J. Martin, Dave C. Swalm School of Chemical Engineering, Mississippi State University, 330 Swalm - President's Circle, Mississippi State, MS 39762 and Kirk H. Schulz, Vice President for Research and Economic Development, Mississippi State University, 617 Allen Hall, Mississippi State, MS 39762.

My research as a graduate student and as a post-doctorate researcher has focused on modifying the surface of titanium to increase the ability to bond a bioactive coating. During my graduate career, two novel reaction schemes were designed which greatly improved the bond between chitosan and commercially pure titanium, grade 4. These two schemes involved the use of two silane molecules in a toluene solvent and two metal treatments. We employed a commonly used silane molecule, 3-aminopropyltriethoxysilane, and a novel silane molecule, triethoxsilylbutyraldehyde, as the linker molecules between the titanium surface and chitosan. We also examined the effects of passivated titanium as compared to piranha treated titanium. X-Ray Photoelectron Spectroscopy (XPS) was used to examine the surface of the titanium following each individual reaction step and to ensure that the chitosan film was unchanged by the four treatment combinations. To further ensure that the mechanical properties of the chitosan film were unchanged, nano-indentation was employed, which demonstrated that the hardness and elastic modulus values were statistically similar to published research. Tensile testing confirmed that the chitosan films were bound to the titanium surface much more strongly than previously published research. I have extended my research as a post-doctorate researcher to include wastewater treatment with a novel chemical, peracetic acid. The hope of this research is to be able to treat wastewater without having to store the dangerous chemicals associated with chlorination and also to reduce or eliminate the production of disinfection by-products which can harm the environment.

As a graduate student and as a post-doctorate researcher, I have had the opportunity to work with many different professors, departments, schools, and companies. I have had the opportunity to work with Plant and Soil Sciences, the Center for Advanced Vehicular Systems, the University of Texas Health Science Center, and Electrical and Computer Engineering. As the primary operator of the XPS machine for most of my graduate career, and now as a post-doctorate researcher, I have been responsible for operating and maintaining the machine, along with teaching others to properly run and analyze their samples. Since the XPS machine is equipped with both a reaction chamber and ion sputtering gun, I have also been responsible for maintaining and operating these two additions to the machine. As part of the maintenance of the machine, I have performed bake-outs, replaced filaments, and performed troubleshooting of the equipment with technicians via phone conversations.

I have also had the opportunity to both create and teach classes at Mississippi State University. As a graduate student, I was in charge of creating and teaching a simulation laboratory to accompany the Process Design class (ChE 4134), a senior level design class. I received teaching evaluations, with an average of 4.7/5.0 for both the Fall and Spring semesters. I was also solely responsible for Unit

Operations I and II (ChE 3222 and 3232), two junior level laboratory classes, which included assigning laboratory projects, grading laboratory reports and reporting final grades at the end of the semester. I again received teaching evaluations, with an average of 4.5/5.0 for both classes during the Spring semester.

As an assistant professor, I plan to continue to perform research in biomedical implants. I also would like to continue working with other departments and other professors, as I have had the ability to do at Mississippi State University. I truly enjoy teaching and would like the opportunity to create a class that extends on the introductory materials class to cover implant issues, problems, and possible solutions. I also would like to teach both undergraduate and graduate level classes.

6ds Liquid Fuel Production Via the Integration of Glycerol Processing Over C-Supported Pt-Re Catalysts with Fischer-Tropsch Synthesis

Dante A. Simonetti[1], Edward L. Kunkes[1], Jeppe Rass-Hansen[2], Ricardo R. Soares[3], and James A. Dumesic[1]. (1) Department of Chemical and Biological Engineering, University of Wisconsin - Madison, 1415 Engineering Drive, Madison, WI 53706, (2) Department of Chemistry, Center for Sustainable and Green Chemistry--Danish Technical University, Lyngby, Denmark, (3) Faculdade de Engenharia Química, Universidade Federal de Uberlândia, Av. João Naves de Ávila 2121, Uberlandia, Brazil

We have studied the direct production of liquid alkanes from glycerol via an integrated process involving catalytic conversion of concentrated glycerol solutions to H_2/CO gas mixtures (synthesis gas) with subsequent Fischer-Tropsch synthesis. We show that the addition of Re to C-supported Pt-based catalysts increases the rate of synthesis gas production by a factor of 5 compared to Pt/C at conversions lower than 20% and with a H_2 co-feed. However, Pt-Re/C is an order of magnitude more active than Pt/C at conditions leading to higher CO pressures (greater than 20% conversion and without H_2 co-feed), suggesting that the primary promotional effect of Re is to weaken the interaction of CO with the surface, thereby decreasing the CO coverage and allowing the catalyst to operate at high rates in the presence of gaseous CO. Temperature-programmed reduction, CO and H_2 chemisorption, and X-ray diffraction studies indicate the formation of a Pt-Re alloy. Importantly, we show that this Pt-Re/C catalyst produces synthesis gas at high rates and selectivities suitable for Fischer-Tropsch synthesis (H_2/CO between 1.0 and 1.6) from concentrated glycerol feed solutions (between 50 wt% and 80 wt%) at low temperatures (548 K) and high pressures (5-17 bar). The oxygenated hydrocarbon byproducts from glycerol conversion do not significantly inhibit Fischer-Tropsch synthesis, and in some cases, have a synergistic effect of participating in hydrocarbon chain growth. Finally, we show that liquid hydrocarbons can be produced from glycerol in a two-bed, single reactor system at 548 K and pressures between 5 and 17 bar. The hydrocarbon products have a selectivity to C_{5+} alkanes of 75% at 17 bar, and the aqueous liquid effluent contains between 5 wt% and 15 wt% methanol, ethanol, and acetone, which can be separated from the water via distillation and used in the chemical industry or recycled for conversion to gaseous products.

6dt First Principles Studies of Energy Materials

Devina Pillay, Naval Research Laboratory, 4555 Overlook Ave. S.W., Washington, DC 20375

The development of novel energy materials depends on the ability to exploit their microscopic chemical and physical properties. My research focuses on attaining a fundamental understanding of the electronic, geometric, and magnetic characteristics of heterogeneous catalysts and battery cathode materials that govern their efficiency and effectiveness. In this poster I will discuss my ongoing research efforts on heterogeneous catalysts and Li metal oxide battery materials.

Oxide Supported 1B Metal Nanoclusters

While Au is known to be chemically inert in its bulk form, nanometer size Au particles dispersed on TiO_2 have been found to exhibit high activities for a variety of catalytic oxidation processes at or below room temperature [1]. Au atoms weakly interact with the TiO_2 surface and become unstable toward sintering in response to changes in the gaseous environment even at moderate temperatures [2-5]. Oxygen adspecies can greatly alter TiO_2 surface properties and Au-TiO_2 interfacial interactions, which may in turn directly influence the nucleation, growth, and sintering of Au particles [6-8]. In this poster, I will present the results of density functional theory calculations on the interactions between oxygen species and small Au particles on TiO_2 (110), with a focus on i) the dynamics of oxygen species on Au/TiO2 and ii) the effect of oxygen species on Au particle nucleation, growth, and sintering.

The Effect of Contaminants on Pt Alloy Catalysts

A critical issue in proton exchange membrane fuel cells (PEMFCs) is the slow rate of oxygen reduction reaction (ORR) on the Pt cathode electrocatalyst. Recent results have shown that alloyed systems such as Pt_3Ni have a higher activity than Pt metals towards the ORR which improves the overall efficiency of the fuel cell [9]. We have analyzed the electronic and geometric properties of Pt, Ni and Pt_3Ni (111) surfaces using first principles calculations. In addition, we have investigated the effect of sulfur contaminants on the chemical reactivity of these systems. Preliminary calculations have shown that S atoms can reduce the adsorption energy and even block adsorption sites between oxygen and hydrogen adsorbates and the (111) metal surfaces. Current results indicate that a single S atom can affect 13 neighboring Pt surface adsorption sites which is in good agreement with experimental data [10]. This effect is due to electrostatic interactions as well as a change in the electronic properties of these surfaces upon S adsorption. I will discuss (i) electronic and geometric properties of the metal surfaces, (ii) the adsorption geometry and bonding mechanism of oxygen and hydrogen gaseous species and (iii) the affect of S contaminants on oxygen and hydrogen surface kinetics.

Improving Phase Stability of Li Metal Oxide Battery Materials

Many properties crucial to good battery electrode performance can be accurately predicted using density functional theory (DFT) methods [11]. New materials can therefore be analyzed computationally and those with potentially superior qualities can be identified prior to synthesis and testing in the laboratory. This can drastically reduce the amount of time and money spent on battery development, while simultaneously providing valuable microscopic-level insight into the physics and chemistry of electrode processes.

Stability against distortion in battery electrodes indicates a longer lifetime and a lower waste of energy during charging and discharging cycles. Preliminary experiments indicate that the addition of Al to $LiCoO_2$ is advantageous in terms of stability [12]. Here we determine computationally the optimal concentration of Al and Mg dopants by calculating the intercalation voltage of $LiCoO_2$. Additionally, we analyze the electronic and geometric changes that these dopants induce, and correlate this to the intercalation voltage. Other results on various Li 3d transition metal oxide materials as well as dopants will be discussed.

References

[1] M. Haruta, Catal. Today, Size- and support-dependency in the catalysis of gold, 36, (1997), 153-166.

[2] **D. Pillay**, Y. Wang, *et al.*, Korean J. Chem. Eng., A comparative theoretical study of Au, Ag and Cu adsorption on TiO_2(110) rutile surfaces, 21, (2004), 537-547.

[3] **D. Pillay**, G.S. Hwang, J. Molec. Struct., Structure of Au, Ag, and Cu Clusters on the TiO_2 (110) surface, (2005).

[4] **D. Pillay**, G.S. Hwang, Phys. Rev. B, Growth and structure of small gold particles on rutile TiO_2 (110), 72, (2005).

[5] **D. Pillay**, Y. Wang, *et al.*, Catal. Today, Nucleation and growth of 1B metal clusters on rutile TiO_2(110): Atomic level understanding from first principles studies, 105, (2005), 78-84.

[6] **D. Pillay**, Y. Wang, *et al.*, J. Am. Chem. Soc., Prediction of tetraoxygen formation on rutile TiO_2 (110), 128, (2006), 14000-14001.

[7] D. Pillay, G.S. Hwang, J. Chem. Phys., O_2-coverage-dependent CO oxidation on reduced TiO_2 (110): A first principles study, 125, (2006).

[8] Y. Wang, **D. Pillay**, *et al.*, Phys. Rev. B, Dynamics of oxygen species on reduced TiO_2(110) rutile, 70, (2004), 193410.

[9] V.R. Stamenkovic, B. Fowler, *et al.*, Science, Improved oxygen reduction activity on Pt_3Ni(111) via increased surface site availability, 315, (2007), 493-497.

[10] **D. Pillay**, M.D. Johannes, J. Phys. Chem. B, Sulfur Poisoning of Pt(111) and Pt_3Ni(111) Surfaces: A First Principles Study, (In Preparation).

[11] G. Ceder, M.K. Aydinol, *et al.*, Comp. Mat. Sci., Application of first-principles calculations to the design of rechargeable Li-batteries, 8, (1997), 161-169.

[12] G. Ceder, Y.M. Chiang, *et al.*, Nature, Identification of cathode materials for lithium batteries guided by first-principles calculations, 392, (1998), 694-696.

22a Real Laboratories At A Distance

Jim Henry, UT-Chattanooga, 615 McCallie Av, Chattanooga, TN 37403

It is recognized how engineering teams can create distance organizations to collaborate on experiments; share information and knowledge; remotely operate instrumentation; dynamically acquire, archive, e-publish, access, mine, analyze, and visualize data; and deliver unique learning, workforce-development, and innovation tools.

These facets of experimentation, sharing, remote operation, data acquiring, analysis and visualization and unique learning tools are available in undergraduate chemical engineering laboratories.

This poster illustrates the techniques and variety of remote operation of unit operations and control systems for student experimentation and learning.

Web Page: chem.engr.utc.edu/labs

22b Fem Using Comsol: Applications For Fuel Cells

Jason M. Keith, Faith A. Morrison, and Julia A. King. Michigan Technological University, Department of Chemical Engineering, 1400 Townsend Drive, Houghton, MI 49931-1295

This presentation will demonstrate the use of COMSOL Multiphysics software for performing finite element analysis for problems in momentum, heat, and mass transport. The problems can be solved by undergraduate chemical engineers in the transport phenomena sequence. The problems presented here have applications to fuel cells.

Introducing new software into the curriculum can sometimes be difficult for students. We advocate the use of a handout with step-by-step instructions for the students to follow. After completion of this problem, the students are then asked to solve a more complicated problem on their own.

22c Superpro Designer: An Interactive Software Tool For Designing And Evaluating Integrated Chemical, Biochemical, And Environmental Processes

Nirupam Pal[1], Demetri Petrides[2], and Charles Siletti[2]. (1) California Polytechnic State University, Dept. of Civil and Envr Engr, San Luis Obispo, CA 93407, (2) INTELLIGEN, Inc., 2326 Morse Avenue, Scotch Plains, NJ 07076

The successful design and evaluation of integrated chemical manufacturing and end-of-pipe treatment processes is a challenging task that can be facilitated by the use of advanced computer-aided process design and simulation tools. Such tools also play an important role in teaching environmental engineering and other design-oriented courses. They are particularly useful in emphasizing the importance of systems approach in the design of integrated processes. In our department we have successfully used SuperPro Designer, a comprehensive simulator from Intelligen, Inc. SuperPro carries out material and energy balances, calculates environmentally significant stream properties, estimates the size and cost of equipment, and performs economic evaluation of integrated processes. Using SuperPro, the student and tomorrow's engineer can readily analyze the effect of changes in operating conditions on the environmental impact and the economics of a process. The functionalities of SuperPro and our experience in using this tool in several courses will be presented.

Web Page: www.intelligen.com/literature

Solving problems in binary batch distillation on the computer using Mathcad® - Part 2

The author has been active in obtaining computer solutions for various separation processes, for example, computer solutions for binary distillation problems using McCabe-Thiele and Ponchon-Savarit methods for steady state operations. Batch distillation has also been examined. Some of these, however, can be obtained from other sources such as commercially licensed software such as Aspen®, Hysys® and Chemcad®. The cost of the license can be expensive and in certain special circumstances the solutions provided may not be totally satisfactory.

Several programs for solving batch distillation problems have been generated using Mathcad®. The purpose of generating these programs was partially for teaching purposes although design usage is possible. Two basic types of distillation were considered. In the first type, the distillate composition is maintained constant and the reflux ratio is varied with time, and in the second type, the reflux ratio is fixed and the distillate composition is allowed to vary with time. For each of these types, a program was generated that could treat 26 different binary systems (systems without azeotropes). Each system was assigned a number that was connected to information in a database. By selecting the proper number, all the necessary information for that system would load. Additional programs for treating systems with azeotropes (maximum and minimum boiling) are currently being developed. In all cases, the number of ideal stages for the column and the feed composition must be specified, although the programs are provided with default values. **For all systems the operating line and the stages may be animated to illustrate the course of events as a_function of time.** The programs will provide numerical output and graphical output as a function of time. The activity coefficients are generated by means of the Wilson equation and the vapor pressures by means of the Antoine equation.

These computations are very difficult to achieve by hand because it involves fitting stages to an operating line that isn't known. This requires graphical trial and error procedures that are very time consuming, particularly if numerous stages are involved. Graphical integration is often needed. Here, the computer does all of this difficult work. The animation of the program is a wonderful aid to the process of teaching batch distillation, but also shows some aspects of the procedure not often considered. For example, consider a batch distillation in which the reflux ratio is held constant. Assuming that the column has five ideal stages, those stages at the start of the process will be doing roughly the same amount of enrichment. However, as time progresses, the lower end of the operating line may begin to approach the equilibrium curve. Thus, the bottom stages may have very small steps, resulting in very little enrichment. For such a situation, it may be that a column with fewer stages can do the same separation more economically.

The systems used for this study were picked in order to obtain a wide range of relative volatility. The fourteen systems examined are as follows:

$System_n :=$

"1-propanol & water"
"2- propanol & water"
"Acetone & chloroform"
"Acetone & methanol"
"Chloroform & methanol"
"Difluorochloromethane & dichlorodifluorometha
"Ethanol & benzene"
"Ethanol & water"
"Ethyl acetate & ethanol"
"Methanol & benzene"
"Methanol & ethyl acetate"
"Methyl acetate & methanol"
"Methyl tertiary butyl ether & methanol"
"Propylene oxide & methanol"
"Tetrahydrofuran and water"

The phase diagrams for the system ethyl acetate and water exhibits a minimum boiling azeotrope. (See figure 1) The program generates both the temperature, composition diagram for the vapor and liquid and it also generates the ordinary xy diagram for this system. (It can do this as well for all the systems shown above). The program can deal with systems showing a maximum boiling azeotrope. Such an example is the system acetone and chloroform. (See figure 2) It is not possible to illustrate on paper the animated McCabe-Thiele diagrams. These must be seen on the computer. However, the programs generate graphics which show the composition and the moles of the feed which remain in the still pot at any given time. Two types of routines are run. One routine assumes a constant distillate composition, while the reflux ratio increases. The second type assumes a constant reflux ratio, while the composition of the distillate decreases with time. The time elapsed is set relative to a time constant.

t, x1, y1 phase diagram

P= 760

System = "Ethyl acetate & ethanol"

Pressures are in mm Hg.

Temperatures are in degrees C.

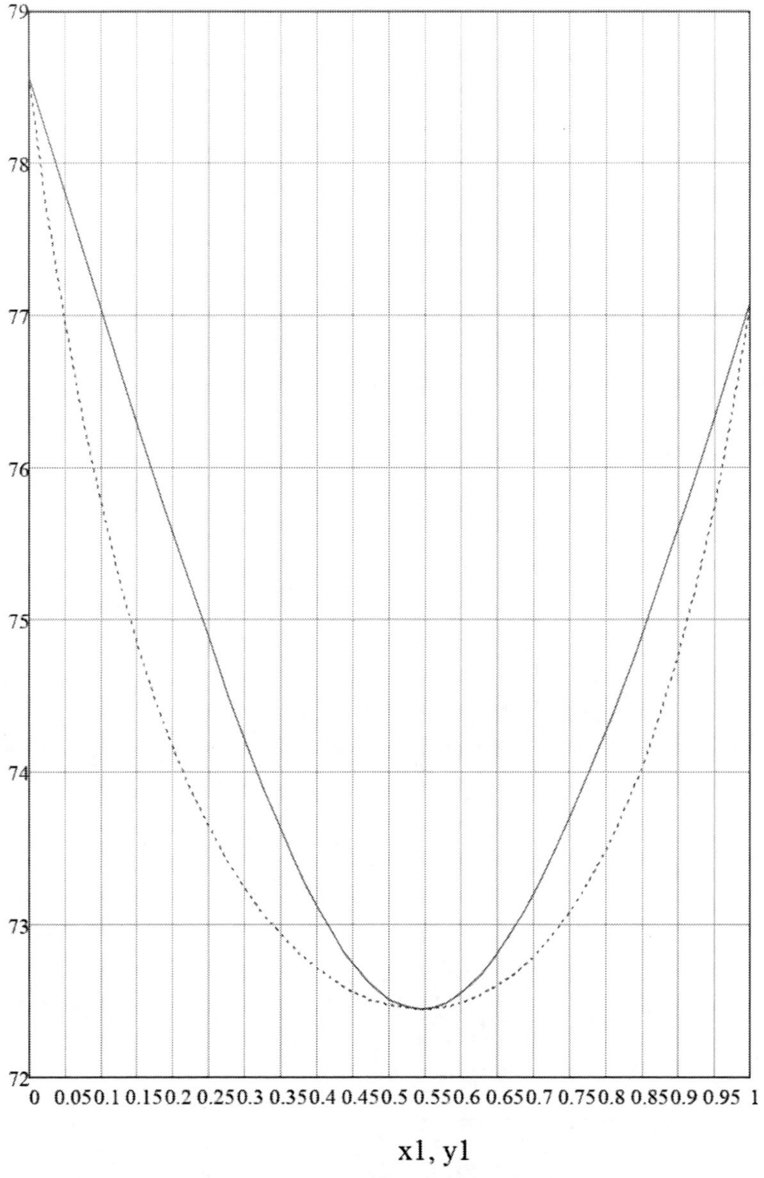

Figure 1

P = 760

t, x1, y1 phase diagram

Pressures are in mm Hg.
Temperatures are in degrees C.

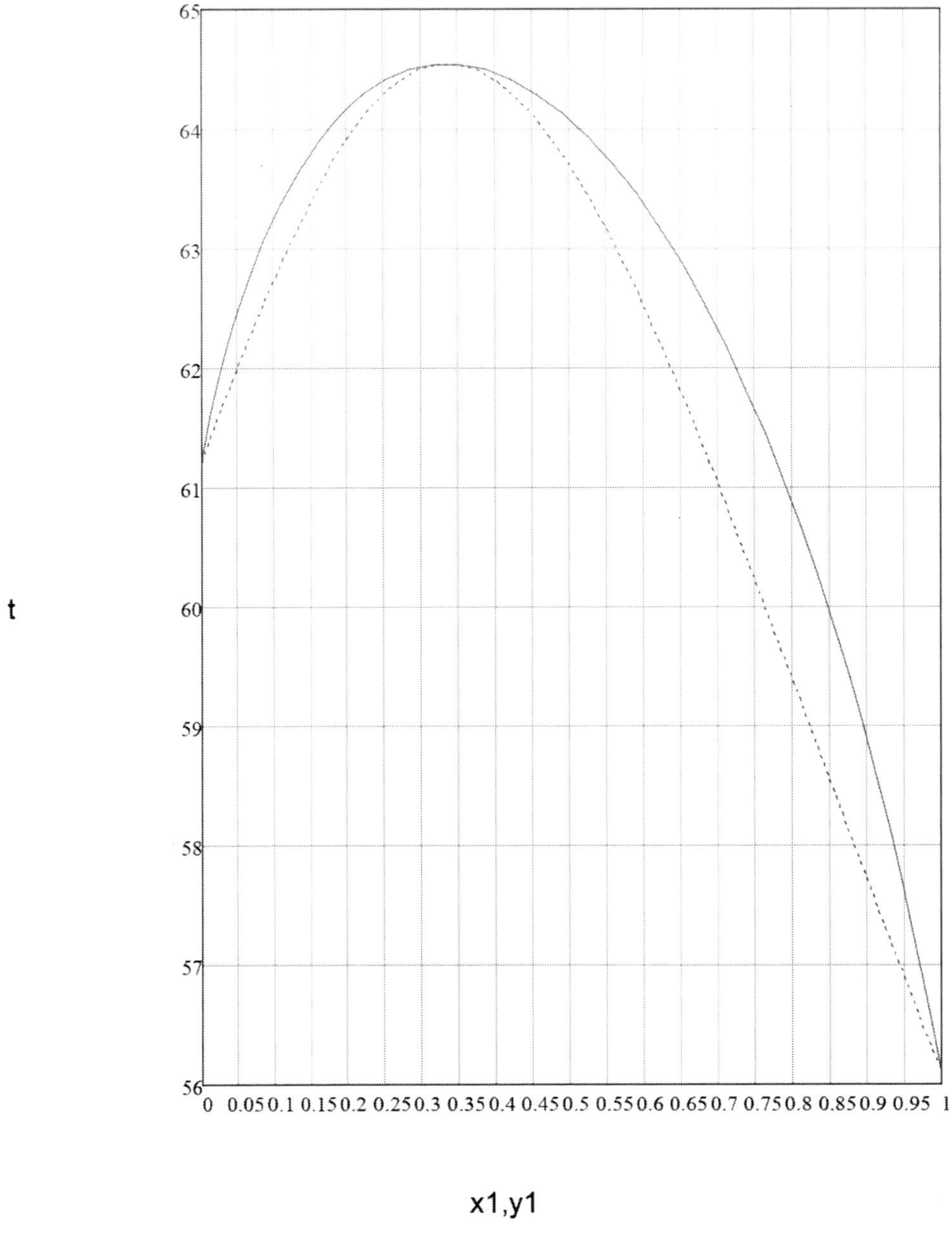

Figure 2

azeotype = "max. boiling'

BIBLIOGRAPHY

1. McCabe, W. L., and E. W. Thiele: *Ind. Eng. Chem.,* **17,** 605 (1925).
2. Ponchon, M.: *Tech. Mod.*, **13,** 20, 55 (1921).
3. Savarit, R.: *Arts Metiers*, **1922,** 65,142,178, 241, 266, 307.
4. Treybal, R. E.: "Mass Transfer Operations," 2nd ed., McGraw-Hill Book Company, New York, 1955.
5. Wankat, P. C.: "Equilibrium Staged Operations," Prentice Hall Inc., Upper Saddle River, New Jersey, 1944.
6. DECHEMA, "Vapor-Liquid Equilibrium Data Collection", Gmehling, J, Onken, U. and Arlt, W.
7. Bart, E. N. and J. Kisutcza: *Chisa Summaries,* **2, Separation Processes,** 639 (2004).
8. Bart, E. N. and J. Kisutcza: *Chisa Summaries,* **2, Separation Processes,** 344 (2006).
9. Bart, E. N. and J. Kisutcza: Meeting of the Mid-Atlantic Section, ASEE, (Fall 2006).
10. Bart, E. N. and J. Kisutcza: Meeting of the National Conference, ASEE, (June 2007) 3113.

22e Polymath - Now Integrates Problem Solving with Excel and Matlab

Michael B. Cutlip, University of Connecticut, 191 Auditorium Rd., Rm. 204, Storrs, CT 06269-3222, Mordechai Shacham, Ben Gurion University of the Negev, Chemical Engineering Department, Beer-Sheva, 84105, Israel, and Michael Elly, Intel Corp., Qiryat Gat, Israel.

POLYMATH Software now supports the integration of problem solving with Excel and MATLAB. This integration allows the export of POLYMATH programs directly to Excel and MATLAB from within POLYMATH by a single key press (Excel)or the proper selection of export option (MATLAB).

The export to Excel automatically generates a complete Excel spreadsheet using the notation and equations from the POLYMATH program. The Excel export automatically reformulates all intrinsic functions and logical variables and conditions of the original problem. Solutions utilize Excel functions, tools such as Goal Seek and Add-Ins such as Solver. POLYMATH provides an Add-In called ODE_Solver for solving systems of ordinary differential equations in Excel.

The option for MATLAB generates code that can be integrated into an m-file for the problem. The MATLAB code is autmatically ordered during the code generation. Intrinsic functions and logical variables and conditions of the original problem are automatically translated. Templates for easy creation of working MATLAB m-files for the problem are provided.

General POLYMATH Software Overview

POLYMATH 6.1 Educational is a proven computational system that supports advanced problem solving in chemical engineering. Capabilities include Linear Equations, Nonlinear Equations, Differential Equations, Data Analysis and Regression. Low-cost site licenses are provided through the CACHE Corporation for academic departments. These licenses enable use in computer labs and distribution of individual copies to all student, faculty, and staff. Individual use educational and professional versions are also available.

POLYMATH enables the user to apply effective numerical analysis techniques during interactive problem solving on personal computers. Graphical and tabular results are provided for easy understanding and for incorporation into papers and reports. POLYMATH has a minimal learning curve and provides extensive error checking during problem entry that leads to great efficiency in problem solution.

Web Page: www.polymath-software.com/

23a Incorporating Food Into The Chemical Engineering Curriculum

Majid Salim, Chris Barr, and Allen Hersel. John J McKetta Department of Chemical and Bioprocess Engineering, Tri-State University, 1 University Ave, Angola, IN 46703

As a university in the Midwest where a large percentage of our graduates work in the food and consumer goods industry, our department has been incorporating food experiments into the sophomore and junior level laboratories. Since a majority of the college level laboratory experiments have quantitative results, we wanted to implement some experiments which have quantitative results as well as subjective results such as taste, appearance, and aroma.

Another factor in designing new subjective experiments had to do with the lack of statistical analysis in our current laboratory experiments. Feedback from our Industrial Advisory Board meetings has shown that a sound background in statistical analysis is a needed skill for the graduating seniors. We already teach a Numerical Methods class and are requiring our students to take Probability and Statistics. However, we noticed that we did not have many experiments which implemented statistics. As a department which prides itself with having experiments that coincide with all of the Chemical Engineering classes taken, we feel these new food-based experiments will add much needed "real world" statistical applications into the labs.

Two food based experiments that have been developed by our department are a coffee experiment and a chocolate experiment. In the coffee experiment, students will explore the quantitative results by performing a two parameter statistical model based on the roasting time and the grind size. The subjective portion of the experiment will be an "expert" taste panel of students and faculty who drink coffee on a regular basis.

Similarly, the chocolate experiment will have a two parameter statistical model as the quantitative result. However, this statistical model will be comprised of the cocoa to sugar ratio and the mold temperature of the chocolate. Once again, the students will present the final product to a taste panel for the subjective results of taste, texture, and appearance.

As a result of the new coffee and chocolate experiments, students will be better prepared for their future careers because of the "real world" statistical applications and the new insights into subjective results. These new insights about subjective results will give students the understanding that sometimes quantitative specifications are not as important if consumers won't use the product. Moreover, it shows that not every consumer looks for the same things in a product.

23b Incorporating Risk Assessment And Inherently Safe Design Into Process Design Education

Jeffrey R. Seay, Process Technology and Engineering Department, Degussa Corporation, 4301 Degussa Road, Mobile, AL 36590 and Mario R. Eden, Auburn University, Department of Chemical Engineering, 222 Ross Hall, Auburn University, AL 36849-5127.

Process safety is a fundamental component of sound process design. Although the chemical industry has demonstrated an excellent safety record over the years, the quantities and hazardous nature of many of the substances typically handled by chemical manufacturers make the potential for large scale disasters a constant concern. Because safety is so critical in industry, it is vital to introduce the concepts of safe process design practices during undergraduate chemical engineering education. Gaining an appreciation of this during a chemical engineer's education can only enhance chemical manufacturing safety in the future. An understanding of how risk is quantified in the chemical process industries will allow future process design engineers to mitigate those risks at the earliest stages of conceptual process development – at the stage where the engineer has the greatest influence on the final process design. This poster will introduce, via a case study example, the benefit of including risk assessment methodology and inherently safer design practices into the process design education of chemical engineering students. This poster will illustrate an industrial case study example that can be used to introduce the concepts of risk assessment and inherently safe process design into the classroom.

The case study to be presented as the classroom example will be based on two alternative processes for breaking an alcohol-water azeotrope. Using risk assessment methodology and inherently safe design practices, the process safety implication of both processes will be illustrated. This design problem will be structured to demonstrate how these types of scenarios are evaluated in industry, and how these techniques can be carried over into the classroom.

One of the responsibilities of every chemical engineer is to ensure that the excellent safety record enjoyed by the chemical process industry is maintained. Therefore it is important to begin introducing the fundamentals of process safety during undergraduate chemical engineering education. The purpose of this poster presentation will be to underscore, by case study example, the natural relationship between inherently safe process design and conceptual process development and how it can be integrated into undergraduate process design education. As will be illustrated by this case study, taking a holistic approach to process safety education can serve to reinforce the benefits of beginning to consider the safety implications of the decisions made during conceptual process development. By reinforcing the benefits of making inherently safe design choices during conceptual process development, students of process engineering will be better prepared for the challenges of meeting the high standards of safety set by today's chemical industry.

23c Nanoscale Studies Course Development At North Carolina A&t State University

Kenneth L. Roberts, Mechanical and Chemical Engineering, North Carolina A&T State University, 1601 E. Market St., 627 McNair Hall, Greensboro, NC 27411-0001

Nanoscale, or atomic-level, studies currently dominate the future interests of national governments, financial institutions, multinational corporations, and leading academic institutions around the globe. A new Nanoscale Studies Initiative (NSI) at North Carolina Agricultural and Technical State University was created in part to connect its seven University academic units by means of an upper-level interdisciplinary elective course, Chemical Engineering (CHEN) 505 - Nanoscale Studies : The Societal Implication of Invention. This senior-level course was open to all Senior-level students at the University was first offered in the Spring 2006 Semester. This course exposed students from two academic disciplines, engineering and technology, in the interdisciplinary field of nanoscale studies. The course provided an overview of topics ranging from the technical to the philosophical in such fields as nanomedicine, nanoethics and public policy, nanobiotechnology, nanobusiness, nanomanufacturing and nanoeletronics.

CHEN 505 provided students a brief overview of technical topics in the pure and applied nanosciences. CHEN 505 students later explored how future inventions and economic opportunities at the nanoscale impacted the greater society in such fields as the social sciences, business, ethics, medicine, education and religion. Topics covered in this course included: Public Policy & Educational Implications of Nanotechnology; Ethical Issues & Unintended Consequences for Nanotechnology; Strategic Impacts on the Future of Nanobusiness & Economics; Challenges & Vision for Nanomedicine and Pharmaceutics; Social Acceptance & Socio-Economic Impacts of Nanotechnology; and Legal Implications & National Security Aspects. This paper will discuss the development and implementation of this new course.

23d Understanding Gibbs Free Energy Through 3D Phase Diagrams For Pure Components

A. Alarcón-García, L. G. Ríos-Casas, D. R. Téllez-Muradás, and J. R. Flores-Tapia. Chemical and Food Engineering Department, Universidad de las Americas, Puebla, Mexico, Ex-hda Sta. Catarina Martir, Sn. Andres Cholula, 72820, Mexico

This work is related to a previous one regarding the representation of Thermodynamic Properties in 3D diagrams. This time it is focused on the analysis of Free Gibbs Energy for pure components to achieve a better understanding of the equilibrium criteria of fluid phases.

The fundamental property relations are obtained from the substitution of the first and second Law of Thermodynamics to define the thermodynamic equilibrium state of a closed system, thus the independent variables are not chosen arbitrarily.

The thermodynamic properties data for water are plotted in 3D to represent the resulting equations for G (Gibbs free energy). The database was taken from Haar et. al. (1984) and introduced in Matlab®. Different colors were used to distinguish the various regions: homogeneous liquid phase, homogeneous gas phase, vapor-liquid region and supercritical fluid. The mesh is constructed with values at constant temperature and pressure and the curves for constant vapor fraction are included inside the vapor-liquid region.

The 3D diagrams obtained can be used to generate projections in 2 dimensions: P-G, G-P, T-G, G-T for fluid homogeneous phases and vapor-liquid equilibrium. The behavior of free Gibbs energy may be associated to other concepts like fugacity and fugacity coefficient to be used as additional equilibrium criteria.

It is evident the complexity the engineering students have to handle when the professor starts the study of thermodynamic properties, but a deeper understanding could be attained by adding a visual approach. In the last few years, this methodology has been applied in the Thermodynamics course for sophomore Chemical Engineering students. The professor has perceived the students show good acceptance of this educational material and the evaluations have revealed a better understanding of these concepts.

The success of this approach relies on the proven fact that the learning experience is more effective when more senses are involved.

23e Force Computation Between Spheres And Flats In An Aqueous 1:1 Symmetric Electrolyte Solution Using Matlab

Xiaoting Hong, R. Eric Berson, and Gerold A. Willing. Chemical Engineering, University of Louisville, 106 Ernst Hall, Louisville, KY 40292

There is no question that discussion of the forces of attraction and repulsion between different surfaces such as flats and spherical particles are of need and importance in both the undergraduate and graduate thermodynamics courses in a chemical engineering or chemistry curriculum. It is also an important issue for students from an environmental engineering or pulp and paper background who need to understand the principles behind flocculation and coagulating during the course of waste water treatment and the paper making process. In practice, numerous products and processes depend on controlling the balance between attractive and repulsive interactions: products such as paints and inks, aerosols and gels, and rubber latexes; processes such as filtration and clarification, emulsification and demulsification. To fully explain the forces acting between different surfaces in a complex fluid, one must introduce the DLVO theory. The basic idea of the theory is that colloidal stability is dependent on the balance between two opposing contributions: a short range repulsive force and a long range attractive force. Different repulsive force equations have been derived based on the solutions of the Poisson-Boltzmann equation under different assumptions such as the Langmuir approximation, Derjaguin approximation and the Debye-Hückel approximation for both the constant surface charge and constant surface potential cases. Previously published work has theoretically demonstrated a numerical integration method for calculating the repulsive interactions when a pair of plates approaches one another to a specific surface separation. Hamaker's molecular model can then be used to calculate the van der Waals force contribution to the total force at this separation. Here we compare the total forces from approximate equations and the theoretical integration for two spheres and two flats under same conditions by using newly developed Matlab programs. Moreover, for sphere-flat systems, we can show the force curve difference by changing system parameters such as zeta potential and electrolyte concentration. These systems will be linked to specific case studies that can be implemented in the classroom to help students understand the role of colloidal scale interactions on suspension stability and discovering how varying the system parameters can influence these interactions.

23g Industrial Collaboration For Improving Undergraduate Hands-On Process Control Education

Atanas Serbezov[1], Ronald Artigue[1], Richard Plapp[2], and Darryl Carpenter[2]. (1) Department of Chemical Engineering, Rose-Hulman Institute of Technology, 5500 Wabash Avenue, Terre Haute, IN 47803, (2) Eli Lilly and Company, Indianapolis, IN 46285

Several nationwide studies on the state of process control education in the Chemical Engineering curricula have pointed out to a growing gap between what students learn as undergraduates and the needs of industry. Process control experts from both industry and academia agree that a problem exists, and that one way to narrow the gap is a closer liaison between engineering programs and industry. Control education is not the sole responsibility of academia. Professional control engineers must be involved in the control education process as well, particularly when one considers that many of the students will work in industry, often as control engineers. Industry has a vested interest in seeing students who are well-educated in the latest control instrumentation and techniques, since that is its recruiting pool. At the same time industry has a wealth of problems which make for interesting and challenging laboratory experiments, and it has the financial resources to help in providing modern laboratory facilities. This presentation describes a successful collaboration between the Department of Chemical Engineering at Rose-Hulman Institute of Technology and Eli Lilly and Company, the result of which is a much improved undergraduate process control experience for the students. Its major achievement is that all chemical engineering students are exposed to industrial-grade process instrumentation, control system architecture, and control and tuning algorithms beyond PID and Ziegler-Nichols. Another achievement is that students are introduced to the industrial practices of data storage, management, retrieval and mining. The backbone of the new process control experience in the Department of Chemical Engineering at Rose-Hulman Institute of Technology is an industrial distributed control system (DCS), namely DeltaV by Emerson, which is incorporated in the undergraduate Unit Operations (UO) laboratory. The new environment in the UO laboratory is very similar to the environment in a typical chemical, petrochemical or pharmaceutical plant. Students learn how to maintain their process under control, take the process safely from one operating condition to another, collect and analyze data using a historian, respond to process alarms and remotely troubleshoot their experiments using limited process information. In addition to financial support, the industrial partner helped with expert advice in the design of new and the revamping of the existing laboratory experiments. It also provided sabbatical opportunities for the faculty involved in process control education. The experience of both parties in this project, namely industry and academia, has been very positive, and we hope that this can provide a model for other academia-industry partnerships for enhancing process control education in the undergraduate curriculum.

23f Experiences Applying Statistical Concepts And Designed Experiments In The Laboratory Courses

David R. Mills, Chemical Engineering, Auburn University, 212 Ross Hall, Auburn, AL 36849-4827

Chemical engineering students are most often asked to solve problems in which input data is assumed to be infinitely precise. Answers to problems are likewise given as exact values. In reality, all real measurements and resultant calculations have an inherent degree of uncertainty. Similarly, industrial product qualities vary as complex functions of process input parameters with associated uncertainty. Since many chemical engineering graduates end up working with some aspect of quality control in which Statistical Process Control (SPC) and Design Of Experiments (DOE) are applied, it is important that students gain experience with data containing uncertainty and to be able to reasonably draw conclusions based on the data. Additionally ABET stipulates that students "have the ability to design and conduct experiments". Transport, Unit Operations and Process Control Laboratory courses are often the first place students encounter "real" data and are ideal places to apply statistical methods. For these reasons, we use laboratory courses to develop the concept of uncertainty and "statistical thinking" through the use error propagation techniques, confidence intervals based on standard deviations of repeat measurements and simple regression analysis. Students end their final laboratory course with a "capstone" DOE project in which they develop their own experiment from g conceptualization, to writing of procedures to achieve experimental goals and minimize systematic error, as well as statistical design and analysis via ANOVA. Materials used to teach error analysis and DOE, including software and tutorials will be discussed. Applicability to specific laboratory experiments and examples of student-designed experiments will be presented. Successes and challenges will also be discussed.

58a Energizing An Introductory Chemical Engineering Course with Biodiesel

Katherine A. Taconi and R. Michael Banish. University of Alabama in Huntsville, 130 Engineering Building, Huntsville, AL 35899

The biorefining industry is experiencing tremendous growth in the United States as well as in the State of Alabama. Therefore, it is important that chemical engineering departments include instruction and education in the area of biofuels and biorefining in the traditional chemical engineering curriculum. Biofuel production is also a timely and relevant way to make chemical engineering "come alive" and stimulate interest and improve retention among students.

The Department of Chemical and Materials Engineering at the University of Alabama in Hunstville is incorporating a hands-on laboratory activity allowing students to produce their own biodiesel into the introductory chemical engineering course. This presentation will describe the development and implementation of this lab activity. Included in the presentation will be the necessary materials and supplies, procedures, cost estimates, examples of supplemental lecture material, and outcomes from student evaluation.

Capstone Design Project for Non-Chemical Engineering Major Students: Production of Biodiesel

Ping Wang *and* Weilu Lin
Department of Bioproducts and Biosystems Engineering
University of Minnesota at Twin Cities
St Paul, MN 55108

Biodiesel made from renewable resources such as vegetable oil, animal fats, and recyclable cooking oils is one major alternative transportation fuel to petroleum products. According to the statistics from National Biodiesel Board, there are presently 105 biodiesel plants operating in the United States with a capability of producing 864 million gallons of biodiesel per year. As the third largest soybean producer in the United States, Minnesota is the first state successfully implemented a statewide initiative that requires blending two percent biodiesel throughout its entire diesel fuel supply.

The biodiesel design project was developed for senior students of the Department of Bioproducts and Biosystems Engineering (BBE) at University of Minnesota. This course promotes the practice and application of fundamental concepts of biobased products engineering. As most of the students enrolled in the class may not have a background with unit operation and other engineering design concepts, teaching engineering design processes and modern design tools and skills has to be conducted at the same time as the design project proceeds.

Biodiesel production from soybean oil was selected as the subject by the students for the BBE Capstone Design Project in Spring 2007. The course was divided into three phases. The first phase was for general instructions on process design concepts and principles. Along with that, the students were also required to conduct literature and data searching for information they will need for market analysis, industrial processing routes and feedstock availability. The second phase involved development of process diagrams using process simulation tools, HySYS in this case, equipment design, and cost and benefit analyses. At the final stage of the project, students were required to prepare a project report, a 20 minutes oral presentation and a poster to peer students and faculty members. In addition to engineering elements, training and practicing of team work ethics, communication and leadership skills are also essential elements of the course.

58c Global Projects in Engineering- a Multidisciplinary Course

Allyson Frankman, Jacob Jones, W. Vincent Wilding, and Randy S. Lewis. Department of Chemical Engineering, Brigham Young University, 350 Clyde Building, Provo, UT 84602

With engineering increasingly becoming an international discipline, engineering programs are enhancing opportunities to prepare students for such a global industry. Being an engineer of tomorrow's world will require students to understand and work with different cultures, peoples, practices, ethics and paradigms. In addition to successful careers, opportunities for engineering graduates to apply their expertise to solve both technical and social problems in the world around them are beneficial.

In order to engage engineering students in global opportunities, a 3-credit hour course was recently developed that involved chemical, mechanical, civil/environmental, and electrical/computer engineering students. The first course was held during the Winter 2007 semester and was followed by a trip to Tonga in May 2007 with 26 students. Students ranged from sophomores to seniors. The trip involved the implementation of a biodiesel process using coconuts. The second course will be held Fall 2007 and will involve several renewable energy/sustainable technical projects that will be implemented in a small village outside of Cuzco, Peru. A major component of the course is working on the technical projects in multi-disciplinary teams.

Having recently completed the first course and the trip, aspects of the course and the associated project will be discussed with regards to the curriculum structure, teamwork, funding, project coordination, project implementation, and student evaluation. The completed assessment will be discussed along with the current changes planned for the next course.

58d Teaching Of "Sustainability, Technology, And Society" At Northwestern University

Harold H. Kung, Chemical and Biological Engineering, Northwestern University, 2145 Sheridan Road, E136, Evanston, IL 60208-3120

Our experience in offering a senior/graduate level elective at Northwestern University, entitled "Sustainability, Technology, and Society" will be presented. The course was offered as a chemical engineering course, but its contents were intended for advanced science and engineering students, and there was the assumption that the students have been exposed to general and organic chemistry. The objective of the course is to discuss the global impact of natural resource utilization and recovery, and the role of science, technology, policy, and business strategies in enhancing sustainability. Because of the advanced science and engineering background of the students, many technical topics can be addressed at a rather sophisticated level.

The course covers topics ranging from effect of human activities on the environment, ranging from concept and some detailed applications of life cycle assessment and life cycle cost analysis of processes, products, and policies, to examples of technology to enhance sustainability, and examples of companies formulating business strategy based on sustainability. Because of the recent attention received, energy related topics, including global warming, form one of the major themes. Other themes are derived from chemically related activities. Other aspects of the course include: (1) emphasis on current events, (2) case studies of companies, and (3) project based student activities.

58e Microbial Fuel Cells as a Multidisciplinary Teaching Tool

Mohammad A. A. Dewan[1], Bernard J. Van Wie[1], Zbigniew Lewandowski[2], and Haluk Beyenal[1]. (1) Chemical Engineering, Washington State University, Spokane street, Pullman, WA 99164, (2) Center for Biofilm Engineering, Montana State University, Bozeman, MT 59717-398

We propose using microbial fuel cells (MFCs) as a teaching tool promoting interdisciplinary interactions among undergraduate students in chemical engineering, bioengineering, environmental engineering and microbiology. MFC experiment embodies principles of thermodynamics, mass transport, electrochemistry, material balance and basic microbiology. We have designed a set of measurements which require that the students apply the fundamentals of their own disciplines and, in addition, learn the scientific principles of other disciplines. This process is designed to help the students to perform in working in environments where employees from various disciplines work together on the same project. For example, chemical engineering students will learn safety measures required for working with microorganisms, culturing cells, counting cells which are normally known by microbiologists. Similarly, microbiology students will learn mass and energy balances, measurements and calculations related to electronics.

Through designing and operating MFCs the students learn principles of energy production, by calculating energy efficiency, coulombic efficiency, power density, and constructing polarization curves. In addition, they learn how to use electrochemical instrumentation, by operating potentiostats connected to computer assisted data acquisition systems. At the same time they also learn principles of microbiology, by culturing bacteria, counting cells, taking images of the biofilms on the surfaces of electrodes. From working with undergraduate students in the past, we know that they are enthusiastic about using alternative sources of energy and are fascinated by producing energy using microorganisms. Setting up a microbial fuel cell and making the electrochemical measurements takes approximately three weeks, for a group of three to four students recruited from chemical engineering and microbiology, working for two hours per day and three days a week.

112a Coaching Students For Improved Team Performance

Pedro E. Arce and Joseph J. Biernacki. Chemical Engineering, Tennessee Technological University, Department of Chemical Engineering, P.O. Box 5013, Cookeville, TN 38505

Teamwork skills remain at the top of the list of important characteristics that employers seek in new graduates. None-the-less, most employers continue to report that new hires are unprepared to work effectively on teams. This should come as no surprise, since most engineering programs have traditionally not offered teamwork training as an explicit activity. At Tennessee Technological University (TTU) the Department of Chemical Engineering has implemented a rigorous program in teamwork training that enables students to discover the importance of teamwork and how successful teams are formed in industry and other environments and offers them numerous opportunities throughout the curriculum to practice teamwork. The TTU team training environments is an integral part of our new integrated laboratory-lecture curriculum wherein the majority of lab activities are integral parts of previously classroom-oriented courses.

112b "High Performance Learning Environments [Hi-Pele]: Role Of Team Functions On The Design, Implementation, And Assessment."

Sharon Sauer, Chemical Engineering, Rose-Hulman Institute of Technology, 5500 Wabash Ave, Terre Haute, IN 47803 and **Pedro E. Arce**, Chemical Engineering, Tennessee Technological University, Department of Chemical Engineering, P.O. Box 5013, Cookeville, TN 38505.

Sound pedagogical and high retention instructional environments must be learned-centered, i.e. they must allow the students to explore, compare, and critically use their thinking in understanding a new piece; they need to have a knowledge-centered characteristic so that students can know not only facts, concepts, principles and values but also when they need to apply the tools. As in research and development in an engineering discipline, effective learning environments must allow for the students to assess and validate their progress, i.e. be assessment-centered in understanding a new piece so that they can eliminate misconceptions and have chances to improve what they know. Finally, these environments need to be community-centered so that the students can actually build their new knowledge, connect with and "engineering culture," and growth professionally.

High Performance Learning Environments or Hi-PeLE are unique in the sense that they capture the four centerness described above not only by using typical learning cycles but also by adding a linear engineering sequence. This important aspect uses functional-team based learning and highlights the use of an experimental prototype that it is not usually present in others environments. Hi-PeLE also promote life long learning by documentation cycles and reduce effectively the number of subjects in a course by employing "principal objects of knowledge" (POK's) . Hi-PeLE has been very successful at the FAMU-FSU College of Engineering, Mississippi State University, Rose-Hulman Institute of Technology, and Tennessee Tech in promoting very high level of critical thinking and student readiness.

In this presentation, the authors will describe the pivotal role of the team-bsed learning and the optimal preparation needed by both student and instructor in achieving the successful outcomes of the approach. Also, avenues for further work in research will be highlighted.

References: Bransford et al., "How People Learn: Mind, Brain, Experience and School", National Academic Press, 1999. Arce, P. and L. Schreiber, "High Performance Learning Environments," Chem. Eng. Education, Fall 2004. Stacey Klein "Legacy Cycle Workshop", Tennessee Tech University, March 2007. Sauer, S. and P. Arce, "Team Member Selection: A Functional-based Approach", ASEE Proceeding, 2003. Arce, P.; J. Biernacki, and P. Melton, "The Experimental Prototype", ASEE National Meeting, 2006. Ralling, R.; S. Allen; and P. Arce, "Personalized Class Binders, PCB". AIChE Annual, 2005. Arce, P. "The Colloquial Approach", Journal of Science Education and Technology (1994).

112c Lego Nxt Robotics--Introducing Che Freshmen To Engineering Skills And Concepts

Bill B. Elmore, Dave C. Swalm School of Chemical Engineering, Mississippi State University, MS 9595, Mississippi State, MS 39762

Abstract

We have completed our first year of using LEGO NXT RoboticsTM for teaching fundamental engineering skills and concepts to second semester freshmen. The goals of this course include establishing collegiality among the students through teams and project-based learning; building concepts of engineering problem analysis and design; and, reinforcing fundamental skills of engineering problem solving with appropriate tools (e.g. software).

Through this past academic year, students have actively participated in the course design and modification, working to strengthen the link between use of the LEGO robotics kits and analogies to chemical engineering problems and processes.

We will report on successes and failures and future improvements to the course with related emphases on ties to chemical engineering.

112d Old Dead Guys: Using Active Learning Techniques To Teach History

Joseph H. Holles, Michigan Technological University, Dept. of Chemical Engineering, 1400 Townsend Drive, Houghton, MI 49931

Old Dead Guys: Using Active Learning Techniques to Teach History

Introduction: Teaching students in the classroom on a regular basis quickly reveals the importance of activity breaks. Active learning is an effective use of activity breaks where, for example, a small group of students work together on a problem with each individual group then discussing their answer with the entire class. An alternative active learning technique is to educate students about the history and personalities of Chemical Engineering. Almost every lecture in an engineering class contains references to the people and their accomplishments that form the foundation for today's students: Gibbs free energy, Arrhenius equation, Reynolds number, McCabe-Thiele plot. While many of these people are at least vaguely familiar to us as instructors, the students are usually completely unfamiliar with them. Since there is usually at least one of these "Old Dead Guys" in every lecture, they provide an opportunity to re-engage the students through an active learning exercise.

Methods: When an "Old Dead Guy" is encountered during a class period, it is usually towards the end of a derivation. This provides an ideal opportunity for an active learning break. The students are first asked to guess when the "Old Dead Guy" lived and did the work that bears their name. After a few inaccurate guesses, a picture of the person is shown to the students. This usually allows a slightly more accurate guess. This is followed by listing birth date, degrees earned, and institutional affiliations. Then the persons major accomplishments are listed (e.g., Nobel Prize), including the accomplishment that is the focus of the ongoing lecture. Finally a short discussion of why the accomplishment occurred at that time, the historical context in which it occurred, what other historical events influenced it, and other related broad societal influences are discussed. The pictures and biographical information are usually obtained from internet or other reference sources.

Results: The historically focused active learning breaks have been used over the last five years in a required kinetics and reaction engineering class and an elective industrial chemical production class. The students respond well to them, especially if the historical context and societal influences create a bit of controversy (e.g., Haber and Bosch developed catalysts to synthesize ammonia so that it could be turned into explosives to support Germany in World War I, and Thomas Midgely who invented chlorofluorocarbons and tetraethyl lead). This also allows the students to gain perspective on the industry they will soon be a part of. The "Old Dead Guys" can and should also contain women and other minorities (e.g., Maud Menten). After awhile, the students come to expect the breaks. Several students have gone out and found pictures and biographical info on their own. In addition the students actively participate by deciding who in the department the Old Dead Guy most closely resembles. Finally, the students have adopted the idea for their own use. When they do classroom presentations, the almost universally incorporate their own "Old Dead Guy" and one class got into a competition to see who could get the oldest "Old Dead Guy". Therefore, a historical active learning break provides a positive way to introduce history into the curriculum.

112e Graduate Student Active Learning in Modeling and Numerical Methods

Karen High and Eric L. Maase. School of Chemical Engineering, Oklahoma State University, 423 Engineering North, Stillwater, OK 74078

Chemical Engineering 5743, "Chemical Engineering Modeling," traditionally has been taught in the lecture format. The author has taught the course as such for seven years. In the fall of 2005, the co-author developed EXCEL/VBA demonstrations, activities, and homework, and the author did the same with MATLAB. These computer tools were used by the students to solve algebraic, ordinary differential equations (initial value and boundary value) and partial differential equations (parabolic and elliptic). Traditional courses will focus on numerical and mathematical methods as necessary skills. In the fall of 2006, the authors decided to add new components to the course exposing students to case studies, problem based learning, team work, and experimentation with the aim of promoting creative and critical thinking as students identify and practice the art of modeling.

The 13 students in the fall 2006 course were tasked with a semester project to "Design an encapsulated drug (of your group's choice) that effectively delivers an appropriate dose for an appropriate number of hours. Select the most cost effective method." The students were introduced to the problem with a case study described in "User's Guide to Engineering" by Jensen, a freshman introductory engineering text. This "Dissolution Case Study" has been developed based on a series of several Chemical Engineering Education articles. In the case study, a lollipop simulates the coating of a new pharmaceutical. A simple first order ODE describes the rate of change of the lollipop radius with respect to time. This rate is dependent on the mass transfer coefficient, the concentration gradient, and the molar density of the lollipop. The case study then tasks the freshman students to develop experimental procedures to test the proposed model.

For the graduate course, the students were put in heterogeneous groups by considering their country of origin, the degree they are pursuing (M.S., Ph.D.), and their background experience in computer applications / programming, experimentation / lab work, and numerical method experience. The graduate students began by creating an initial experimental protocol for their physical model (equipment/ supplies needed, data collection procedure), their initial computer modeling/numerical methods appropriate for solving the mathematical model, necessary simplifications made to begin the research, and expectations of future considerations that more appropriately model the real system. Throughout the semester the students improved their physical and mathematical modeling complexity to reach the goal of encapsulated drug design. Teams gave final presentations of the results of their efforts.

Assessments are presented examining the effectiveness of our team-based, active learning experience. Modeling, experimental design, experimental procedure and competence, and proficiency in applied numerical methods and computer programming are evaluated. This project proved invaluable for beginning graduate students to identify the connection between modeling, mathematical descriptions and experimental reality. It was also intriguing for the students to not only consider mathematical but physical models where experimentation with the real system is, at best impractical, as is the case in biological systems. The students gained valuable teamwork skills not common for many graduate students as well as gained improvements in communication skills.

170a Advancing Green Engineering through Partnerships between University and Pharmaceutical Industry

C. Stewart Slater, Mariano J. Savelski, Brian G. Lefebvre, and Robert P. Hesketh. Department of Chemical Engineering, Rowan University, 201 Mullica Hill Road, Glassboro, NJ 08028-1701

Rowan University student projects have examined how to apply green engineering strategies to both R&D and manufacturing in several pharmaceutical and specialty chemical industries. This has been accomplished through industry-university partnerships with pharmaceutical companies in the New Jersey area. Several grants from the U.S. Environmental Protection Agency and industry have supported initiatives in green chemistry, engineering and design. These projects have the broader goal of supporting sustainability in the industry. We have sought to improve on our concept of green engineering education integration in a way that not only enhances student learning, but improves the environment. This is through real-world projects in our engineering clinic sequence. The green engineering interventions use a vertically integrated student team of junior, senior and masters level engineering students who work with the industrial partner on approaches to pollution prevention. These projects provide an illustration of how companies can take advantage of universities in their region to assist in sustainable development and manufacturing. Students are able to learn about green engineering design and apply their knowledge directly to a real industrial case. The industries involved partner by providing support in terms of senior staff, scientific/engineering process data, access to plant facilities, and take initiatives recommended forward to process improvements on the plant scale. These projects have the potential to yield the following green engineering outcomes: solvent reduction, increase efficiency in use of raw materials, waste reduction, and energy reduction. Projects examine specific process improvements and use life cycle assessment to quantify the broader impact. Overviews of representative projects are presented. Partnering with industry on the 2008 AIChE Topical Conference on Green Engineering and Sustainability and in the Pharmaceutical Industry will be described.

170b Incorporating Semester-Long Projects In Thermodynamics To Improve Critical Thinking / Real Word Problem Solving

Donald P. Visco, Tennessee Technological University, Box 5013, Cookeville, TN 38505

A common response from external assessment measures indicates that during the time students are on campus to earn their B.S. degree, the skills of critical thinking and real-world problem solving are not addressed in a satisfactory manner. To this end, we present the use of critical thinking questions and semester-long projects in a thermodynamics class as a backdrop for students to practice and develop these skills. Pre and post-assessment results over the course of two semesters of thermodynamics indicate statistically significant progress in these areas. This presentation will provide the mechanism by which this technique was implemented as well as examples of questions and projects used during these courses.

170c Introducing Labview Alongside Controls-First Semester Findings

Gregory E. Ogden, Chemical & Environmental Engineering, University of Arizona, 1133 E. James E Rogers Way, Tucson, AZ 85721 and Anthony J. Muscat, Chemical and Environmental Engineering, University of Arizona, P.O. Box 210011, Tucson, AZ 85721.

This paper presents initial findings and results of the department's ongoing effort to modernize experimental facilities and incorporate data acquisition systems into the undergraduate laboratory. Labview programming was used to teach the process control course that students take in the second semester of their senior year. This was likely their first introduction to Labview. In the past Matlab with Simulink was used to model process control systems. Although the Matlab environment is powerful and flexible, Labview was chosen because of the capability to communicate with external equipment via a USB based interface. Measurement Computing USB devices were utilized for both temperature (USB-TC) and analog (1208FS) signal processing. The students learned basic Labview programming along side controls. Beginning with simple computer-based assignments, the students developed and debugged virtual instruments (vi's) of increasing complexity until they wrote complete data collection and control vi's that were used to control a two-tank in series experiment. The Labveiw assignments were all geared around controlling liquid levels and flow into two tanks in series. Early projects were designed to provide experience in data manipulation, using the front panel and annotating the block diagram to explain the process. Students created a vi to graphically indicate the height of water in a tank as a function of volume. These vi's were expanded to include subvi's designed to correlate pressure sensor input data to liquid level which were tested in the lab. Students gathered voltage vs. flow data and developed correlations that were used in Labview to control pump flow. Finally, students used tank height and pump flow correlations in a dynamic vi to control the liquid level in a second tank to a desired setpoint using a PID algorithm. Labview provided students with a means to acquire data, perform analysis, and output results to control a system. This experience should give students a better insight into the complexities of effective controller design. The faculty intend to introduce Labview programming earlier in the ChE curricula so that students can take full advantage of its DAQ and control functions when they take process controls. This will allow faculty to develop more involved computer and experiment-based activities that will help students to better relate control theory to actual practice.

170d Effective Use of E-Books in Chemical Engineering Classroom

Sasha Gurke, Knovel Corp., 20 Hawley Street, Binghampton, NY 13901

The use of technical publications available online in academic libraries is below the level expected in the Internet-dominated society. The severity of the problem depends on the type of the publication with journals doing the best, especially in research universities and among graduate students, and e-books the worst. There are several reasons for poor performance of e-books, including difficulty of finding on library portals and domination of the Web by general search engines like Google. This paper explores the ways to remedy this situation by including e-books in the curricula. Successful integration of e-books available from www.knovel.com in chemical and chemical engineering curricula at leading US universities is discussed.

Web Page: www.knovel.com

170e Podcasting In The Introductory Materials And Energy Balance Course

Jessica O. Winter, Chemical and Biomolecular Engineering/Biomedical Engineering, Ohio State University, 140 W 19th Ave, Columbus, OH 43210

The influence of podcasting in an introductory materials and energy balance course was assessed. The course evaluated was a quarter-length class focused on steady state energy balances and transient mass and energy balances. The class consisted of 18 lectures, with weekly quizzes and homework sets, 1 midterm and 1 final. All lectures were posted on an internal website using audio-only podcasting technology. Course slides containing all lecture materials were also available on the internal site. Out of 50 students enrolled in the class, 44 chose to participate in the study (conducted under IRB exemption, granted by OSU IRB committee), and 13 (30%) reported using podcasts throughout the semester. The perceived impact of podcasts on learning, general interest in podcasting, and course attendence were correlated. Although not compared with a control, the availability of podcasts is not believed to have influenced course attendance with 1-2 students (2-4%) absent on average per class. Podcasting represents a new technology that may enhance learning by providing the ability for students to repeat lecture material that was unclear during initial presentation and also provides a mechanism for students to make-up missed classes.

229a Session Introduction

John Corn, Chemical & Biomolecular Engineering, Ohio State University, 416 Koffolt Lab, 140 W. 19th Ave., Columbus, OH 43210 and Marina Miletic, Department of Chemical and Biomolecular Engineering, University of Illinois Urbana-Champaign, 206 Roger Adams Lab, Urbana, IL 61801.

Session Introduction - 5 minutes

229b A Chemical Engineering Fundamentals Laboratory

J. M. Caruthers[1], D. Ramkrishna[1], F. Ribeiro[1], W.N. Delgass[1], Y. Zvinevich[1], D.S. Corti[1], and R. Chhabra[2]. (1) Department of Chemical Engineering, Purdue University, 480 Stadium Mall Drive, West Lafayette, IN 47906, (2) Depart. of Chemical Engineering, IIT Kanpur, Kanpur, India

It is well known that hands-on experience is a critical component of effective education for most undergraduate students. However, in many large state universities the majority of the education is in a classroom format, where an engineering laboratory experience is typically limited to one or two semesters towards the end of the student's academic career. The reasons for limited undergraduate laboratory experience, especially in universities that have significant research activities, is the considerable expense in both instructor time and space necessary for the operation of teaching laboratories. In order to address this situation we have developed a new Fundamentals Laboratory to provide undergraduate chemical engineering students hands-on experience during the sophomore and junior years in a manner that is both cost effective to operate and requires only a modest amount of laboratory space. A series of 12 modular experiments have been developed for three ChE core courses: fluid mechanics, heat & mass transfer and kinetic & reaction engineering, where at Purdue we have 8 replicates of each experiment. Each experiment is self contained and can be completed in a pair of two hour lab session scheduled in a single week. At the end of the week a technician disconnects the 8 experiments and then connects a new set of 8 experiments for the following week's class; thus, over a 12 week period each of the three core undergraduate courses have 4 different experiments – all in the same laboratory with only one technician and with no additional faculty instructors. With the 8 replicate experiments we can provide a hands-on experience in these three courses to over 120 students per year per class. The experiments reinforce basic principles (e.g. heat conduction, Dittus-Boelter correlation, pressure drop vs. flow rate in a pipe, pump characteristic, plug flow reactor, etc.) just as the students are being taught that concept in the lecture course. We will report on how the addition of hands-on experiments improves the educational effectiveness in teaching the core chemical engineering principles.

229c Incorporation of An Industrial Distributed Control System in the Chemical Engineering Unit Operations Laboratory

Atanas Serbezov[1], Ronald Artigue[1], Richard Plapp[2], and Darryl Carpenter[2]. (1) Department of Chemical Engineering, Rose-Hulman Institute of Technology, 5500 Wabash Avenue, Terre Haute, IN 47803, (2) Eli Lilly and Company, Indianapolis, IN 46285

The field of Process Control in the chemical, petrochemical and pharmaceutical industries has entered a new stage marked by enormous advances in computers and instrumentation. In order to respond to the new reality and to better prepare our graduates for industry, we are revising the process control component in the Chemical Engineering curriculum at Rose-Hulman Institute of Technology. The backbone of the new process control experience is an industrial distributed control system (DCS), namely DeltaV by Emerson. This presentation describes the deployment and the incorporation of the DCS in the undergraduate Unit Operations (UO) laboratory. The Unit Operations laboratory in the Chemical Engineering curricula is well recognized for its educational value. The UO laboratory is often the only place where exposure to industrial practice and new technologies occur within the typical four-year engineering curriculum. This is especially relevant in the process control area, where the gap between classroom-based training and the real world is a core issue. One of the major goals of the redesigned process control experience in the Chemical Engineering curriculum at Rose-Hulman Institute of Technology is to expose all chemical engineering students to industrial-grade process instrumentation, control system architecture, and control and tuning algorithms beyond PID and Ziegler-Nichols. Another goal is to introduce students to the industrial practices of data storage, management, retrieval and mining. The UO laboratory in the Chemical Engineering department at Rose-Hulman Institute of Technology has over a dozen pilot-scale process units (skids), such as heat exchangers, pumps, distillation columns, chemical reactors filtration units and fermentors. Most of them have been designed for manual operation and had only local gauges. Over the past two years six of the existing experiments have been re-designed and converted to run under the DCS. Each of the redesigned UO skid is treated as a plant area and is functionally isolated from the others. Students can control the experiments and access the data historian through operator stations in a centralized control room or through remote operator stations located in close proximity to each experiment. Closed circuit TV cameras monitor critical process areas and are accessible through the school network. The new environment in the UO laboratory is very similar to the environment in a typical chemical, petrochemical or pharmaceutical plant. Students learn how to maintain their process under control, take the process safely from one operating condition to another, collect and analyze data using a historian, respond to process alarms and remotely troubleshoot their experiments using limited process information. The DCS incorporation in the UO laboratory was accomplished in close collaboration with industrial partners who provided significant financial and technical support. The re-design of the experiments and the documentation associated with it followed the best industrial practices. Most of the configuration, commissioning and troubleshooting work was performed by the undergraduate students as part of special project-based courses.

When Things Go Wrong

William E. Josephson, Chem. Eng. Dept., Auburn University, Auburn, AL, USA
Jaya Krishnagopalan, Chem. Eng. Dept., Tuskegee University, Tuskegee, AL, USA
David R. Mills, Chem. Eng. Dept., Auburn University, Auburn, AL, USA

Introduction

The junior level unit operations laboratory course usually has objectives that include development of technical writing abilities, reinforcement of concepts learned in previous lecture courses and the acquisition of hands-on experience. Students (singularly or in groups) operate and analyze small-scale industrial equipment (e.g., a centrifugal pump) or measure some physical properties (e.g., thermal conductivity), do some number-crunching, put together a formal report and go on to the next experiment on the schedule. When the experiments all go as expected then the life of the student is relatively stress-free and the above objectives are routinely met. However, equipment problems sometimes intrude on the normal course of events and prevent the "correct" data from being gathered. Rather than present a barrier to learning, such occasions can provide new insights into the issue under investigation. This paper documents four such cases in which lab "problems" were used to aid student learning. The cases cover the areas of conduction heat transfer, observation of fluid flow, determination of friction factors in piping and determination of viscosity of a shear-thickening fluid.

Course Setup

The examples that follow come from our experiences with the junior level lab course at two universities: Auburn University's CHEN 3820 course is titled 'Chemical Engineering Laboratory I' and Tuskegee University's CENG 320 course is titled 'Unit Operations Laboratory I'. Although differences exist (e.g., in number of experiments covered, in the use of a one hour weekly lecture component in addition to the three hour weekly lab session) the basic structure of the courses is the same. The students are organized into groups and work as a team on various experiments during the semester. For each experiment they are given objectives and instructions on use of the appropriate experimental apparatus. A full-length written report is due one week after the completion of the experiment. At the end of the semester each group gives a presentation covering one of their completed experiments.

The pre- and co-requisites for the lab courses include courses in fluid dynamics and heat transfer. The typical experiments reflect this with an experimental selection that includes experiments in which the student groups analyze centrifugal pumps, determine frictional losses in pipes and fittings and determine heat transfer coefficients for various systems. In the normal course of events the students conduct the experiment and write their reports by drawing upon the knowledge they have acquired in lecture courses. However, when there are problems with an experiment then the students are forced to intensely consider the situation and utilize their lecture-gained knowledge to a much greater extent. Having to actually think about why an experiment is not producing the results that it 'should' is sometimes a novel experience for students in a lab course. As a result, student learning may be enhanced when there are difficulties in the proper running of an experiment. The four cases presented below are examples of instances in which such problems were turned into opportunities.

Conduction Heat Transfer

The Conduction Heat Transfer experiment is a fairly simple one in which the students are asked to investigate the one-dimensional conduction of heat along a metallic sample. The students use an apparatus in which thermocouples are attached to the sample at regular intervals and a temperature profile is generated. Construction of a best fit line through a plot of thermocouple position versus temperature and use of Fourier's Law enables the calculation of the material's thermal conductivity.

During one semester one of the thermocouples began malfunctioning. It would report temperature values at least 5 C° above the true value. Unavailability of parts meant that the problem could not be fixed before the groups were scheduled to perform the experiment. The instructor elected not to inform the groups of the exact nature of the problem. Instead, the groups were simply told that one of the temperature sensors was malfunctioning. The groups did not know which sensor was involved nor the exact nature of the errors (i.e., readings too high or too low). The groups were then responsible for taking their data normally, analyzing it and determining a workaround that would allow them to determine the sample's thermal conductivity.

Most groups figured out on their own a solution to the dilemma involving the generation of multiple values of conductivity using selected pairs of temperature readings. Comparison of the various conductivities resulted in the singling out of the suspicious value and thus determination of which thermocouple was malfunctioning.

The Reynolds Experiment

In a second case students were attempting to determine the Reynolds number range over which a flow transitioned from laminar to turbulent. The groups used the popular Osborne Reynolds apparatus in which a controlled flow of water is drawn through a vertical glass tube. A fine filament of dye is introduced into the water stream in order to visualize the water flow pattern. Various flow rates of water are used in order to run over a range of Reynolds numbers.

The two primary objectives of this experiment are for the students to refresh their capabilities in computing Reynolds numbers and for the students to observe and quantify the transitional flow regime. Accordingly, the students were expecting to observe the transition from laminar to turbulent flow somewhere in the range of Re = 2,100 to 4,000. The instructor was thus surprised to find that the students were consistently reporting that the transition from laminar to turbulent flow occurred in the range of Re = 700 – 900. This was the case even after all numbers and unit conversions had been double checked. Eventually (after the semester was over) the instructor determined the source of the problem: at some time in the past a modification had been made to the dye injector tubing. The modification caused the dye to flow at a much higher velocity than the apparatus was designed for and, as a result, the dye stream would disperse at relatively low water flow rates. Removal of the modification restored the apparatus to its expected operation.

The value of having this problem occur did not lie in a direct enhancement of student learning. Although the students were diligent in repeating their experiment without complaint and rechecking their calculations they did not learn anything new by doing so. Instead, the benefit of this equipment malfunction was that it indicated to the instructor that many students were unclear as to the real definitions of terms like laminar and turbulent flow. When the problem was occurring students commonly disconnected what their eyes were telling them ('the flow is turbulent') from what their calculations were telling them ('the Re is 800, thus the flow is laminar'). Upon noticing this discrepancy the instructor took advantage of the opportunity to lecture the students on the true meaning of these terms and to warn them about excessive faith in sensor readouts.

Friction in Pipes and Fittings

The Friction in Pipes and Fittings experiment is a common one in which water flows in a loop through pipes of various nominal size, schedule and material. There are four pipes made of PVC and one pipe made of stainless steel. Pipe sizes and schedules are given to the students along with other material in an experimental handout. One of the objectives of this experiment is for the students to determine the relationship between friction factor and Reynolds number/relative roughness. The generated data should be compared to that of the literature (e.g., the Moody plot).

All students have found that use of the stainless steel pipe results in the lowest friction factor at a given Reynolds number. This is despite the fact that its relative roughness is calculated to be the highest of all the pipes. The reason for this discrepancy between anticipated and actual results is unclear at this time.

Despite our lack of knowledge regarding why the discrepancy exists, knowing that it exists and seeing how the students deal with this unexpected result is quite valuable from the instructional point of view. Some students gloss right over the discrepancy (they are called to task when their reports are graded) while others attempt to explain the results as being somehow related to some change in the stainless steel pipe. Surprisingly few students take the time to verify that the information given in the handout corresponds to the actual experimental apparatus (it does). The most thoughtful students, after verifying all their data, compare their numbers to those on a standard Moody plot. When this is done the students see that their stainless steel pipe data matches up well; it's the PVC data that is out of whack. The problem is not that the stainless steel pipe friction factors are too low; it's that the PVC friction factors are too high.

Determination of Viscosity

Finally, in a Determination of Viscosity experiment, students attempting to quantify the behavior of a shear thickening fluid (55% cornstarch solution) found instead that the experimental apparatus was indicating shear thinning behavior. This occurred during the first semester that a new rotating spindle dial viscometer was used. The instructor was unfamiliar with the apparatus and set up an experiment in which the supplied standard cylindrical spindle was used for all fluids including the cornstarch in water solution. Use of such a spindle for dilatant fluids can be problematic due to separation of the solid materials (cornstarch) from the water. This segregation of materials is dependent on the rotational speed of the spindle and will increase with increasing spindle speed. At high speeds the spindle is encountering more water and less corn starch and the drag on the spindle decreases. As a result, the viscosity of the fluid appears to decrease with increasing shear rate (i.e., it's a shear thinning fluid) instead of increasing.

After careful analysis of their data most student groups performing the viscosity experiment with the cylindrical spindle in corn starch realized what was going on. A more appropriate vane spindle was used in subsequent semesters and produced qualitatively accurate results.

Conclusions

In the above cases various problems with experimental equipment posed challenges to students. Some of the difficulties were simple enough to understand and could be overcome via reinterpretation of the data (e.g., Conduction Heat Transfer) while in other cases the problems were more complex and required careful analysis (e.g., Determination of Viscosity). All of the cases resulted in enhanced student comprehension of the experiment and demonstrated that a disproportionate amount of learning can take place when experiments don't go smoothly.

229e The Chemical Engineering Laboratory Experience At The University Of South Carolina

James A. Ritter and Charles E. Holland. Department of Chemical Engineering, University of South Carolina, Columbia, SC 29208

Two of the most significant highlights of my teaching career have been the modernization of the Chemical Engineering Laboratory facilities and the redesign of the Chemical Engineering Laboratory two course (6 credit) sequence, ECHE 460 and ECHE 461. I have not only taught this two-course sequence for well over a decade, I have humbly placed it on an equal basis with my extensive and productive research program as being a significant and defining part of my career at USC.

I believe USC has a great chemical engineering undergraduate laboratory. It is not just me saying this. The students, the faculty and even two sets of ABET evaluators have all agreed that the Chemical Engineering Laboratory and its associated courses are suburb. I am also very proud of the fact that the Chemical Engineering Laboratory and its associated two-course sequence survived two ABET assessments, the one in 2000 and the one in 2006, receiving outstanding, essentially raving reviews in both cases.

I began this journey in the fall of 1994, when I was asked to take over the two-semester Chemical Engineering Laboratory course. I accepted this challenge because the two-course sequence and the laboratories needed some help. I would like to emphasize that I did not simply teach the same laboratory course that was developed by my predecessors. I redesigned the entire course. Highlights of the improvements I made early on in its development included: hiring an excellent full-time laboratory manager (Charles E. Holland), developing new and/or significantly improving the existing experiments, implementing as a standard, a two teaching assistant (TA) minimum assigned to each experiment, writing or rewriting all of the laboratory experimental handouts and procedures, and ensuring that rigorous theory and supporting computational analyses were part of each experiment, and

The effort Chuck and I put into this endeavor paid off. USC has had for many years now seven to eight experiments running each semester with state-of-the-art data acquisition and control systems based on LabView software from National Instruments and a "statistical design of experiments" approach where the students carry out full factorial designs to assist them in analyzing their data. Experiments in ECHE 460 include: Plate and Frame Filtration; Adsorption and Diffusion in Microporous Zeolites using Gas Chromatography; Heat Transfer Analysis in Solids; Analysis of Centrifugal Pumps and Pipe Network Flow; Viscometry and Rheology; Ethanol/Water Vapor-Liquid Equilibrium using an Othmer Still; Electrochemical Methods: Voltammetry; and Packed Column Hydrodynamics

Experiments in ECHE 461 include: Hollow Fiber Membrane for Gas Separation: N2 Production from Air; Fundamentals of Fixed Bed Adsorption Processes: Analysis of Adsorption Breakthrough and Desorption Elution Curves; Batch Distillation with Rectification: Separation of Ethanol/Water using an

Oldershaw Distillation Column; Performance and Characterization of a Proton Exchange Membrane (PEM) Fuel Cell; Hollow Fiber Membrane Contactor: Gas and Vacuum Stripping of O2 from Water; Kinetics and Reactor Engineering: Study of a Batch Reactor; Absorption in Packed Columns: Sulfur Dioxide Absorption from Air using Water; and Drying of Process Materials: Combined Heat and Mass Transfer

Chuck and I also have been and will continue to develop a new experiment each year, or to take an existing experiment and to modify it so that it can be used over a period of two to three years without repeating the same experiment or analysis. This makes the experiment far more interesting to teach and minimizes the sharing of information by students from different classes. We are currently developing four new experiments. These include concepts on: agitation and aeration, gas phase catalytic reactor, photocatalysis, and reverse osmosis.

My teaching philosophy in the Chemical Engineering Laboratory two-course sequence has always been to have the students: run well planned experiments that allow for the collection of interpretable and analyzable data; analyze the data using statistical and experimental design techniques; correlate the data with simple but fundamental models; present the results in a logical fashion; and write a comprehensive and well organized report on each experiment

Typical results from this approach show a consistent improvement in both the style and content of the presentations and reports as the semester progresses, especially during the second semester. I have been very pleased with this kind of tangible result that I believe demonstrates that I have been effective at teaching some of the most important areas in engineering education, i.e., summative knowledge, analysis, presentation and writing.

I have also been very receptive to comments and suggestions from the laboratory TAs (i.e., graduate students), undergraduate students, faculty, and staff. As a result, several significant evolutionary changes have been made over the years. These changes include.

Pre-Experiment Quiz (PEQ): The PEQ was implemented the first time in ECHE 461 in fall 2001 and in ECHE 460 in spring 2002 based on feedback from the TAs that the students were not reading the experimental handout prior to coming to the laboratory.

Outcome: very few students have failed the PEQ; students now read the entire experimental handout prior to going to the laboratory. No more complaints have been received from the TAs on this point.

Final Exam (FE): The FE was implemented the first time in ECHE 461 in fall 2001 and in ECHE 460 in spring 2002 based on students requesting less of their grade to be a group grade, and more importantly, to determine if the students are retaining any information about the experiments they ran during the semester.

Outcome: most students receive 100 points; a few receive 75 points, one or two receive 50, 25 and even

0 points during a semester; in effect, the FE provides for a more even distribution between an individual and group grade and it signifies that the greater majority of the class actually learned how to interpret or analyze results from their experiments; and it challenges the student's ability to know how well they think they know a topic; I now get first hand knowledge of this ability.

Reduction from Six to Four Tasks (i.e., Experiments): this task reduction was implemented the first time in ECHE 461 in fall 2002 and in ECHE 460 in spring 2003 to provide time for a student within a group or for the group to delve more deeply into one or more of the experiments and to provide a more realistic (possibly industrial type) experience where report writing tends to be an interactive process between colleagues and management, at least in the early stages of a career.

Outcome: Up to one-third of the reports are now turned in for re-grading after a major revision has been done, usually by a group member that was not pleased with the original group grade they received; to me it is obvious that a student has more opportunity to learn about a given experimental topic.

New Experiments: new experiments were implemented and old ones eliminated continually over the past 12 years to provide sufficiently accurate and precise data for the undergraduates to analyze, to stay current with the latest trends in the chemical engineering community, and to provide the students with a broader and more challenging exposure to different chemical engineering property measurement techniques and unit operations or processes.

Outcome: the students spend a great deal of time analyzing data that is indeed analyzable, and correlating the results with significant, curriculum-reinforcing theoretical developments, as evidenced partly by the frequent visits of the students to my office with relevant technical questions.

20 Questions: this new assessment tool was implemented the first time in ECHE 460 in spring 2005 and in ECHE 461 in fall 2005 as a means to encourage the students to learn more about a given topical experiment, either from gaining a fundamental understanding of underlying phenomena or from gaining a more practical understanding of the apparatus and procedure, and of the equipment including measurement devices.

Outcomes: Most of the students generally take 20 Questions very seriously and are probably prepared well enough to answer all 20 questions if called upon to do so during their presentation, which is the goal; a few students take a chance on the random nature of the draw of the two questions they are required to answer in 5 minutes and hope to get only easy ones; based on the scores and feedback from the students, it is clear to me that the students are learning more about each experiment and its underlying theoretical and practical developments.

Overall, the students fully realize and appreciate the fact that they receive a first rate chemical engineering laboratory experience at USC. I know this because Chuck and I repeatedly receive accolades from them about the rewarding experiences they had with the course, especially a few years after they graduated and used some of their skills in industry or graduate research. This kind of feedback is very gratifying as it provides a unique assessment tool for which we use to judge success.

229f Competition Between Student Groups In The Protein Production Challenge

Brian G. Lefebvre, Loren E. Connell, and Kevin D. Dahm. Chemical Engineering, Rowan University, 201 Mullica Hill Road, Chemical Engineering, Glassboro, NJ 08028-1701

Converting biological discoveries into commercial-scale processes requires that graduating chemical engineers obtain an understanding of modern bioprocess principles. This paper describes the development and implementation of a five week long protein production project into a multidisciplinary upper level elective course on bioprocess engineering.

The protein production project was developed and implemented during the fall 2006 semester. A key element of the protein production project was the competition between student groups. Group performance was rated using a production rubric, and the team with the highest score was guaranteed an "A" on the project. The rubric included equipment rental costs and production bonuses for producing large quantities of protein of high purity. The equipment rental costs were carefully determined to encourage continued student experimentation in the laboratory. The protein production project was mentioned as the "best feature of the course" on over half of the student course evaluations.

254a Introduction

Robert Wellek, The National Science Foundation, Division of Chemical & Transport Systems (ENG/CTS), 4201 Wilson Boulevard, Arlington, VA 22230

Presenter will provide overview of NSF Workshop Session I.

254b NSF Cbet Overview

Judy A. Raper, Division of Chemical, Bioengineering, Environmental and Transport Systems, National Science Foundation, Arlington, VA 22230

The Division Director of Engineering/NSF will discuss the new organization of the NSF CBET, the current budget projections, a new general emphases and trends, dates for the call for proposals, and personnel changes at NSF/CBET.

254c Highlights of Cbet Cluster on Chemical, Biochemical & Biotechnology Systems

John Regalbuto, Division of Chemical, Bioengineering, Environmental & Transport Systems, National Science Foundation, 4201 Wilson Boulevard, Arlington, VA 22230

The authors will very briefly review the current objectives and discuss the trends for the NSF program activities in Cluster A - in the Process and Reaction Engineering, Biochemical Engineering, Biotechnology, Chemical and Biological Separations, and Catalysis and Bio-Catalysis Programs

254d Highlights of Cbet Cluster on Transport and Thermal Fluids Phenomena

Phillip R. Westmoreland, Division of Chemical, Bioengineering, Environmental and Transport Systems, National Science Foundation, Arlington, VA 22230

The presenter will very briefly review the current objectives and discuss the trends for the NSF program activities in Cluster B-- in the Particulate & Multiphase, Thermal Transport Processes, Fluid Dynamics & Hydraulics, Interfacial Transport & Thermodynamics, and the Combustion, Fire, & Plasma Systems programs. Other Program Directors will be introduced.

254e Highlights of Cbet Cluster on Environmental Engineering & Sustainability

Trung Van Nguyen, The National Science Foundation, Division of Chemical, Bioengineering, Environmental and Transport Systems (ENG/CBET), 4201 Wilson Boulevard, Arlington, VA 22230

The authors will very briefly review the current objectives and discuss the trends for the NSF program activities in Cluster C-- in the Environmental Engineering and Technology, Energy for Sustainability.

Chemical, Bioengineering, Environmental, and Transport Systems Division (CBET)

Biomedical Engineering and Engineering Healthcare Cluster

AIChE Update – November 2007

Cluster Mission

- Provide opportunities to develop novel ideas into projects that integrate engineering and life science principles in solving biomedical problems that serve humanity.
- Programs in cluster focus on:
  - High impact transforming technologies for deriving information from cells, tissues, organs, and organ systems,
  - Extraction of useful information from complex biomedical signals,
  - New approaches to the design of structures and materials for eventual medical use,
  - Biophotonics,
  - New methods of controlling living systems, and
  - Characterization, restoration, and/or substitution of normal functions in humans.

Programs in Cluster

- Biomedical Engineering (Demir)
- Biophotonics, Advanced Imaging, and Sensing for Human Health (Esterowitz)
- Research to Aid Persons with Disabilities (Jaeger)

Biomedical Engineering

- Contact Information
  - Semahat Demir, Ph.D.
  - sdemir@nsf.gov
- Mission
  - To provide opportunities to develop novel ideas into projects that integrate engineering and life science principles in solving biomedical problems that serve humanity

Current Program Guidelines PD 08-5345

- Focus on high impact transforming methods and technologies;
- Models and tools for understanding and control of biological systems;
- Fundamental improvements in deriving information from cells, tissues, organs, and organ systems;
- New approaches to the design of structures and materials for eventual medical use;
- New methods of understanding and controlling living systems;
- Information technology relevant to biotechnology including bioinformatics;

Current Program Guidelines
PD 08-5345

- New novel methods of reducing health care costs through new technologies;
- Emphasize the advancement of fundamental engineering knowledge, possibly leading to the development of new methods and technologies;
- Emphasize novel application of existing technologies to advance fundamental knowledge of both engineering and life sciences;
- Encourage initial evaluation of discovery-level research in a clinical setting but not supporting clinical trials;
- Highlight multi-disciplinary nature, integrating engineering and the life sciences; and
- Balance theory, mathematical modeling, and experiment.

Biophotonics, Advanced Imaging, and Sensing for Human Health

- Contact Information
  - Leon Esterowitz, Ph.D.
  - lestero@nsf.gov
- Mission
  - Innovative basic research in photonics, imaging, and sensing that is very fundamental in science and engineering is needed to lay the foundation for new technologies beyond those that are mature and ready for application in medical diagnostics and therapies.

Current Program Guidelines PD 08-7236

- Low Coherence Sensing at the Nanoscale - - Low coherence enhanced backscattering (LEBS), n-dimensional elastic light scattering, and angle- resolved low coherence interferometry for early cancer detection (dysplasia)

- Macromolecule Markers - - Innovative methods for labeling of macromolecules, new compositions of matter/methods of fabrication of multi-color probes such as might be used for in-vitro marking and detection of specific pathological cells and or pathogens

- Multi-probe Molecular Manipulations - - New optical approaches that permit specific molecular action on cells which conjointly bind two or more different probes with specificity for different macromolecular markers

Current Program Guidelines
PD 08-7236

- Neuro-photonics - - Development of new biocompatible detection technologies that could serve as massively parallel interfaces for communicating with networks of cells such as brain tissue slices. Studies of photon activation of neurons at the interface of nanomaterials attached to cells

- Biomimetic and Miniaturized Devices - - Innovative miniaturized optical tools or devices for the interrogation and manipulation or creation of specific reactions in complex cell or organ culture

- Photon-cell Interactions - - Fundamental studies of novel photonic properties of nanoparticles and optical reporters and their interaction with cells and their internal organelles

Research to Aid Persons with Disabilities

- Contact Information
  - Robert Jaeger, Ph.D.
  - rjaeger@nsf.gov
- Mission: Support research that will lead to the development of:
  - New technologies, devices, or software for persons with disabilities.
  - Characterization, restoration, and/or substitution of human functional ability or cognition.
  - Interaction of persons with disabilities and their environment.

Current Program Guidelines PD 08-5342

- Areas of particular recent interest include
  - Neuroscience/Neuroengineering.
  - Rehabilitation Robotics.
  - Telerehabilitation and Rehabilitation in the home environment.
- Emphasis is placed on significant advancement of fundamental engineering and scientific knowledge and not on incremental improvements.
- Proposals should advance discovery or innovation beyond the frontiers of current knowledge in disability-related research.

Current Program Guidelines
PD 08-5342
RAPD Undergraduate Design Projects

- The primary goal of this thrust is to provide a meaningful design experience for the engineering student who will work with a specific disabled individual to design a prototype "custom-designed" device or software.
- Each PI is expected to implement a high percentage of projects each year. It is also expected that the projects will contain appropriate levels of quantitative engineering analysis.
- The proposal must include a short description of ten possible design projects. These projects should be suitable for an undergraduate student, or a small team of students, to complete in one or two semesters.

Closing Thought…

This is the crew of the *B'Quest.* They completed the 2,500 mile Transpac yacht race in July 2005. Each member of the crew, with the exception of the skipper, was a person with a disability. NSF-funded engineering students at San Diego State University created specially designed equipment for the yacht to allow each crew member to participate in the race. A very successful undergraduate design project.

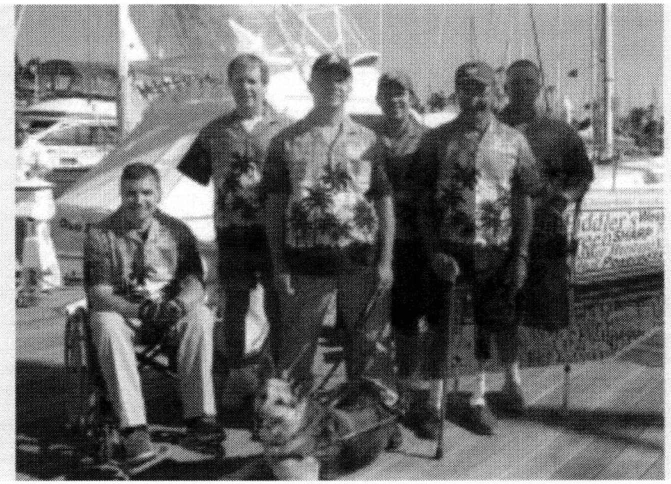

(NSF Research to Aid Persons with Disabilities Program, Award #0314095)

254g NSF Special Programs Overview

Judy A. Raper, National Science Foundation, Arlington, VA 22230

The Division Director of Engineering / NSF will briefly give an overview of special NSF programs such as IGERT, REU, RET, SBIR, ERC, etc).

254h Q And A Session

Vijay T. John, Chemical and Biomolecular Engineering, Tulane University, 300 Lindy Boggs Bldg., New Orleans, LA 70118 and **William B. Krantz**, Chemical and Biomolecular Engineering, National University of Singapore, Room E5-02-11, Block E5, 4 Engineering Drive 4, Singapore 117576, Singapore.

Q and A session

287a Finite Element Simulations In The Unit Operations Laboratory

William M. Clark and David DiBiasio. Chemical Engineering Department, Worcester Polytechnic Institute, 100 Institute Road, Worcester, MA 01609

The unit operations laboratory aims to reinforce fundamental principles by connecting theory and equations from text books with data from real equipment. Providing an additional connection between experiment and theory, through computer simulation, might result in more complete physical understanding. COMSOL Multiphysics(TM) is a user friendly, finite-element based, software package that allows students to solve partial differential equations within a specified geometry. We are testing the hypothesis that using this software to set up and solve momentum, energy, and mass balances and also to visualize the velocity, pressure, temperature, and concentration profiles within the equipment will help "bring to life" not only the experiments but also the mathematics. Use of simulations for a heat exchanger, a gas permeation unit, and a fluid flow apparatus will be discussed. Performance and attitudes of students who used the simulations will be compared to those who did not.

287b Understanding The Fundamental Phenomena That Influence Physical Properties And Fluid Thermodynamics

David A. Gallagher, CAChe Research, 13690 SW Otter Lane, Beaverton, OR 97008 and Andreas Klamt, COSMOlogic GmbH&CoKG, Burscheider Str., Leverkusen, 51381, Germany.

The easy-to-use software package, COSMOtherm, allows students to quickly explore and predict a wide range of physical properties and fluid thermodynamics such as vapor pressure, boiling points, activity coefficients, Henry's constant, gas-solubility, azeotropes, miscibility-gaps, solubility in mixtures, partition coefficients, vapor-liquid equilibria, liquid-liquid equilibria, solid-liquid equilibria, pKa, etc., etc.

However, more than just numbers, COSMOtherm provides an intuitive understanding of the fundamental phenomena that influence physical properties and fluid thermodynamics through graphical visualization of, for example, electronic properties superimposed upon the chemical structure. Visualization of electrostatic potentials on the molecule surface and sigma profiles, etc., illustrate the potential for hydrogen bonding, dipole-dipole, dipole-induced-dipole and Van der Waals interactions.

As the underlying methodology is based on first principles quantum mechanics (DFT), COSMOtherm does not suffer from the limitations of the empirical based methods, and so it can deal with an almost limitless range of compounds. To speed the student's learning, the educational COSMOtherm package comes with a large library of pre-calculated solvents and chemicals, enabling most property calculations to be performed in just seconds on a laptop or desktop Windows PC.

The author will explain the underlying methodology in COSMOtherm and illustrate some typical student exercises.

287c Molecular Simulation Modules In Undergraduate And Graduate Education: Examples From Molecular Engineering

Christopher R. Iacovella, Aaron S. Keys, Michael J. Solomon, Mark A. Burns, and Sharon C. Glotzer. Department of Chemical Engineering, University of Michigan, 2300 Hayward Street, Ann Arbor, MI 48109-2136

Computer simulation is a critical tool to aid in the understanding of complex phenomena in chemical engineering, yet its inclusion in the curriculum has often been limited. As the field becomes increasingly more diverse, simulation becomes even more important because it can effectively communicate new and difficult material to students. Technology, and in particular cyberinfrastructure, has now advanced sufficiently to allow students to investigate complex phenomena simulationally in short periods of time. As such, virtual laboratory experiments utilizing simulation can now play a powerful role in the learning process. Here we discuss the use of molecular simulation modules developed using the Glotzilla simulation API in both undergraduate and graduate chemical engineering courses. The simulation API provides for the easy creation of Monte Carlo and molecular dynamics simulations, realtime 2-d and 3-d rendering tools, as well as a host of analysis methods. Virtual laboratory modules developed to date focus on such concepts as phase separation, crystallization, liquid-crystalline ordering, and polymer dynamics. The modules are coordinated with a Wiki environment and link to the Materials Digitial Library (MATDL) Soft Matter Wiki [1] and digital library repository [2]. This provides students with a wide array of relevant-linked material ranging from definitions of key terms, explanations of algorithms, research examples, and links to relevant literature. As examples, we show virtual lab modules designed for an undergraduate course in molecular engineering developed at the University of Michigan.

The MatDL project is a collaboration between the University of Michigan and Kent State University, and is sponsored by the National Science Foundation under grant number DUE-0532831.

[1] http://matdl.org/matdlwiki/

[2] http://matdl.org/repository/

287d Fascination And Frustration With Excel And The Peng-Robinson Equation For Mixtures

J. Richard Elliott, Chemical and Biomolecular Engineering, University of Akron, Akron, OH 44325-3906

If your students are like ours, they are happy to program in any computer "language," as long as it is Excel. Even applying the VBA utility within Excel is a hard sell with the majority of our students.

One possible response is to teach only methods that can be computed by hand. This certainly forces a more fundamental understanding of the equations, but it limits the types of models that can be considered and it can be tedious, especially for a test question.

On the other hand, modern engineering rarely involves that kind of hand calculation and the models are much more sophisticated. Process simulators or thermodynamics packages present the workings of the model equations as black boxes, with little insight as to what is happening within the box. The pitfall is that the students cannot understand what is wrong when something goes awry, or even recognize that something might be wrong.

These observations suggest the need for some kind of middle ground. We illustrate this middle ground with the Peng-Robinson (PR) equation for mixtures as applied to fluid phase equilibria. The PR model is sufficiently sophisticated for most modern process simulators. The analytical solution of this cubic equation circumvents the need for calling the Excel Solver. The PR model and its multiple roots are familiar to students from their prior study of pure fluids. The expressions for the fugacity coefficient have been derived in class and similar derivations are familiar from homework covering other equations of state. In this context, providing a spreadsheet to students comes very close to simply saving them some tedious typing. It is up to the students to assemble the fugacity coefficients to solve practical problems.

The spreadsheet to be illustrated in this presentation provides the capability for students to solve flash problems by selecting the correct assembly of fugacity coefficients based on the appropriate roots. But beware of oversimplifying by simply calling the Excel Solver or applying Excel's iteration feature! Those tricks may work at low pressure if your initial guess is very good, but mixing gaseous components and raising the pressure causes the solution to vary from the three-root region to the one-root region, forcing the student to guide the convergence with intelligence. In this way, the students can appreciate the difficulty of some types of convergence and have an alternative to simply stating "stupid computer" when the black box goes awry. Assessing the students' capability to learn and apply this methodology is demonstrated with ConcepTests, a project that simulates a distillation train, and exam performance.

Nano/bio-Systems Modeling In Undergraduate and Graduate Education Using Gaussian '03 and Material Studio

Katherine L. Keeton[1] and Daniela S. Mainardi[1], (1) Louisiana Tech University, Ruston, Louisiana

Summary

Nanosystems Modeling is a course developed by Dr. Daniela Mainardi, Assistant Professor of Chemical Engineering and Nanosystems Engineering at Louisiana Tech University as part of her NSF Career Grant (CTS-0449046). The course was originally designed to complement existing standard courses with this computer-based modeling course in the College of Engineering and Science, where students learn different state-of-the-art computational modeling techniques and apply them to assigned activities and in a comprehensive final project. There is an outreach component associated to the course in which students design learning kits that can be taken directly into the schools or to professional development workshops. The goal is to educate others about a variety of research topics and the use of molecular modeling simulations.

Nanosystems Modeling Coursework Material

The uniqueness of the Nanosystems Modeling course is that it allows all students in the Engineering and Science curricula at Louisiana Tech to be exposed to and familiar with the use of state-of-the-art computational modeling techniques, and to prepare learning kits to educate several audiences. Hence, the personal development of students increases as they are taught to convey material to persons of all ages. Students also establish a better understanding of previously taught chemistry and related topics where only abstract topics are given to them.

Initially the Nanosystems Modeling students are introduced to the concept of molecular modeling and the types of modeling techniques that will be discussed in the course. They learn how molecular modeling is used to predict and explain experiments and how each of the techniques is to be applied to a molecular system. Students are introduced to the software used in the course and the basics of LINUX and UNIX. The software utilized in this course are Gaussian '03 and Material Studio [1,2]. The computational techniques introduced in the course include Density Functional Theory, Molecular Dynamics, and Monte Carlo Methods. The students are first exposed to the importance, development, and mathematical basis of each technique. Weekly activities are assigned to complete a simulation on a simple molecular system and summarize the results in report form. A summarized timeline for the course is shown in Figure 1.

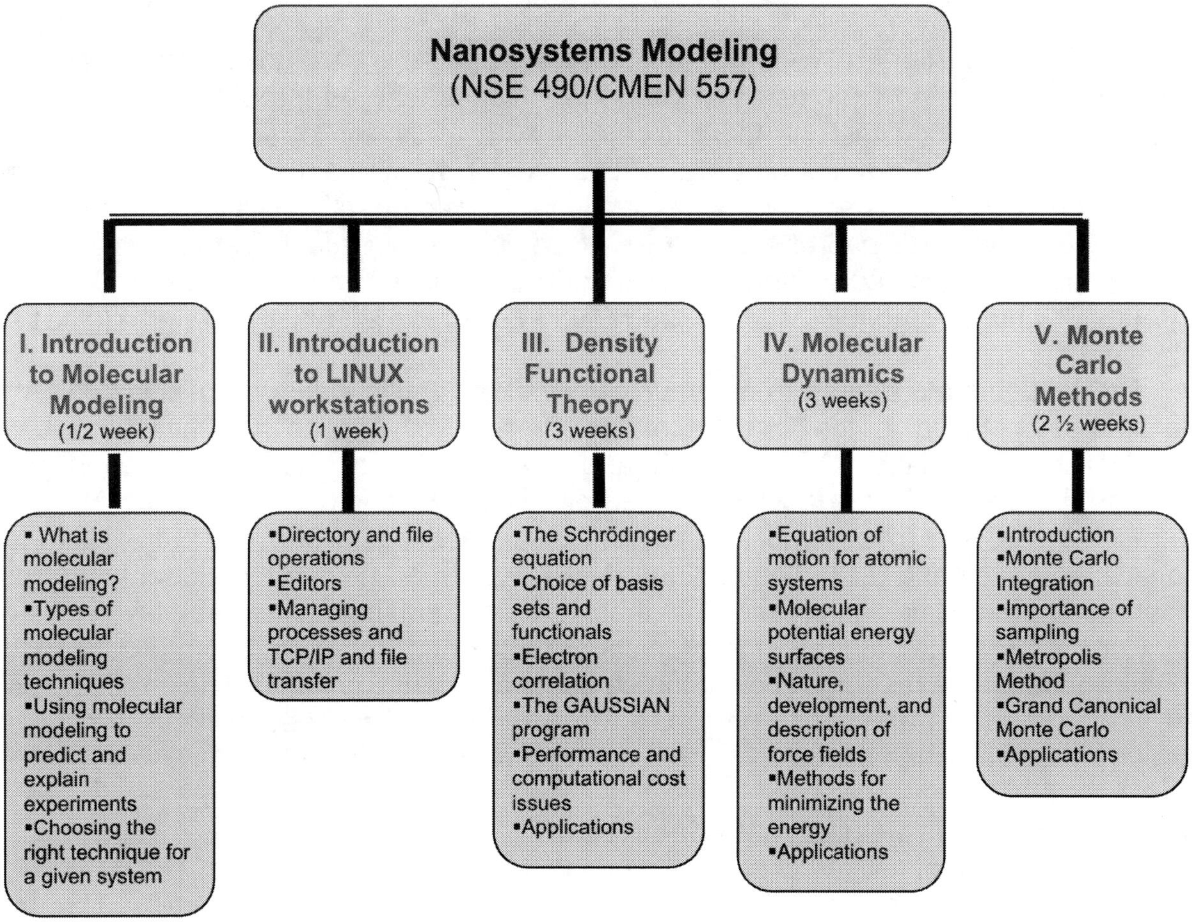

Figure 1. Outline of the *Nanosystems Modeling* course for the Louisiana Tech quarter (10 weeks).

There are many results that can be obtained by the modeling techniques used. In the weekly activities (both in class and assigned) and the final project, the following results that can be obtained through the application of the techniques are:

- III. Density Functional Theory : bond lengths in hydrocarbons, dipole moments and molecular polarity, formal charge vs. atomic and localized vs. delocalized charge, structure of hydrogen-bonded complexes, electronic structures and transition states, determining rates of reactions.
- IV. Molecular Dynamics: simulated annealing (high temperature dynamics followed by slow cooling) to search for low- energy minima, reaction mechanisms with dynamics trajectories using potential surfaces, calculations on isolated biomolecules or on systems in solution.
- V. Monte Carlo Methods: time dependent systems (voltammetry, temperature programmed), adsorption and reactions on surfaces.

Nanosystems Modeling Final Projects

The final project of the Nanosystems Modeling course is used to completely evaluate an assigned fuel cell-related topic using all of the techniques learned throughout the coursework. The students are guided through the assignment with a division of subtopics that requested specific information. In the last part of the final project the students are asked to design kits that could be brought into schools or professional development workshops that will convey their assigned topic and address the State of Louisiana's Grade Level Expectation(GLEs). Each student is assigned an audience to present their topic to. The audiences range from K – 4th grade, 5th and 6th grade, 7th and 8th grade, 9th -12th grade and lastly, general (all ages and educational backgrounds).

The students are trained to communicate their scientific ideas to specific audiences by Linda Ramsey. Linda is the Director of the Center For Applied Teaching and Learning to Yield Scientific Thinking (CATALyST) and the Director of the Center for Educational Excellence, Louisiana Tech's faculty and staff professional development center. Ms. Ramsey prepares the students to accomplish this task by presenting in one class the National and State science, math, and technology standards [3]. This effort helped the students address the appropriate GLEs and design learning kits that were appropriate for the grade levels assigned.

For the graduate students, the projects are assigned based on information provided to Dr. Mainardi about the research project the students are participating in. The projects assigned to the undergraduates are based on topics that are being discussed in their curriculum and other current event issues in society such as energy-related topics.

Research and Final Project of Katherine Keeton

Katherine Keeton is a PhD student working under the advisement of Dr. Mainardi. She participated in the Nanosystems Modeling course in the Spring 2007 quarter. Through this course she furthered her knowledge of computational modeling techniques and addressed a specific idea in the final project component of the course. At the conclusion of the quarter she was selected to be a GK-12 Fellow (supported by NSF grant 0638730) and has used the educational component of this course in professional development workshops and at neighboring cities' school.

The basis of her research with Dr. Mainaridi's group currently involves investigating limitations on electron transfer in an enzymatic fuel cell that uses the methanol dehydrogenase (MDH) enzyme. Many graduate students currently working in Dr. Mainardi's group address various aspects of the enzyme and its possibility of providing biochemical energy. MDH has practical applications in chemical fuel cells (enzymatic fuel cells).

Ms. Keeton's research is focused on effectively transferring an electron from the enzyme's cofactor, PQQ, to the substrate of the anodic chamber of the fuel cell. Current theories propose that the location of PQQ limits the electron transfer and the addition of mediator molecules may aid in the transport process. The composition of the mediator molecule affects the ionization potential and electron affinity. In studying these molecules, it is her goal to show that size and composition of various molecules will aid in electron transfer from the active site of the molecule to the substrate. Ms. Keeton was given the task of investigating the molecules N,N,N,N-tetramethyl-phenylenediamine (TMPD) and Phenazine Ethosulfate (PES) as part of her Nanosystems Modeling final project.

Audience

The assigned target audience for Ms. Keeton's final project was kindergarten through 4th grade. The GLEs were examined from the Department of Education for the state of Louisiana. For example, in Kindergarten one learns letters and the formation of words. Two scientific expectations for this level of schooling that are relative to this project include learning to predict and anticipate possible outcomes (SI-E-A1) and recognizing the use of objects to examine different degrees of magnification (SI-E-B3) [3].

In the fourth grade the expectations only have a few additions to the ability of the student to do scientific inquiry, but the expectations to the understanding of scientific inquiry are expanded significantly. In addition to the previously mentioned expectations, fourth graders are expected to base explanations and logical inferences on scientific knowledge and observations (SI-E-B4). They are also expected to describe procedures and be able to communicate the results and procedure to others (SI-E-B5). Two other applicable expectations are the ability to use evidence to propose additional experimentations (SI-E-B6) and the ability to give examples of how scientific discoveries can affect society (SI-E-B6) [3].

Components of the Learning Kits

Ms. Keeton decided to create a two component activity to explain the effective use of a mediator in an enzymatic fuel cell to kindergarteners. An animated cartoon is used which includes an interactive component. Because kindergarteners usually can not read, vocal support is added to guide the students through the learning process. Secondly, a physical exercise is conducted using volunteers from the class.

The animated, or flash, applet first shows a toy car driving around. The narrator explains that the car cannot run forever without replacing batteries. The animation then zooms inward on the toy car and shows the inner workings. The animation zooms in by following a path through the lens of a microscope to convey that a microscope was used to "look inside". Another magnification shows the fuel "sparking" which will signal to the students that a form of power is present. The narrator asks the children if they wish they could carry toys around everywhere and never have the batteries die. The animation then shows a picture of a toy and then magnify the inner workings. The narrator explains that if a fuel is given to the toy (from a supply) that the fuel needs something to bring it to the metal to "spark". Further explanation is given to explain that the mediator is what brings the electron to the electrode.

The last part of the animation has a toy that is in need of fuel. The fuel is in the cup but it is only slowly leaking out of the storage. The initial sketch for this concept is shown in Figure 2. The student will then have to use the computer's mouse to "drag and drop" the mediator from the storage to the electrode. The comparison is made as if a water balloon is slowly leaking water and if you use a straw it will all funnel out fast. This is analogous to a mediator allowing an electron to move quickly instead of leaking out. The animation lights up and blink when the student correctly places the mediator, signaling that the "toy" now has power to run "forever".

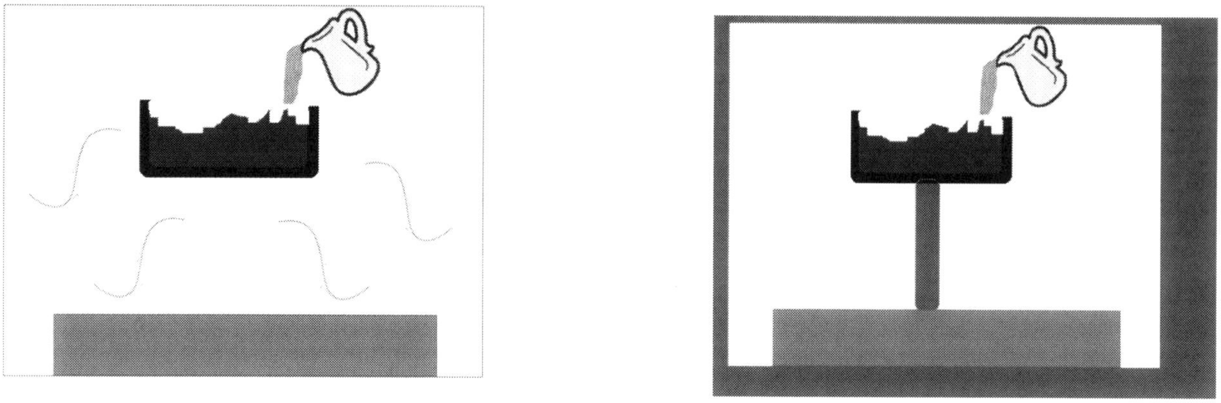

1. Toy is unanimated. The electricity (electrons) cannot get to where they need to go (substrate).

2. Toy is animated (blinking red) because a mediator has been correctly placed to allow the electricity (electrons) to get where they need to go (substrate).

Figure 2. Initial sketches for an interactive animation implying the use of a mediator (red tube) to allow for effective transfer of electrons.

The second part of the students' learning is presented in the form of a physically interactive component. A group of five to ten students (depending on the class size) is asked to stand in a circle. It is explained that they are representing the enzyme. Another group of five to ten students is asked to stand in a straight line to represent the base (the electrode). A last group of five to ten students is asked to join hands to act as the mediator. One student (or perhaps the teacher) is asked to represent the electron. At the beginning of the exercise the person representing the electron is asked begin in the middle of the enzyme circle and then to wander around aimlessly. It is then communicated that the electron may not know where to go and that there is not a strong attraction to the substrate. Then the mediator group is asked to first join hands with someone in the enzyme circle and on the other end someone from the substrate group. The electron person moves from their initial position within the center of the enzyme and the person follows the chain down to the electrode as the mediator group "pulls" the person towards the base. A diagram of a typical setup is shown in Figure 3. The overall goal of the activity is to communicate to the students is that with the use of a mediator molecule (or chain), the electron transfer is faster, and therefore more effective.

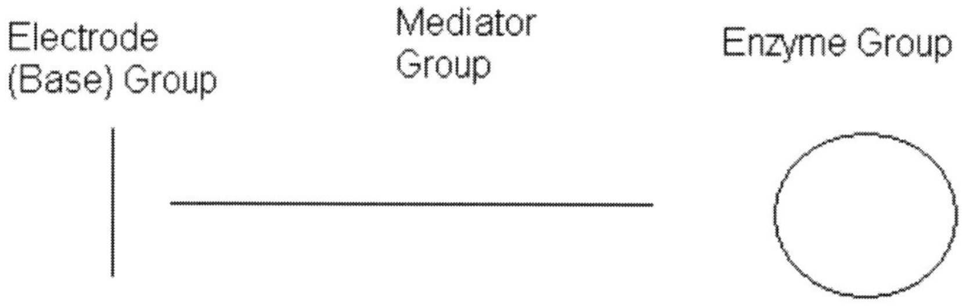

Figure 3. Diagram representing the positioning of groups for students representing the enzyme, mediator, and electrode.

The fourth grade students are more scholastically mature and a separate concept is used to address the more advanced GLEs and introduce age appropriate vocabulary. An animation is still used as an effective way of presenting the material to the students. The reason for using an animation rather than just a verbal explanation is because an animation can appeal to the student's everyday life. Most students at that age watch cartoons. Presenting an educational element in something familiar and enjoyable to the students may result in better absorption of the material. The use of an animation also varies from a teacher lecturing and the use of a computer animation may spark more interest in the student because of the novelty of the characters and graphics presented.

The theme that is proposed to be used with the fourth graders is CHEESE. In the beginning of the animation a camera flashes and the narrator says "cheese." Then the educational component then begins by showing the toy car that never has to be refilled with gas. The inner workings of the car are shown to explain a difference in magnification. The use of a fuel cell is then presented and the use of an enzyme is explained. The word 'Cheese' is used as a mnemonic for remembering that there are three "e"'s in an enzymatic fuel cell. The three major components are the Enzyme, Electrode, and the Electron. The students are asked to repeat "The mediator allows the electron to go from the enzyme to the electrode." This allows the students to affectively communicate the idea that a mediator is responsible for the electron transferring from the enzyme to the electrode. The students are also informed of the social impacts of this technology.

Lastly, the physical exercise component of this presentation is once again conducted, but with an investigating component addition. The group that represents the mediator is asked to spread out with their arms completely stretched out. The student playing the role of the electron is asked to move from the enzyme to the electrode at the same pace as previously done. This demonstrates to the students that the electron will transfer but may take longer.

The mediator group is then asked to reconfigure to the original position of closely lined up. Some students are asked to switch places and then the person representing the electron is asked to traverse down the mediator once again but the student is asked to walk with an increased pace. This conveys to the students that a different arrangement of molecules (i.e. a different mediator) may have properties that accelerate or decelerate the transferring of the electron. The students are then asked what other modifications of the mediator will affect the transfer or the electron.

Through the learning instruments of informative computer animations and physical demonstrations, the necessity of the mediator is conveyed. The societal benefits from the use of enzymatic fuel cells for the use of small portable devices are presented. The grade level expectations previously noted are explored in the use of these activities. The overall importance of this research is then effectively conveyed.

Implementation of Outreach

There are many sources available for the implementation of this outreach at Louisiana Tech University and in the surrounding community. Within the Engineering and Sciences Department, Louisiana Tech has an outreach program entitled GK-12 Fellows Program. This NSF supported program provides assistantships to students to serve as science laboratory teachers in rural and at risks schools. Ms. Keeton was one of the seven chosen fellows and she currently teaches inquiry based science laboratories to 7th and 8th grade students at Simsboro High School (Simsboro, Louisiana) and at Carroll Magnet Jr. High School (Monroe, Louisiana). Two full days out of each week she facilitates laboratories designed around the

State of Louisiana's curriculum that address appropriate GLEs. She has also been a facilitator and guest speaker at a professional development workshop conducted by the Nanoscience Educational Research Outreach (NERO) program at Louisiana Tech University (NSF grant 0602029). The presentation entitled "What makes a beautiful model?" explained the general concept of modeling and how it is appropriate for various age levels and can be incorporated into different subject material. She explained that there were many learning kits for all grade levels designed and developed by the students who participated in *Nanosystems Modeling*. Lastly, Ms. Keeton spoke specifically about the computational studies that could be performed in the classroom.

Tanya Culligan, a teacher at Caddo Parish Magnet High School in Shreveport, Louisiana, was a participant in the NERO workshop and was paired with Dr. Mainardi as a partner teacher. Through her month long stay in this program she learned basics of the mathematical operations performed by the software (Material Studios). She also conducted a research project under the direction of Dr. Mainardi. As part of the program, she was asked to explain how she would implement what she learned into her classroom and was provided monies to do so. Mrs. Culligan is an example of a teacher that brought computational molecular modeling into her classroom from material she learned as a result of the educational outreach from NERO and student developed project from *Nanosystems Modeling*.

Conclusions

Students taking Nanosystems Modeling are able to i) Identify the proper computational techniques for modeling a given system ii) interpret and describe the outputs of simulations iii) apply different computational techniques on a variety of applications iv) derive and compute calculations combining the outputs of several simulations v) design a computer – based experiment using the material covered in the class that can be easily explained to an audience, a vi) evaluate, select, and justify the steps to follow in the design of learning materials based on molecular simulations. Students learn how to effectively communicate research ideas to all age groups with different educational backgrounds. The learning kits designed in the course have provided a tool that has been implanted in outreach programs of Louisiana Tech. Students, professors, and teachers from professional development workshops have all benefited from the knowledge gained by learning computational modeling techniques.

References

1. Gaussian 03, Revision C.02, M. J. Frisch, G. W. Trucks, H. B. Schlegel, G. E. Scuseria, M. A. Robb, J. R. Cheeseman, J. A. Montgomery, Jr., T. Vreven, K. N. Kudin, J. C. Burant, J. M. Millam, S. S. Iyengar, J. Tomasi, V. Barone, B. Mennucci, M. Cossi, G. Scalmani, N. Rega, G. A. Petersson, H. Nakatsuji, M. Hada, M. Ehara, K. Toyota, R. Fukuda, J. Hasegawa, M. Ishida, T. Nakajima, Y. Honda, O. Kitao, H. Nakai, M. Klene, X. Li, J. E. Knox, H. P. Hratchian, J. B. Cross, V. Bakken, C. Adamo, J. Jaramillo, R. Gomperts, R. E. Stratmann, O. Yazyev, A. J. Austin, R. Cammi, C. Pomelli, J. W. Ochterski, P. Y. Ayala, K. Morokuma, G. A. Voth, P. Salvador, J. J. Dannenberg, V. G. Zakrzewski, S. Dapprich, A. D. Daniels, M. C. Strain, O. Farkas, D. K. Malick, A. D. Rabuck, K. Raghavachari, J. B. Foresman, J. V. Ortiz, Q. Cui, A. G. Baboul, S. Clifford, J. Cioslowski, B. B. Stefanov, G. Liu, A. Liashenko, P. Piskorz, I. Komaromi, R. L. Martin, D. J. Fox, T. Keith, M. A. Al-Laham, C. Y. Peng, A. Nanayakkara, M.

Challacombe, P. M. W. Gill, B. Johnson, W. Chen, M. W. Wong, C. Gonzalez, and J. A. Pople, Gaussian, Inc., Wallingford CT, 2004.
2. Accelrys Software, Material Studio http://www.accelrys.com/products/mstudio/
3. Division of Student Standards & Assessments, "Grade-Level Expectation," 2007: http://www.doe.state.la.us/lde/saa/1819.html

287f Simulation-Based Projects In Biotransport - An Elective Course In The Chemical Engineering Curriculum

Laurent Simon, Otto H. York Department of Chemical Engineering, New Jersey Institute of Technology, University Heights, Newark, NJ 07102

Biotransport (ChE427/BME427) is a course offered at the New Jersey Institute of Technology to undergraduate chemical and biomedical engineering students. Specific areas covered include transport within cells, basics of vector and tensors, conservation relations and momentum balance. Two weeks of the semester are dedicated to introducing concepts out of which simulation-based projects are developed. These topics involve pharmacokinetic analysis, oxygen delivery to tissues and hemodialysis. Because most of the students participating in the class are new to modeling and simulation of biological systems, nearly a month is devoted to teaching analytical methods and to help with the transition from modeling to simulation, data interpretation and presentation. The students write and present a final report to the rest of the class at the end of the semester. This contribution focuses on student performance and feedback, strategies developed to improve the computing component of the course and challenges, such as, the choice of simulation platforms.

EFFICIENT SOLUTION OF MULTIPLE-MODEL, MULTIPLE-ALGORITHM PROBLEMS IN UNDERGRADUATE AND GRADUATE EDUCATION

Mordechai Shacham, Ben-Gurion University, Beer-Sheva, Israel
Michael B. Cutlip, University of Connecticut, Storrs, CT 06269, USA
Michael Elly, Intel Corp., Qiryat Gat, Israel

Introduction

Mathematical software packages, such as Excel®, Maple™, Mathcad, MATLAB®, Mathematica® and POLYMATH, are currently utilized for numerical problem solving in engineering education (Cutlip et al., 1998, Shacham and Cutlip, 1999). From the numerical solution perspective, it is convenient to characterize the various problems as Single Model-Single Algorithm (SMSA) problems and complex problems with some combination of Multiple Models and Multiple Algorithms. Typical examples of SMSA type problems include the following:

1. Steady state operation of a tubular reactor where the model consists of a system of ordinary differential equations and explicit algebraic equations. One numerical integration algorithm (such as the 4th order Runge-Kutta) can be used to solve this model.
2. Calculation of the bubble-point temperature for a non-ideal liquid mixture where the model includes a system of implicit and explicit algebraic equations. A non-linear equation solver algorithm (such as the Newton-Raphson technique) can be used to solve this problem .
3. Fitting the Wagner equation to vapor pressure data using a linear regression algorithm.

The application of the mathematical software packages for solving SMSA problems has essentially replaced all other solution techniques, as can be seen in many recent textbooks (see, for example, Fogler, 2005). However, for complex problem types, the solution process is often more involved. The types of models included in the "complex" category are:

1. Multiple Model Single Algorithm (MMSA) Problem.

A typical example of such a problem is the cyclic operation of a semi-batch bioreactor (Cutlip and Shacham, 2007). The three modes of operation of the bioreactor (initialization, processing, and harvesting) are represented by different models comprising of ordinary differential equations and explicit algebraic equations. All models can be solved by one numerical integration algorithm (such as the 4th order Runge-Kutta).

2. Single Model Multiple Algorithm (SMMA) Problem.

An example of such a problem is the problem of parameter estimation in dynamic systems. In this case there is a model comprising of ordinary differential equations and explicit algebraic equations, with parameters that should be fitted to experimental data using nonlinear regression techniques. One option is to solve this system by integrating the differential equations with specified parameter values in an internal loop, and minimizing the sum of squares of the difference between the calculated and the experimental values using an optimization algorithm, in the outer loop .

Additional examples for SMMA problems include: the solution of two point boundary value problems where the integration of the model is carried out in the inside loop and a nonlinear equation solver algorithm adjusts the boundary values in an outer loop, and the solution of differential-algebraic systems of equations where the same algorithms are used but in an opposite hierarchy.

3. Multiple Model Multiple Algorithm (MMMA) problem.

A typical example of such a problem is the optimization of the semi-batch bioreactor, described earlier, with respect with some of its operational parameters. An additional example is the modeling of an exothermic batch reactor, where the two stages of operation (heating and cooling) require different models and different integration algorithms (stiff and non-stiff).

The solution of such complex problems can be rather cumbersome and time consuming even if mathematical software packages are used, as manual transfer of data from one model to another and consecutive manual reruns may be frequently required. However, combining the use of several software packages of various levels of complexity, flexibility and user friendliness, can considerably reduce the time and effort required for solving complex models. Following this premise, the models representing the various stages of the problems are coded and tested using a software package (for example, POLYMATH, a product of Polymath Software, http://www.polymath-software.com) that requires very little technical coding effort. After testing each of the modules separately, they are combined into one program using a programming language, or a mathematical software package that supports programming (for example, MATLAB, a trademark of The Math Works, Inc. http://www.mathworks.com). To minimize the probability of introducing errors into the model equations, an option within POLYMATH for automatic conversion of the programs representing the various modules can be converted to MATLAB functions. MATLAB can then be programmed to handle the consecutive and repetitive calls to the various models, applying the appropriate solution algorithms, and assigning the hierarchy of the computations .

In this paper, the solution of a problem involving the biokinetic modeling of a chemostat with imperfect mixing will be carried out to demonstrate the solution of a SMSA problem, the generation of parametric runs of an SMSA problem, and the solution of a SMMA problem.

Modeling a chemostat with POLYMATH and solution for one set of parameter values

The chemostat problem is presented by Cutlip and Shacham (2008). The complete details of the problem are provided in Appendix 1. The chemostat may be considered to be modeled as a reactor with a completely-mixed volume V_1 that interacts with another completely-mixed but stagnant volume V_2. This is shown in Figure 13–13 of Appendix 1. Volume V_2 represents the poorly mixed regions within a production fermenter. The microbial system to be modeled involves substrate S going to product P only under the action of cells X. The separate balances on the substrate, cells, and product in each reactor utilize Monod kinetics and a cell death rate constant given by k_d. The four balance equations are shown in Appendix 1. The assignment involves solution of the model equations for one set of parameter values, plotting the substrate (S), cell (X), and product (P) concentrations versus the dilution rate: $D = F_1/V_1$ (where F_1 is the feed flow rate to the well mixed volume: dm^3/hr, and V_1 is the well mixed part of the reactor volume: dm^3). Additionally, the cell production rate, DX_1, and product production rate, DP_1, should be plotted as a function of the dilution rate. Finally the dilution rate which will maximize the production rate for the cells and the dilution rate which will maximize the production rate for the product are to be determined.

The POLYMATH model for the chemostat, given in Table 1, includes four implicit nonlinear algebraic equations which are obtained from the balance equations. The POLYMATH model (including the "comments" which start with the # sign) provide complete documentation of the equations, the values of the constants and the initial estimates used for the four unknowns: S_1, S_2, X_1, and X_2. Statements 1 through 4 of Table 1 present the implicit equations for obtaining the substrate concentration in the well mixed volume (S_1) and in the stagnant volume (S_2), and the cell concentration in the well mixed volume (X_1) and in the stagnant volume (X_2). Explicit variables and constants are described in statements 5-18. Initial estimates for the unknowns in the nonlinear equations are provided in lines 19 to 22.

The results for the case where F_1 = 0.17 L/hr are given In Table 2. For this case, the dilution rate D = 0.085 hr^{-1}; the cell production rate DX_1 = 0.0119717 g/hr and the product production rate DP_1 = 0.0066612 g/hr.

Exporting the chemostat model to MATLAB for parametric runs

Parametric runs can be carried out with POLYMATH by manually changing the parameter values. However, carrying out such runs by programming within MATLAB can be more efficient. The MATLAB function representing the operation of the chemostat is generated automatically by POLYMATH (Table 3). Note that MATLAB requires input of the variable values into the function in a single array (**x**, in this case), and return of the function values in a single array (**fx**, lines 18-21 in Table 3). The variable values are put back into variables with the same names as used in the POLYMATH model (lines 2-3) to make the MATLAB code more meaningful. POLYMATH orders the equations sequentially as required by MATLAB and converts any needed intrinsic functions and logical expressions.

In order to carry out parametric runs while changing the feed flow rate F_1 this variable has to be added as an input parameter to the *NLEfun* function (Table 3) and a main program has to be prepared which changes the value of F_1, solves the system of nonlinear equations, collects the pertinent data, and plots the results of the parametric runs. Part of this main program is shown in Table 4. The value of F_1, is changed starting at F_1 =0.1 up to F_1 = 0.25 with steps of 0.01. The MATLAB library function *fsolve* is used to solve the system of algebraic equations in line 7 of Table 4. The variable values needed for preparing the various plots are calculated and stored in lines 8 through 10.

A plot of S_1, X_1, and P_1 as function of the dilution rate is shown in Figure 1 and the cell and product production rates are plotted in Figure 2. Observe that there are clear maximum points for the two production rates in the vicinity of D = 0.08 hr^{-1}. Determination of the exact points of the maximum involves the solution of a SMMA problem.

Maximization of the production rates – solving an SMMA problem

The two optimization problems can be defined as minimization problems (as only minimum searching library functions available in MATLAB) as follows:

$$\min_{F_1} -DX_1 \quad \text{and} \quad \min_{F_1} -DP_1 \quad \text{where } D = F_1 / V_1 . \tag{1}$$

The calculation of D, X_1 and P_1 associated with a particular value of F_1 involves the solution of a system of nonlinear algebraic equations (NLEs) while a minimization algorithm is required in order to find the values of F_1 that satisfy Eq.1. Thus this is a single model (the chemostat) and multiple algorithm (one for solution of NLEs and one for minimization).

The MATLAB library function *fminbnd* for single value minimization can be used for finding the minimum of the functions in 1. In order to carry out the minimization two new functions should be prepared. The first one (shown in Table 5) obtains F_1 as input, uses the *fsolve* library function to solve the chemostat model, and returns – DX_1 to the calling function. The second function does the same except that it returns the value of – DP_1. Two calls to the library function *fminbnd* identify the highest production rate for cells DX_1 = 0.01207 g/hr at dilution rate of D = 0.079932 hr^{-1} and the highest production rate for product DP_1 = 0.0066643 g/hr at dilution rate of D = 0.083729 hr^{-1}.

Conclusions

This work demonstrates that the effort involved in solving complex problems which involve multiple models and or multiple algorithms can be considerably reduced by using a combination of software packages. This is achieved by utilization of the most effective software tools for the various tasks. Use of a "user-friendly" mathematical software package such as POLYMATH for the preparation of the individual models reduces the technical effort involved in coding and minimizes the potential errors that may enter the final completed mathematical model. Conversion of the verified separated modules into a more versatile program within a package such as MATLAB requires essentially the addition of a main executive program that controls the interaction with the user, the sequencing and integration of the results of the various modules, and the reporting of the results.

In addition to reducing the time and effort required for building the process models, the proposed approach yields modular programs that are well documented and easily modified. It also ensures that the models presented in the documentation are almost identical to the models used in the program.

The POLYMATH and MATLAB programs used in this study are available at the site: ftp://ftp.bgu.ac.il/shacham/chemostat/.

References

1. Cutlip, M., Hwalek, J.J., Nuttall, H.E., Shacham, M., Brule, J., Widman, J., Han, T., Finlayson, B., Rosen, E. M. and Taylor, R., "A Collection of Ten Numerical Problems in Chemical Engineering Solved by Various Mathematical Software Packages." *Comput. Appl. Eng. Educ.*, 6(3), 169-180(1998).
2. Cutlip, M. B. and Shacham, M. *Problem Solving In Chemical and Biochemical Engineering with Polymath, Excel and MATLAB*. Prentice-Hall, Upper Saddle River, New-Jersey, 2008.
3. Fogler, H. S., *Elements of Chemical Reaction Engineering*, 4th Ed, Prentice-Hall, Upper Saddle River, New-Jersey, 2005.
4. Shacham, M. and M.B. Cutlip, "A Comparison of Six Numerical Software Packages for Educational Use in the Chemical Engineering Curriculum", Computers in Education Journal, IX(3), 9-15 (1999).

Table 1. POLYMATH model for the chemostat with imperfect mixing

No.	Equation # Comment
1	f(S1) = F1*S0+F2*S2-(1/Yxs)*(mum*S1/(Ks+S1))*X1*V1-F1*S1-F2*S1 # Substrate balance on the well mixed volume
2	f(S2) = F2*S1-(1/Yxs)*(mum*S2/(Ks+S2))*X2*V2-F2*S2 # Substrate balance on the stagnant volume
3	f(X1)=F2*X2+(mum*S1/(Ks+S1)-kd)*X1*V1-F1*X1-F2*X2 # Cell balance on the well mixed volume
4	f(X2)=F2*X1+(mum*S2/(Ks+S2)-kd)*X2*V2-F2*X2 # Cell balance on the stagnant volume
5	P1=Yps*(S0-S1) # Production (g/dm^3)
6	D=F1/(V1+V2) # Dilution rate (1/hr)
7	S0 = 0.6 # Feed substrate concentration (g/dm^3)
8	kd = 0.002 # Cell death rate (1/hr)
9	Yxs=0.4 # Yield coefficient (g cells/g substrate)
10	Yps=.2 # Yield coefficeint (g product/g substrate)

11	Ks = 0.2 # Monod constant (g substrate/L)
12	mum = 0.2 #Maximal specific growth rate (1/hr)
13	V1=1.7 # Well mixed voume (dm^3)
14	V2=0.3 # Stagnant volume (dm^3)
15	F1=.17 # Feed flow rate to the well mixed volume (dm^3/hr)
16	F2=0.2*F1 # Feed flow rate to the stagnant volume (dm^3/hr)
17	PR_DX1=D*X1 # Cell production rate (g/hr)
18	PR_DP1=D*P1 # Product production rate (g/hr)
19	S1(0) = 0
20	S2(0)=0
21	X1(0)=0.2
22	X2(0)=0.4

Table 2. Chemostat results for F_1 = 0.17 L/hr

Variable	Value	f(x)	Initial Guess
S_1 (g/L)	0.208163	4.32E-13	0
S_2 (g/L)	0.044739	6.83E-14	0
X_1 (g/L)	0.140844	-1.73E-13	0.2
X_2 (g/L)	0.202638	-2.74E-14	0.4
D (1/hr)	0.085		
F_1 (L/hr)	0.17		
F_2 (L/hr)	0.034		
PR_DP1 (g/hr)	0.006661		
PR_DX1 (g/hr)	0.011972		

Table 3. MATLAB function (model) for the chemostat with imperfect mixing

No.	Equation % Comment
1	function fx = MNLEfun(x);
2	S1 = x(1); S2 = x(2);
3	X1 = x(3); X2 = x(4);
4	Yps = .2; %Yield coefficeint (g product/g substrate)
5	F1 = .17; %Feed flow rate to the well mixed volume (dm^3/hr)
6	S0 = .6; %Feed substrate concentration (g/dm^3)
7	kd = .002; % Cell death rate (1/hr)
8	Yxs = .4; %Yield coefficient (g cells/g substrate)
9	P1 = Yps * (S0 - S1); %Production (g/dm^3)
10	Ks = .2; %Monod constant (g substrate/L)
11	mum = .2; %Maximal specific growth rate (1/hr)
12	V1 = 1.7; %Well mixed volume (dm^3)
13	V2 = .3; %Stagnant volume (dm^3)
14	D = F1 / (V1 + V2); %Dilution rate (1/hr)
15	F2 = .2 * F1; %Feed flow rate to the stagnant volume (dm^3/hr)
16	PR_DX1 = D * X1; %Cell production rate (g/hr)
17	PR_DP1 = D * P1; %Product production rate (g/hr)
18	fx(1,1) = F1 * S0 + F2 * S2 - (1 / Yxs * mum * S1 / (Ks + S1) * X1 * V1) - (F1 * S1) - (F2 * S1);
19	fx(2,1) = F2 * S1 - (1 / Yxs * mum * S2 / (Ks + S2) * X2 * V2) - (F2 * S2);
20	fx(3,1) = F2 * X2 + (mum * S1 / (Ks + S1) - kd) * X1 * V1 - (F1 * X1) - (F2 * X2);
21	fx(4,1) = F2 * X1 + (mum * S2 / (Ks + S2) - kd) * X2 * V2 - (F2 * X2);

Table 4. Part of the MATLAB "main program" for parametric studies with the chemostat

No.	Equation % Comment
1	options = optimset('Diagnostics',['off'],'TolFun',[1e-9],'TolX',[1e-9]);
2	Yps = 0.2; S0 = 0.6; kd = 0.002; Yxs = 0.4; Ks = 0.2;
3	mum = 0.2; V1 = 1.7; V2 = 0.3;
4	F1=0.1;
5	xguess = [0 0 0.2 0.4]; % initial guess vector
6	for k=1:16
7	xsolv=fsolve(@MNLEfun,xguess,options,F1);
8	S1(k)=xsolv(1); S2(k)=xsolv(2); X1(k)=xsolv(3); X2(k)=xsolv(4);
9	F1list(k)=F1; D(k) = F1 / (V1 + V2); P1(k)= Yps * (S0 - S1(k));
10	PR_DX1(k) = D(k) * X1(k); PR_DP1(k) = D(k) * P1(k);
11	F1=F1+0.01;
12	end

Table 5. A function for calculating the cell production rate for a single value of F_1

No.	Equation % Comment
1	function PR_DX=ProdRateCell(F1)
2	V1 = 1.7;
3	V2 = 0.3;
4	xguess = [0 0 0.2 0.4]; % initial guess vector
5	options = optimset('Diagnostics',['off'],'TolFun',[1e-9],'TolX',[1e-9]);
6	xsolv=fsolve(@MNLEfun,xguess,options,F1);
7	X1=xsolv(3);
8	D = F1 / (V1 + V2); PR_DX = -D* X1;

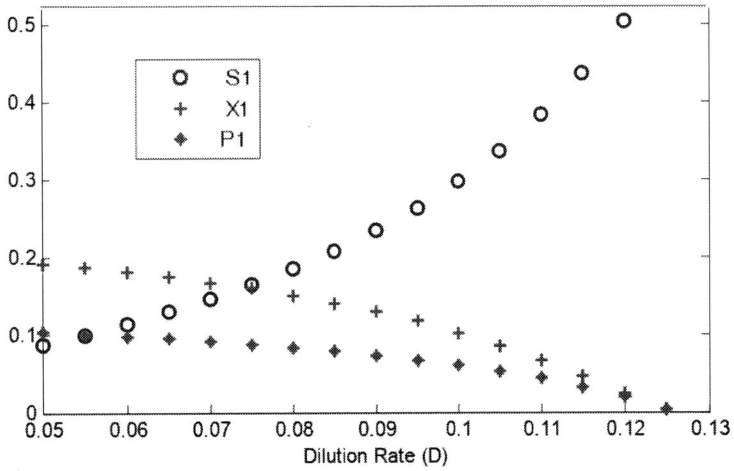

Figure 1. Plot of S_1, X_1 and P_1 as function of dilution rate

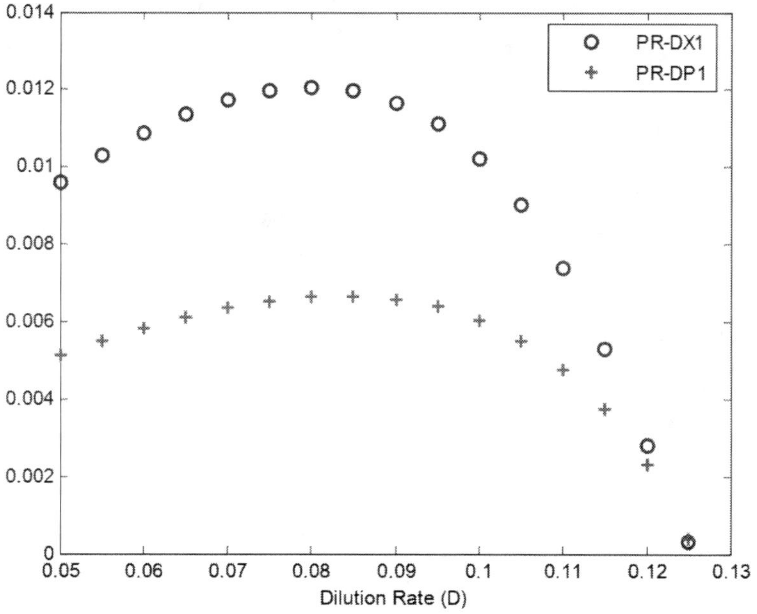

Figure 2. Cell production rate (PR-DX1) and product production rate (PR-DP1) as function of dilution rate

Appendix 1 - Problem Statement for Biokinetic Modeling Imperfect Mixing in a Chemostat

A chemostat is usually considered to be a completely mixed reactor; however, this is not always the case. Consider the situation where the chemostat may be considered to be modeled as a reactor with a completely-mixed volume V_1 that interacts with another completely-mixed but stagnant volume V_2 as shown in Figure A-1. Volume V_2 may be considered to model the poorly mixed regions within a production fermenter.

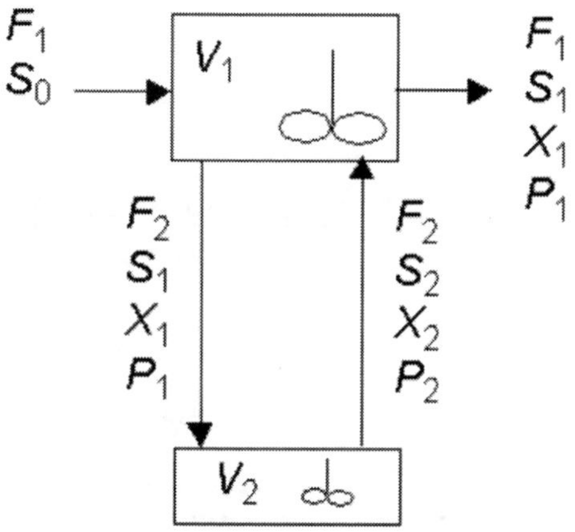

Figure A-1 Chemostat Reactor Model

The microbial system to be modeled involves substrate S going to product P only under the action of cells X. The following separate balances on the substrate, cells, and product in each reactor utilize Monod kinetics and a cell death rate constant given by k_d.

Steady-State Substrate Balance on Volume V_1

$$F_1S_0 + F_2S_2 + \frac{1}{Y_{X/S}}\left(\frac{\mu_m S_1}{K_S + S_1}\right)X_1V_1 = F_1S_1 + F_2S_1 \quad \textbf{(A-1)}$$

Steady-State Substrate Balance on Volume V_2

$$F_2 S_1 + \frac{1}{Y_{X/S}}\left(\frac{\mu_m S_2}{K_S + S_2}\right) X_2 V_2 = F_2 S_2 \qquad \textbf{(A-2)}$$

Steady-State Cell Balance on Volume V_1

$$F_2 X_2 + \left(\frac{\mu_m S_1}{K_S + S_1} - k_d\right) X_1 V_1 = F_1 X_1 + F_2 X_2 \qquad \textbf{(A-3)}$$

Steady-State Cell Balance on Volume V_2

$$F_2 X_1 + \left(\frac{\mu_m S_2}{K_S + S_2} - k_d\right) X_2 V_2 = F_2 X_2 \qquad \textbf{(A-4)}$$

Overall Steady-State Material Balance for Product

$$P_1 = Y_{P/S}(S_0 - S_1) \qquad \textbf{(A-5)}$$

Microbial growth has been studied in a continuous culture and the following parameters were obtained: μ_m = 0.2 h^{-1}, K_S=0.2 g/dm^3, k_d = 0.002 h^{-1}, $Y_{X/S}$ = 0.4 g cells/g substrate, and $Y_{P/S}$ = 0.2 g product/g substrate. Tracer studies have indicated that the incomplete mixing can be described by a well-mixed volume V_1 = 1.7 dm^3 and a stagnant volume of V_2 = 0.3 dm^3 with a constant flow rate relationship of F_2 = 0.2 F_1 in dm^3/h. Chemostat operation is such that X_0 = 0 and S_0 = 0.6 g/dm^3, and the endogenous metabolism can be neglected.

(a) Create a single graph of S, X, and P versus the dilution rate defined by $D = F_1/V_1$.

(b) Plot the cell production rate, the product $D*X_1$, and the product production rate, the product of $D*P_1$, as a function of the dilution rate between 0.05 and 0.130 h^{-1}.

(c) Estimate the dilution rate that will maximize the production rate, DX_1, for the cells and the dilution rate that will maximize the production rate, DP_1, for the product.

312a Introduction - Workshop 2

William B. Krantz, Chemical and Biomolecular Engineering, National University of Singapore, Room E5-02-11, Block E5, 4 Engineering Drive 4, Singapore 117576, Singapore

Introduction to Workshop II

312b Proposal Writing Tutorial

John Regalbuto, Division of Chemical and Transport Systems, National Science Foundation, 4201 Wilson Boulevard, Arlington, VA 22230, James M. Lee, Washington State University, PO Box 642710, Pullman, WA 99164-2710, and Geoffrey A. Prentice, Division of Chemical, Bioengineering, Environmental & Transport Systems, National Science Foundation, 4201 Wilson Blvd., Rm. 525, Arlington, VA 22230.

The speakers will present some general principles for writing successful proposals, in part based of experiences serving as grantees and as reviewer on CBET review panels. (30 minutes). This should stimulate more questions during the remainder of the Session II's interactive panel.

312c Introduction and Overview of the Format for Breakout Sessions

Robert Wellek, The National Science Foundation, Division of Chemical & Transport Systems (ENG/CTS), 4201 Wilson Boulevard, Arlington, VA 22230

The Speaker will introduce the participating NSF PDs and overview the format of the informal breakout sessions which will include one-on one discussions regarding some CBET programs. It will describe the dynamics of a review panel with a mock panel.

312d Interactive Breakout Panels Coordinated by NSF Program Directors

Robert Wellek, The National Science Foundation, Division of Chemical & Transport Systems (ENG/CTS), 4201 Wilson Boulevard, Arlington, VA 22230

The Program Directors & Division Director for CBET will answer questions about the new organization, its scope and goals, trends, review panels, and new Administrative procedures in small interactive groups based on CBET programs. One interactive sub-panel will describe the dynamics of a review panel with an interesting mock panel. Discussions will continue in the NSF Hospitality Suites in small groups and individually.

351a Critical Thinking In A Senior-Level Capstone Lab: Construction Of Biodiesel Reactors

Holly A. Stretz, Department of Chemical Engineering, Tennessee Technological University, Prescott Hall, Box 5013, Cookeville, TN 38501 and Mario A. Oyanader, Universidad Catolica del Norte, Avenida Angamos, Antofagasta, 0610, Chile.

The chemical engineering faculty at Tennessee Technological University have been involved in a curriculum redesign to integrate lab and lecture such that students see an immediate need for the concepts they are learning, and thereby anchor those concepts. Some of the goals of this effort are retention, long-term recall, integrative thinking skills, and improvement of critical thinking. Here we define critical thinking in a narrow sense…the ability to effectively critique the experimental process and modify or improve it appropriately. Nine senior chemical engineering students from a capstone laboratory class have been involved in this study, whereby they were asked to investigate and construct certain projects associated with biodiesel production (a technology which drew significant student interest). The capstone experience thus became a direct, theme-oriented extension of concepts introduced in core coursework. In other words, student engineering competencies were challenged by a project that needed to be implemented from fundamentals to actual operation. In particular, the students had to critique and rewrite an existing proposal for building a reactor experiment, identify and/or justify the best location for the experiment, plan the operation for safety assurance, apply team work strategies, order the components, build the reactor, benchmark reactor performance, collect data, communicate results, and identify future efforts. The educational strategy was based on the Hi-Pele (High Performance Learning Environment) approach proposed by Arce [1,2]. We will discuss test results measuring improvements in critical thinking in this setting, as measured by a critical thinking instrument developed by researchers at TTU [3,4]. Using this instrument, ten specific skills were identified as central to critical thinking, and development in these skill sets is assessed using short-answer essay questions and graded by independent faculty from across campus.

1. Arce, P., The Colloquial Approach: An Active Learning Technique, Jr. of Science, Education & Technology, 3(3), and 145 (1994). 2. Mario A. Oyanader, P. Arce, And Whitaker S., "The Catalytic Pellet: A Rich Prototype for Engineering Scaling", Chem. Eng. Education, 2007. (In Press) 3. Stein, B., Haynes, A., and Understein, J. (2003). Assessing Critical Thinking.. Paper, SACS Annual Meeting in Nashville, Tennessee, 12- 2003. 4. Bransford, J. D., Brown, A. L. & Cocking, R.. R. (Eds.) 2000. How people learn: Brain,mind, experience, and school. Washington, D.C., National Academy Press.

351b Use Of The Aiche National Student Design Competition Problem As A Capstone Design Project

Richard L. Long, New Mexico State University, Chemical Engineering Department, P.O. Box 30001, MSC 3805, Las Cruces, NM 88003-8001

The AIChE National Student Design Competition is the oldest and most prestigious competition for undergraduates in AIChE. It has been offered annually since 1932. Problems are sought from industry, government, and academia. It is furnished to all student chapter advisors and department heads of departments of chemical engineering in the fall for use in the following spring competition or in design classes. For the national competition a set of rules is published with the problem. Of about 190 departments of chemical engineering in the U.S. about 30-50 departments submit entries to the competition. The problem is furnished by the Design Subcommittee, Student Chapters Committee, AIChE. This group works with sponsors from industry, government, and academia to provide the problem and to perform the grading at the national level. Each department is allowed to submit the two best individual and/or group solutions to the competition. A number of departments use the problem in their design classes and as a capstone design experience for their students. One approach to improving performance is to compare the national winning solution with the work done by students at a particular school. The first place winning solution is available through the national office once the grading of the problem is complete. A comparison can show the difference between what a given class of students is doing and what the national competition winner did. This allows a department to improve its desgn instruction in accord with ABET wishes.

351c Design Of A Partial Oxidation Methanol Production Plant

Mark Bricka, Dave C. Swalm School of Chemical Engineering, Mississippi State University, MS 9595, Mississippi State, MS 39762

At Mississippi State University (MSU) our Capstone Design courses consist of a series of two courses, Process Design and Plant Design. Process design introduces the students to design fundamentals and cost estimation techniques while Plant Design allows the student to utilize skills developed in their previous classes and is target to assist the students in applying these stills towards solving a single design problem.

During the spring term of 2007, the students were tasked with designing plant to product methanol using the partial oxidation of petroleum vacuum residue. The design problem was adapted from a previous 1989 AIChE student contest design problem. This problem was modified to reflect the increase in demand for methanol as a result of biodiesel production and the escalating methanol costs. While it was desired to have the students work with a biomass feed it was decided that the data base for vacuum residue. Mush of the data from the 1989 problem was incorporated into the assignment but the problem was vastly expanded to cover the semester time frame of the design. This presentation will discuss the group selection process, the technique of design from site selection to the generation of the cash flow diagram, and the post class assessment (after action report). Pitfalls of such a design course will also be presented.

351d Product And Process Design In Emerging Fields With Industrial By-In

Aydin Sunol, Department of Chemical Engineering, University of South Florida, 4202 East Fowler Avenue, ENB 118, Tampa, FL 33620

This presentation provides an overview of the transformation of Capstone Design Projects over the past two decades at institutions the author have been affiliated with and focus on the challenges associated with various elements of design and student assessment. The projects involve both group and individual performance elements; both the product and the associated manufacturing systems are designed; the safety and environmental issues are addressed explicitly; the assessment of written reports, oral presentations, and poster presentations involve several stake holders and ABET criteria friendly; the mentors include industrial sponsors; and projects from diverse emerging areas in microelectronics, fuels cells, targeted delivery products, biomedical devices, and novel material are used. The impact of the capstone projects on the curriculum including the efforts in recent NSF Spiral Curriculum Reform Implementation grant at the University of South Florida.

351e Papernol -- Recycle Paper to Ethanol

Jim Henry, James Grant, Jennifer Johnson, Aquila Hughley, and Blake Nida. UT-Chattanooga, 615 McCallie Av, Chattanooga, TN 37403

This paper describes the group design of a process facility to produce ethanol from waste paper. The technology is one integrating traditional and emerging chemical engineering topics. The proposed site and operating partner is a recycle paper processing plant at a facility that employs over 100 developmentally disabled citizens. The final plant is proposed to be operable with some of these same individuals. The preliminary economics are favorable. This process is shown to be one of the more thermodynamically sensible routes to ethanol. The presentation will present measures of student improvement.

Web Page: chem.engr.utc.edu/ench430/2007/Teams/Whole.htm

351f Capstone Objective: Develop the Skill of Time, Scope and Depth of Analysis in Complex Projects

Miguel J. Bagajewicz, Chemical, Biological and Materials Engineering, University of Oklahoma, 100 E. Boyd St., T-335, Norman, OK 73019

At the University of Oklahoma, the capstone experience consists of projects conducted in groups of two students, all projects being different.

In this presentation, we will explain why we found two students to be the optimal group size, despite the obvious instructor effort.

We will also explain in what form we challenge students to develop the skill of time, depth and scope management in projects that are required to provide realistic answers to a variety of problems: process design with investment strategies, process operations with economical objectives and product design using microeconomics.

We will finally illustrate how this strategy enabled us to be able to submit papers to this conference out of work done during the capstone experience and finally, we will discuss the confidence building associated to going through our class experience.

414a Opportunities and Challenges Mingling Chemical Engineering into a Graduate Course of Food Colloids

Qixin Zhong, Department of Food Science and Technology, University of Tennessee, 2509 River Drive, 115B McLeod Building, Knoxville, TN 37996-4539

Excellent opportunities are present to integrate chemical engineering with food science curricula. This is especially important in the context that food science has evolved from commodity-based to fundamental research. Food Colloids is such a course because numerous food systems are colloidal in nature and understanding fundamentals is important for improving food quality. However, food science curricula are significantly different from chemical engineering. Instructing such a Food Colloids course needs to consider the shortage of students' preparation in physics, mathematics and other factors. This paper describes the approaches used in a one-credit, introductory graduate-level Food Colloids course. One of the key emphases was to link fundamentals with specific foods. Students' feedbacks were mostly positive in that they experienced the connection between fundamental theories with realistic products. The approaches used may also be adaptable to educate chemical engineering students since food systems are familiar examples offering excellent opportunities to elucidate fundamental principles.

414b Popcorn! - - High School and First Year College Process and Product Design Experiences

Karen High and Sundararajan V. Madihally. School of Chemical Engineering, Oklahoma State University, 423 Engineering North, Stillwater, OK 74078

This paper will compare and contrast the use of a short "popcorn" project as part of a week-long summer program for high school students and the use of the same project as a multi-week project for an orientation class. Pre and post survey assessments were done and will be discussed. In the summers of 2007, a week-long academy, Reaching Engineering and Architecture Career Heights (REACH), was hosted by the College of Engineering, Architecture and Technology at Oklahoma State University, Stillwater. A Chemical Engineering/Industrial Engineering nine-hour module was a part of this academy. During this module, the students worked on product and process design concepts related to the manufacturing of popcorn. Popcorn was used as a mechanism to show chemical reactions (in a microwave), separation (using various filter sizes), and recycle. The students worked in teams to develop the most cost effective, "best" popcorn process to provide the customer with a satisfactory product (based on consumer surveys). The fall freshman orientation course was focused on chemical engineering. The paper will compare student surveys and outcomes for the popcorn project for high school students versus college freshmen.

414c A Project-Based Introduction to Data Analysis for Freshmen

Dr. Stephanie Farrell, Zenaida Otero-Gephardt, and Robert P. Hesketh. Chemical Engineering, Rowan University, Department of Chemical Engineering, 201 Mullica Hill Road, Glassboro, NJ 08028-1701

Hands-on experiences are an excellent vehicle to introduce students to design of experiments (DOE) and data analysis. DOE and data analysis techniques are cornerstones in the ISO and Six-Sigma requirements so prevalent in the chemical process industry. By linking hands-on experiences with concepts in experimental design and data analysis, students benefit from actively designing experiments, operating the equipment and analyzing the results to obtain a conclusion. This process strengthens students' understanding of engineering concepts and of how experimental design and data analysis are used to optimize experimentation and answer engineering questions. An automatic drip coffee maker is used to demonstrate the principles of experimental design and data analysis to Freshman Engineering students.

Food Experiments In The Unit Operations Lab

Mohammad Biswas, Chem. Eng. Dept. Auburn University, Auburn, AL, USA
Laura Kelly, Chem. Eng. Dept. Auburn University, Auburn, AL, USA
William E. Josephson, Chem. Eng. Dept. Auburn University, Auburn, AL, USA
David R. Mills, Chem. Eng. Dept. Auburn University, Auburn, AL, USA

Introduction

The junior level 'Unit Operations 1' laboratory course is found in most undergraduate chemical engineering curriculums. The objectives of this lab usually include development of technical writing abilities, reinforcement of concepts learned in previous lecture courses and the acquisition of hands-on experience. Experiments normally involve operation and analysis of small-scale industrial equipment (e.g., a centrifugal pump, a fluidized bed, an evaporator system) or analysis of some basic physical phenomenon (e.g., heat transfer via conduction, viscosity of Newtonian and non-Newtonian fluids). Food related experiments are typically not included in the curriculum. This can prove to be disadvantageous to students who later pursue a career in a food processing field.

Recently Auburn University's Chemical Engineering Department incorporated a food experiment into the schedule of the junior level lab course. The experiment was not pre-planned by the instructional staff; rather, it was presented to the students as an opportunity in experimental design. Student groups were told to "make up an experiment dealing with food that will teach you something". They were presented with equipment that they could use (e.g., an EASY-BAKE Oven) but were not put under any limitations other than that of practicality. A variety of experiments resulted from this open-ended assignment including a look into the relationship between boiling time and spaghetti stickiness, the design of a method to determine fat content of liquids and an analysis of cookie baking. One proposed experiment delved into the mysteries of making butter. This proposal was refined and used in later terms of the laboratory course. This paper describes the experiment proposals and reports on our experiences in incorporating the 'butter experiment' into the unit operations lab course.

Course Setup

Auburn University's CHEN 3820 course is titled 'Chemical Engineering Laboratory I' and is the junior level lab course found in most chemical engineering departments across the country. The two credit hour course meets each week for a one hour lecture and (on a different day) a three hour lab session. The lecture topics include safety issues, discussions of instrumentation and an extensive look at data analysis techniques. For the laboratory portion of the course the students are organized into groups and work as a team on about six experiments during the semester. Each experiment takes up two three hour lab sessions. The first is a pre-lab session in which the students are shown the relevant experimental apparatus, given instructions in its use and given objectives for the experiment (e.g., develop pump performance curves, verify affinity laws, etc.). The students may operate the equipment and, depending on the experiment, take some preliminary data (e.g., calibrate flowmeters). The student groups then produce a memo in which they describe their proposed experimental procedures and data analysis. The memos are graded and returned to the students before the next lab session, in which the groups will actually perform the experiment. A full-length report is due one week after the completion of the experiment. At the end of the semester each group gives a presentation covering one of their completed experiments.

The pre-requisites for the lab course include courses in transport phenomena. The experiments typically scheduled for CHEN 3820 reflect this, with an emphasis on fluid flow (e.g., an experiment in which students determine frictional losses in pipes and fittings) and heat transfer (e.g., an evaporator experiment in which students determine steam economies and analyze multiple effect systems).

The 'Food Engineering' Proposal

During the Fall 2006 semester there was an opportunity to introduce a new experiment into the CHEN 3820 mix. The general area of 'food engineering' was decided upon primarily because although food processing activities occupy a significant number of chemical engineering graduates nationwide there is usually little formal instruction in the area. After some internal discussion it was decided that the students should propose an experiment dealing with some aspect of food. The three groups taking CHEN 3820 that semester were told only to develop a food-related experiment that would be instructive. Each group was to produce an "Equipment & Objectives" handout that would mimic the ones they were given for established experiments. The groups also needed to determine the range of expected results for their proposed experiment.

The groups were presented with 2006 model EASY-BAKE Ovens but were under no obligation to use them. They were also pointed to the instructor's favorite food engineering references: The Joy of Cooking by Irma S. Rombauer and Marion Rombauer Becker and What Einstein Told His Cook: Kitchen Science Explained by Robert L. Wolke. The three groups (named Alpha, Bravo and Charlie) produced the experiments outlined below.

Group Alpha's Proposal

Group Alpha elected to investigate the college student's method of determining if spaghetti is completely cooked: throw a sample against a wall and, if it sticks, it's done. Their proposal was to boil spaghetti and remove some sample noodles from the water at regular time intervals. After recording their observations as to the noodles' color and firmness, Group Alpha initially proposed to mimic the college student practice and throw the noodles against a lab wall. They would eventually develop a graph depicting the relationship between noodle boiling time and probability of sticking to the wall.

Upon reflection, the members of Group Alpha realized that the likelihood of having poor repeatability of throws would prove to be a major obstacle to minimizing experimental error. Accordingly, they modified their method so that each 'throw' was, in fact, very repeatable. The new method involved simply dropping the spaghetti noodles from a standard height of four feet onto a piece of acrylic glass. The glass was turned onto its side immediately after the noodle was dropped on it. If the noodle slide off then it was judged to be a 'no-stick' noodle.

Group Bravo's Proposal

Group Bravo investigated the correlations between heat capacity and fat content of certain liquid foods. The initial idea for this experiment came after it was noticed from a table in a required text[1] that liquid foods high in fat were reported to have lower heat capacities. For example, the reported heat capacities for skim milk, whole milk and cream of 45-60% fat content were 3.98 – 4.02 kJ/kg·K, 3.85 kJ/kg·K and 3.06 – 3.27 kJ/kg·K, respectively. Group Bravo came up with the idea of using the Easy-Bake Oven as a fat content determining device for liquid foods.

They proposed that small amounts of various liquid foods of known fat content be placed in a pre-warmed oven for ten minutes. The increase in sample temperature could be determined and then used to determine relative heat capacities; samples with a greater temperature increase were judged to have lower heat capacities. A calibration curve could then be developed which related the temperature increases to food fat content.

When developing this experiment Group Bravo used a variety of liquids including coffee and tea (zero fat content), various milks and milk-like products (e.g., smoothies, fat content about 2-5%) and a sample of pure olive oil. In general the resulting data was as expected: the higher fat content liquids showed greater temperature increases. However, there was a problem with use of the developed calibration curve due to lack of sample data between the high fat content olive oil and the other materials. The group did recommend that future work include investigation of this intermediate area.

Group Charlie's Proposal

Group Charlie was the only group to propose an experiment using the Easy-Bake Oven in a traditional manner. The members of this group decided to use the cookie baking capabilities of the Easy-Bake to produce drying-rate curves. Group Charlie's proposal involved weighing cookies during the baking process to determine the amount of free moisture present as a function of time. Various combinations of cookie mix and water were used. It was found that the best results (i.e., production of drying-rate curves showing the various periods of drying) were obtained with samples that had the highest initial water content.

Another Proposal

The three experiments above were proposed during the Fall 2006 semester. Although a variety of factors prevented implementation of any of them during the following semester, it was felt that the exercise was still worthwhile and the students did, in fact, learn something about food engineering.

---

[1] Table A.4-1 in Transportation Processes and Separation Process Principles, 4th ed., Christie John Geankoplis, Prentice Hall, 2003.

During the Spring 2007 semester there was no scheduled food engineering experiment in the rotation of CHEN 3820. However, having heard of the notion from students who took the course in the Fall, one group (Group Foxtrot) asked for permission to propose a food engineering experiment and present it during their end of semester presentation activity. Permission was granted and the group ended up proposing the following butter experiment.

Butter can easily be made at home (or in a lab!) by vigorously agitating whipping cream. Just shaking some cream in a small container will also work and is a popular demonstration in elementary schools (this will be referred to as the elementary school method below). Group Foxtrot's idea was to investigate the effects that various co-solvent additives have on the time needed to make butter. The addition of salt, for instance, will result in a reduction of butter manufacturing time in addition to imparting flavor to the butter.

Group Foxtrot proposed using a pint of heavy whipping cream in a standard 4.5 quart KitchenAid mixer. After addition of various amounts of salt (½ tsp up to 3 tbsp) the butter making process could be monitored. The addition of higher amounts of salt were predicted to lower both the overall time for butter making and the time needed to achieve each of the various stages in the process (e.g., production of whipped cream).

There were variations to the above butter experiment that were also proposed by Group Foxtrot. Co-solvents other than salt such as sugar, cornstarch and calcium sulfate were suggested as being worthy of investigation. Additionally, Group Foxtrot recommended exploring the role of temperature in the making of butter; there should be a temperature effect on both the time needed to make butter and on the yield of butter.

Implementation of The Butter Experiment

A version of the butter experiment was scheduled as one of the regular CHEN 3820 experiments during the Summer 2007 semester. Initially we decided to go with a simple experiment: determine how addition of salt affects the time needed to make butter. The procedure described above was followed using heavy whipping cream and common table salt amounts of zero, ½ tsp, 1 tsp, 1 tbsp, 2 tbsp and 3 tbsp. The first groups that did this experiment in the summer found that butter making time could be reduced by about a third if the highest amount of salt was added.

Once some familiarity was gained with the experiment the procedures were tweaked and modified in order to investigate different phenomena. A group was instructed to use the elementary school method in addition to the stand mixer method and attempt to correlate the two. They were not able to do so; the variations in manually shaking apparently prevented any conclusions from being made. Other groups were asked to compare the use of regular and heavy (higher fat content) whipping creams. Those groups found that, as expected, the heavy cream would produce butter faster at all levels of salt addition. However, one group found that the difference narrowed with the addition of salt and the type of cream used was almost immaterial at the highest salt level. The other group doing this version of the experiment found no such effect.

Further Discussion and Conclusions

The incorporation of a food engineering experiment into the junior level lab course has proved to be popular with the students. This was true for both the "develop a food-related experiment" task undertaken by groups Alpha, Bravo and Charlie and the more scripted butter experiment done by later groups. The novelty factor of both experiments is high and that is an aid to popularity. But, even though they were having fun at the time, the students did take the experiments seriously. Groups Alpha, Bravo and Charlie came up with some interesting ideas while developing their food-related experiments. While the butter experiment was structurally similar to the other experiments in CHEN 3820, most groups did a superb job when presenting the background sections of their reports. Certainly all the students that performed the butter experiment gained knowledge in a subject area that they would not have otherwise been exposed to.

CHEN 3820 will have some version of a food experiment for the foreseeable future. During the course of the current semester (Fall 2007) the groups in CHEN 3820 will be performing a version of the butter experiment. They will most likely be investigating the effect of different co-solvents (salt vs sugar vs cornstarch) on the butter making process. The "develop a food-related experiment" experiment will also return, probably in 2008. In its next implementation the students will be asked to concentrate more on conversion processes such as baking. Consideration of questions such as "what's going on when bread is being made?" will be sure to pique the students' interest and increase their understanding of processes related to food.

414e Hot Potato: Edible Thermodynamics And Heat Transfer

Margot A.-S. Vigeant, Chemical Engineering, Bucknell University, Dana Engineering Building, Lewisburg, PA 17837

While most of our students easily see the connection between their chemical engineering education and their professional lives, many of them find it harder to connect their work to familiar experiences. The traditional solution to this problem is to give students more chemical experience through laboratories, but learning can also be aided by tying class material to everyday experiences. In this talk, I will share some of my ideas and experiences explaining thermodynamics and heat transfer ideas in terms of food, including the best way to cool bottles, the thermal efficiency of popcorn, and getting burnt (or not) by an oven. These ideas can work as demos or thought experiments, and most produce edible results.

Characterizing Rheological Properties of Newtonian and Non-Newtonian Fluid Food Products with a statistical method

Kyung C. Kwon, Floyd Tamara Smith, Nader Vahdat, Claudell Burnell, Tikia Allen, and Paul Jones*, and Yoonkook Park**

Department of Chemical Engineering, Tuskegee University, Tuskegee, Alabama 36088

kwonk@tuskegee.edu, Ph: (334) 727-8976, Fax: (334) 724-4188
*P&G, Cincinnati, Ohio, ** Chemical System Engineering Department, Hongik University, South Korea

A tank-tube viscometer and its novel Newtonian and non-Newtonian viscosity equations were developed to determine flow characteristics of Newtonian and non-Newtonian fluid food products. Rheological behaviors of 60 w% sucrose aqueous solution as a reference/calibration fluid solution for the tank-tube viscometer were characterized with the novel viscosity equations of the tank-tube viscometer.

Several series of experimental data of amounts of 60-w% sucrose aqueous solution drained from the reservoir at various drain durations were applied to the non-Newtonian viscosity equation to calibrate the tank-tube viscometer, and then the same experimental data were applied to the Newtonian viscosity equation to obtain dynamic viscosity values of the solution with the linear least squares method. Dynamic viscosity values of 60 w% sucrose aqueous solution, determined with the calibrated tank-tube viscometer and its Newtonian viscosity equation, were compared with the literature values.

The objectives of this presentation are for our students to be familiar with statistical analysis of experimental data, to understand relative accuracies of experimental data with correlation coefficients obtained from statistical analysis of experimental data, to be able to recognize and omit some bad experimental data deviated noticeably from the rest of experimental data set, and to identify possible experimental error sources.

KEYWORDS: linear least-squares method, Unit Operations Laboratory class, tank-tube viscometer, Newtonian viscosity equation, non-Newtonian viscosity equation, sucrose aqueous solution

INTRODUCTION

Rheological data are indispensable in the design of pumps, compressors, heat exchangers, packed columns, fluidized beds, distillation columns, pipelines, decanters and many others. Rheological properties are utilized in the quality control of consumer products. Certain materials are more acceptable to consumers if they have specific rheological properties [1].

In order to fully determine viscosity values of a fluid, careful experimentation is required. Temperature and pressure affect viscosity of the fluid. Effects of pressure on viscosity of a fluid are negligible at pressures less than 40 atm. However, viscosity of nongaseous fluids is not significantly affected by pressure [2]. Viscosity of Newtonian fluids is determined by measuring the shear stress at one known shear rate, whereas that of all non-Newtonian fluids is determined at more than one shear rate.

There are two main types of viscometer: One type of a viscometer is designed so that a fluid flows through a tube, channel or orifice, while the other type of a viscometer is constructed so that a fluid is sheared between moving surfaces [3 - 5]. Several forms of viscometers have been constructed for the determination of viscosity from the knowledge of the pressure drop-flow rate relationship for a tube. The flow may be achieved simply as a result of gravitational head. This type of viscometer is not very suitable for measurements with non-Newtonian fluids, not only because of the non-uniform shear stress across the vertical tube section, but also because the rate of flow through the vertical tube varies with time as the level falls. To overcome these shortcomings, a constant velocity of a fluid through a vertical tube is maintained by pressure from a gas reservoir. For this type of a viscometer, the length to the diameter ratio of a tube normally exceeds 50 to eliminate the end effects of a tube [5].

Although a few capillary tube viscometers (sometimes called extrusion rheometers) are commercially available, it is common practice to construct them in the laboratory. Several capillary tubes with a range of lengths and diameters are needed. Tubes with internal diameters from 0.794 mm to 6.35 mm should probably cover normal needs [6].

Many researchers used various types of viscometers to measure viscosity of fluids. James et al [7] reported a consistent set of high precision viscosity data for water at various temperatures and for 20, 30 and 40 per cent sucrose aqueous solutions. These were obtained using a glass capillary viscometer with an extensively flared capillary to neglect the kinetic energy correction term for this viscometer. These data are recommended as reliable standards for capillary viscometer calibration.

This paper describes a tank-tube viscometer and its novel viscosity equations developed for application to Newtonian fluids as well as non-Newtonian fluids. The test materials include a sucrose aqueous solution that exhibits Newtonian behavior, and well-known Hershey's syrup that exhibit shear-thinning non-Newtonian behavior. The 60-w% sucrose aqueous solution was used as a calibration/reference fluid for the viscometer.

THEORY

Newtonian Fluids

Rheological characteristics of Newtonian fluids can be determined with the Newtonian and Non-Newtonian viscosity equations of the tank-tube viscometer. The detailed development of Newtonian and Non-Newtonian viscosity equations was described in the papers (Kwon and Pallerla, 2000; kwon et al., 2006). Dynamic viscosity values of Newtonian fluids are determined with the Newtonian viscosity equation, as shown in Equations (1) and (2). Rheological behaviors of Non-Newtonian fluids are characterized with the non-Newtonian viscosity equation, as shown in Equation (3).

$$-\ln\left(1-\frac{m}{(H+L)\pi R^2\rho}\right)=\left(\frac{gR_0{}^4\rho}{8\mu R^2 L}\right)(t) \tag{1}$$

Because the length of a vertical tube attached to the reservoir is much longer than the initial height of a fluid in the reservoir and the amounts of the fluid drained from the reservoir are relatively small, Eq. (1) can be simplified to obtain Eq. (2).

$$\left(\frac{m}{(H+L)\pi\rho}\right)\cong\left(\frac{gR_o{}^4\rho}{8\mu L}\right)(t) \tag{2}$$

$$\ln\left(\frac{m}{\pi R_o^3\rho t}\right)\cong\frac{1}{n}\ln\left[\frac{\rho g R_o(H+L)}{2L}\right]+\ln\left[\frac{n}{1+3n}\left(\frac{1}{k}\right)^{\frac{1}{n}}\right] \tag{3}$$

The left-hand side of Eq. (3) contains the amount of a fluid drained from the tank-tube viscometer at a given drain duration, while the right-hand side of the same equation has a term containing the length of a drain tube. The left-hand side of Eq. (3) is denoted as *Y*, as shown in Eq. (3-1), and a part of the first term on the right-hand side of Eq. (3) is denoted as *X*, as in Eq. (3-2).

$$Y = \ln\left(\frac{m}{\pi R_o^3 \rho t}\right) \qquad (3\text{-}1)$$

$$X = \ln\left[\frac{\rho g R_o (H + L)}{2L}\right] \qquad (3\text{-}2)$$

Plotting Y against X allows flow behavior index (n) and consistency coefficient (k) of a fluid to be determined with the slope and the intercept of the linear regression line, respectively. The non-Newtonian viscosity equation, Eq. (3), is reduced to the Newtonian equation, Eq. (2), when the non-Newtonian viscosity equation is applied to Newtonian fluids. The non-Newtonian viscosity equation, Eq. (3), can be applicable to non-Newtonian fluids as well as Newtonian fluids

where

μ	viscosity of a fluid, Pa-s
L	length of a vertical tube, m
ρ	density of a fluid, kg/m^3
g	acceleration of gravity, 9.81 m/s^2
H	initial level of a fluid in a reservoir tank or height of a reservoir tank, when the reservoir tank is filled up, m
R	inside radius of a cylindrical reservoir tank, m
R_o	inside radius of a vertical tube, m
t	drain duration, s
m	accumulated amount of a fluid drained at t, kg
k	consistency index of shear-thinning fluids, Pa-s^n
n	flow behavior index of shear-thinning fluids

EXPERIMENTAL SETUP

A wide variety of viscometers such as capillary type, glass tube type, rotational type, falling ball type, cup-type, and oscillatory type [5 – 6] are available for the measurement of viscosity. An inexpensive viscometer [13] was fabricated, as shown in Figs. 1 and 2. The viscometer consists of a rectangular reservoir and a set of long vertical drain tubes. The equivalent radius and the height of the reservoir are 3.082 cm and 24.377 cm, respectively. The radius of the vertical drain tube is 0.1051 cm, and the lengths of the vertical drain tubes are 20, 25, 30, 35, and 40 cm. The ratios of the length to the diameter of the vertical drain tubes of the fabricated tank-tube viscometer are 93 - 186 to neglect the end effects on completion of final velocity distribution of a laminar flow in the entrance region of the vertical drain tubes.

The reservoir is made of transparent Plexiglas, whereas the drain tubes of the viscometer are made of stainless steel. The vertical drain tubes are connected at the bottom of the reservoir. Sucrose aqueous solution containing 60-w% sucrose is chosen as a reference/calibration fluid to test both the fabricated tank-tube viscometer and its viscosity equations, since sucrose is soluble in water, and does not inflict any health hazards. Sucrose is also used widely for various food products. The concentration of sucrose in water is 60 w%. An electronic balance placed beneath the end of the vertical drain tube is used to measure accumulated amounts of aqueous solutions drained from the reservoir at a given drain duration. A circulator [3] is used to control temperatures of both the test fluid and the vertical drain tubes of the viscometer

EXPERIMENTAL PROCEDURE

An electronic balance is placed beneath the end of the vertical drain tube of the tank-tube viscometer, where temperatures of both the viscometer and a test fluid are regulated with a circulator, as shown in Fig. 1. The ends of the vertical drain tubes are closed with a plastic cap to stop the flow of the test fluid, after the drain tubes are filled with the test fluid by using a syringe. Thereafter, the drain tubes filled with the test fluid are attached to the rectangular reservoir tank, and then the reservoir tank is filled with the test fluid up to the desired initial height H of a level of the test fluid in the reservoir tank.

An empty receiving beaker is placed on the electronic balance, and tared to read accumulated amounts of a test fluid out of the viscometer. Reading accumulated amounts of the test fluid drained from the reservoir tank at given drain durations starts with an electronic balance, a data acquisition software and EXCEL spreadsheet, when the plastic cap from the end of the drain tube is removed to allow the test fluid to flow into the beaker on the electronic balance. Several series of experimental data of amounts of the test fluid drained from the reservoir at various drain durations are obtained with various lengths of the drain tube of the viscometer.

After the tank-tube viscometer is rinsed with distilled water for the next experiment, the reservoir tank is dried with paper towel, whereas the vertical drain tube is dried with acetone.

Figure 1. A schematic diagram of a tank-tube viscometer

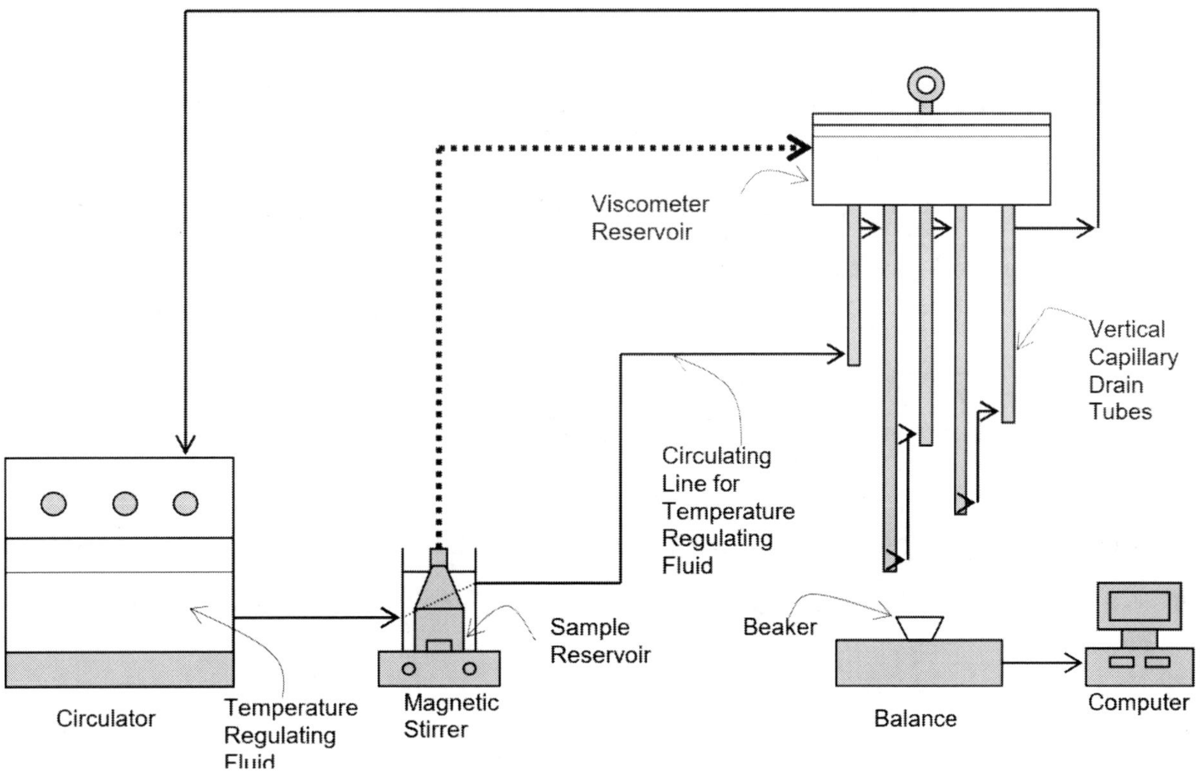

Figure 2. Single vertical capillary viscometer

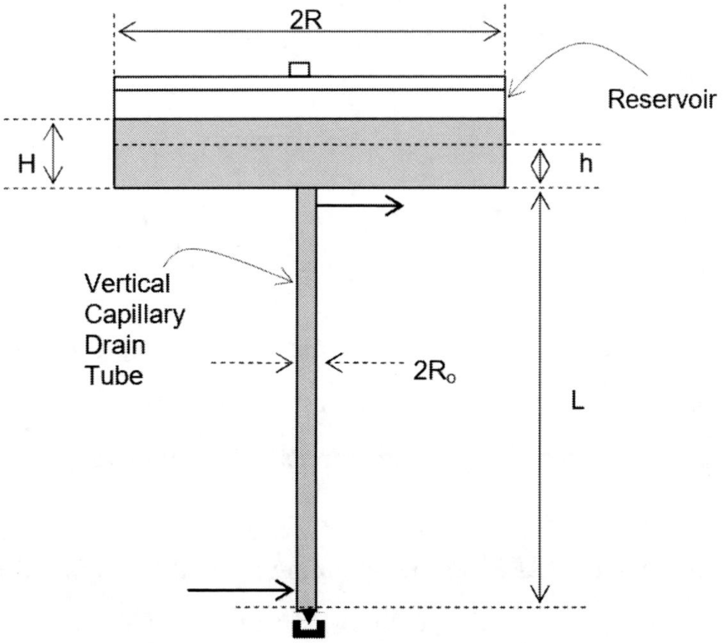

ANALYSIS OF EXPERIMENTAL DATA

Experimental data of accumulated amounts of a test fluid drained from the reservoir of a tank-tube viscometer at various drain durations are obtained, using an electronic balance, a data acquisition software, and EXCEL spreadsheet. The viscometer was kept at a desired temperature with a circulator, which was equipped with a cooling and heating element. Several series of experimental data were obtained using the test fluid under controlled experimental conditions such as fluid compositions and temperatures of the test fluid.

Several series of experimental data of amounts of a Newtonian test fluid drained from the reservoir at various drain durations are applied to the Newtonian viscosity equation (NVE) (Eq. 1). Left-hand side (LHS) values of the NVE are plotted against drain durations. A slope of the best-fit straight line passing through the origin of the rectangular coordinates is obtained through the linear least squares method. A viscosity value is calculated by substituting a density value of the test fluid, the diameters of both the reservoir tank and the vertical drain tube, and the length of the vertical drain tube into the slope value (see Eq. 1).

Several series of experimental data of amounts of a shear-thinning test fluid drained from the reservoir at various drain durations are applied to the non-Newtonian viscosity equation (NNVE) (Eq. 3). LHS values against the right-hand-side (RHS) term containing a variable length of the drain tube in the NNVE are plotted. A slope and an intercept of the best-fit straight line are obtained through the linear least squares method. A flow index n value and a consistency k value of the shear-thinning test fluid are calculated with the slope value and the intercept value, respectively.

RESULTS AND DISCUSSION

Sucrose Aqueous Solution

Figures 3 through 8 are obtained with experimental data at atmospheric pressure. Quintuple experimental data of accumulated drain amounts of 60-w% sucrose aqueous solution at various drain durations, as shown in Fig. 3, are applied to the non-Newtonian viscosity equation (NNVE) (Eq. 3) to calibrate the viscometer, as shown in Fig. 4. Left-hand side (LHS) values against the right-hand-side (RHS) term containing a variable length of the drain tube in Eq. 3 are plotted, as shown in Fig. 4. The average LHS values of the NNVE and their standard deviations for the 20-cm drain tube, the 25-cm drain tube, the 30-cm drain tube, the 35-cm drain tube, and the 40-cm drain tube are shown in Tab. 1. The slope value and the intercept value of the plot are 1.00001 and -0.42101, respectively. The flow behavior index n and the flow consistency *k* of 60-w% sucrose aqueous solution at 25.97°C are obtained from the slope value and the intercept value. The flow behavior index *n* and the flow consistency *k* of 60-w% sucrose aqueous solution at 25.97°C are 1.000 and 38.088 mPa-s with the calibrated viscometer and its NNVE.

Table 1. Average left-hand side (LHS) values of non-Newtonian viscosity equation (NNVE), and their standard deviations and standard deviation percentages, using quintuple experimental data of 60-w% sucrose aqueous solution at 25.97°C.

Drain Tube Length, cm	Average LHS Value, 1/ln(s)	Standard Deviation for LHS Value, 1/ln(s)	Standard Deviation %
20	4.5658	0.0069	0.1520
25	4.2456	0.0033	0.0767
30	4.3732	0.0094	0.2139
35	4.3008	0.0044	0.1030
40	4.4591	0.0044	0.0976

Quintuple experimental data of accumulated drain amounts of 60-w% sucrose aqueous solution at various drain durations, as shown in Fig. 3, are also applied to the Newtonian viscosity equation (NVE) (Eq. 1), after the viscometer is calibrated with the NNVE and the 60-w% sucrose aqueous solution as a calibration fluid. Left-hand side (LHS) values of the NVE are computed with accumulated amounts of 60-w% sucrose aqueous solution drained from the reservoir at various drain durations. Average LHS values of the NVE are plotted against drain durations (see Fig. 5). The slope of the best-fit line is obtained through the linear least squares method. The slope of this plot increases with decreased lengths of the vertical drain tube of the viscometer. The viscosity value is calculated from the slope of the best-fit line by substituting the density value of the 60-w% sucrose aqueous solution as well as the diameter and the lengths of the drain tube of the viscometer into the slope value.

Figure 3. Average drain amounts of 60-w% sucrose aqueous solution against drain durations at 25.97°C, using quintuple experimental data.

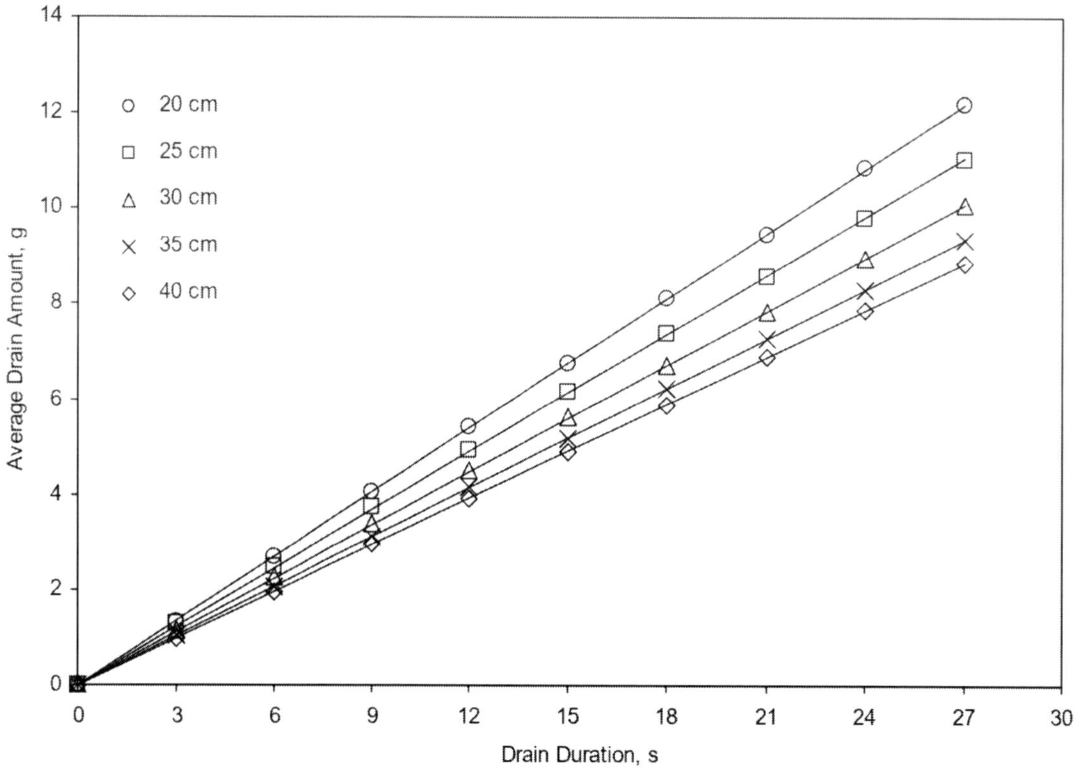

Viscosity values of 60-w% sucrose aqueous solution obtained with quintuple experimental data at 25.97°C are compared with that from the literature [14] (see Fig. 6). The error percentages of the experimental dynamic viscosity values of the 60-w% sucrose aqueous solution obtained from the calibrated viscometer in comparison with the literature viscosity value for the 20-cm drain tube, the 25-cm drain tube, the 30-cm drain tube, the 35-cm drain tube, and the 40-cm drain tube are shown in Tab. 2. The viscosity values obtained from this study are very agreeable with that from the literature viscosity value at 25.97°C with an average error of -0.48 % (see Fig. 6) in comparison with the literature accuracy ±1 % [4]. The average viscosity values and their standard deviations of quintuple experimental data for the 20-cm drain tube, the 25-cm drain tube, the 30-cm drain tube, the 35-cm drain tube, and the 40-cm drain tube are shown in comparison with the literature value 38.0897 mPa-s, as shown in Tab. 2 and Fig. 6.

Standard deviation % of quintuple amounts of 60-w% sucrose aqueous solution drained from each vertical drain tube of the viscometer were against various drain durations, as shown in Fig. 7. Standard deviations of drain amounts decrease with increased drain durations. Initially, high standard deviations of drain amounts were observed for the 20-cm drain tube, the 25-cm drain tube, and the 30-cm drain tube.

Table 2. Average dynamic viscosity (DV) values of quintuple experimental DV values, their standard deviations (SDs) and percentages of SDs, and error % of average experimental DV values in comparison of the literature DV value 38.0897mPa-s of 60-w% sucrose aqueous solution at 25.97°C.

Drain Tube Length, m	Average Experimental DV, mPa-s	SD of Average Experimental DV, mPa-s	SD % of Average Experimental DV	Error % of Average Experimental DV from Literature DV
20	38.1930	0.2561	0.6706	0.2710
25	37.4642	0.5699	1.5211	-1.6422
30	37.6619	0.7019	1.8637	-1.1233
35	38.0566	0.1416	0.3720	-0.0871
40	38.1516	0.1311	0.3438	0.1625

Figure 4. Quintuple experimental data of drain amounts at various drain durations, applied to non-Newtonian viscosity equation (NNVE) for the calibration of the viscometer to determine behavior index *n* and consistency index *k* of 60-w% sucrose aqueous solution at 25.97°C.

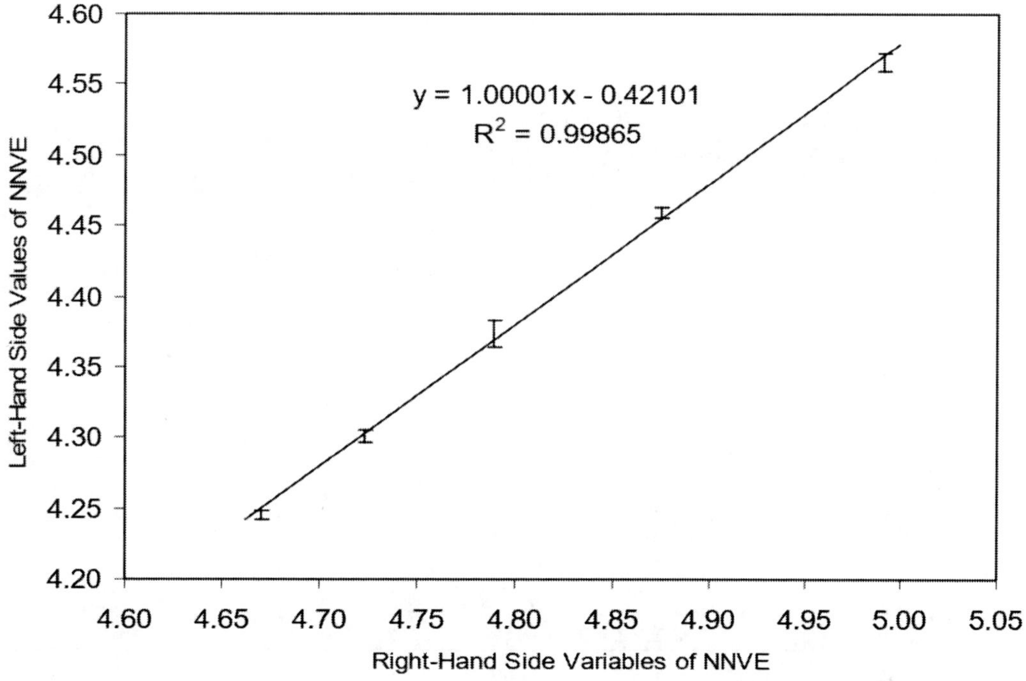

Figure 5. Left-hand side (LHS) values of Newtonian viscosity equation (NVE) for various drain tubes and durations with 60-w% sucrose aqueous solution at 25.97°C.

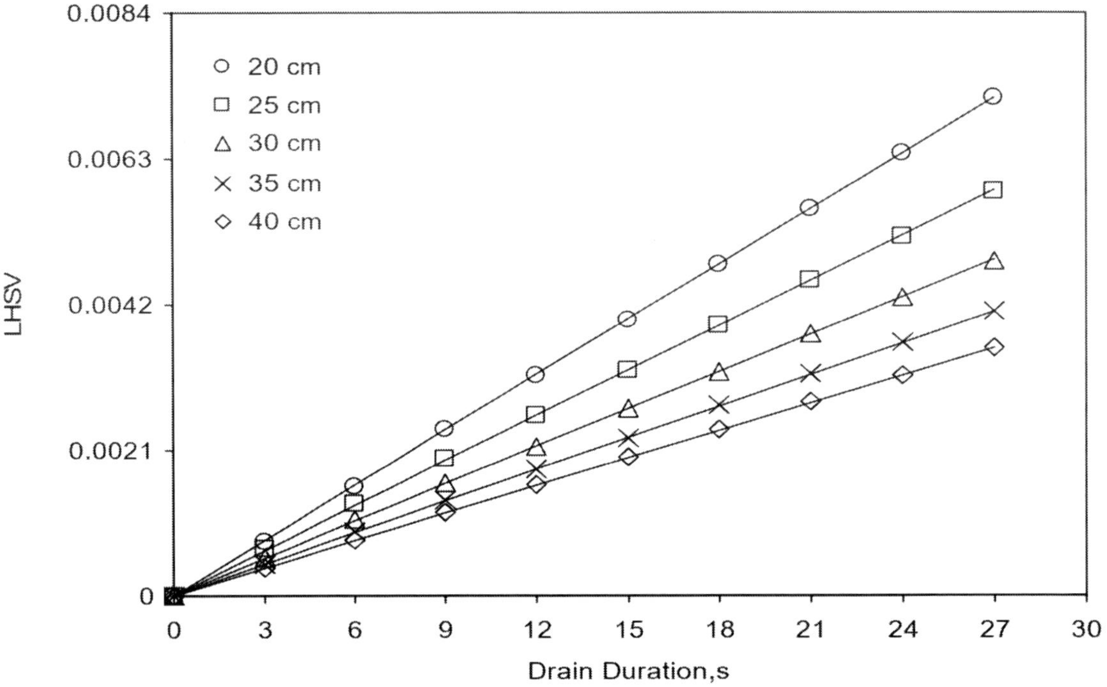

Figure 6. Comparison of average quintuple experimental dynamic viscosity values with literature viscosity value of 60-w% sucrose aqueous solution at 25.97°C.

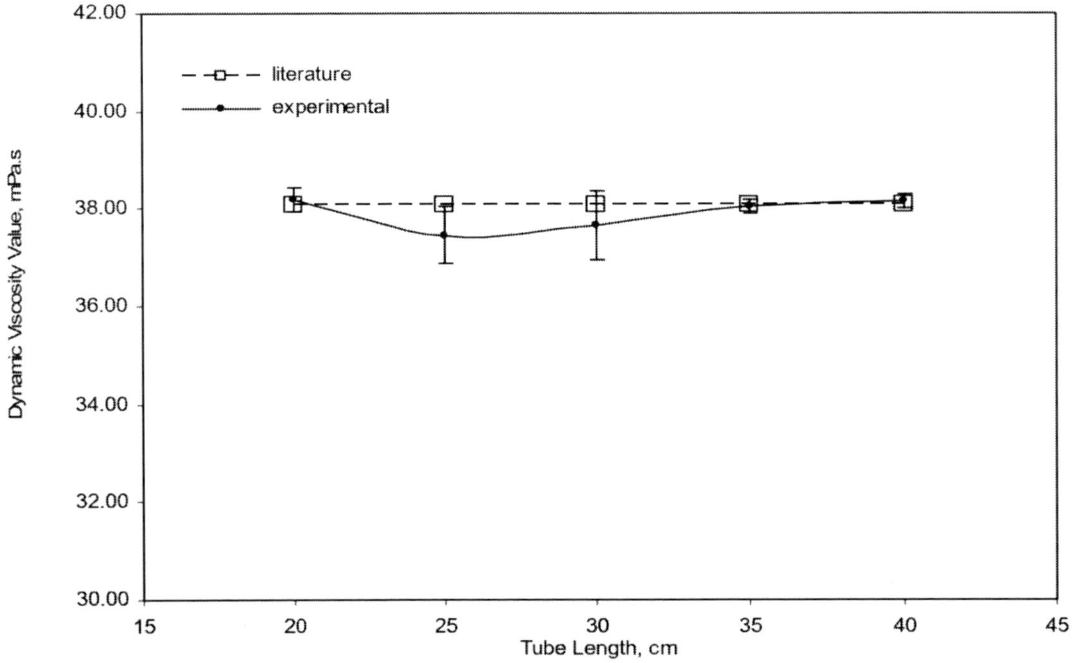

Figure 7. Standard deviation % (STDEVP) of drain amounts of 60-w% sucrose aqueous solution against drain durations at 25.97°C, using quintuple experimental data.

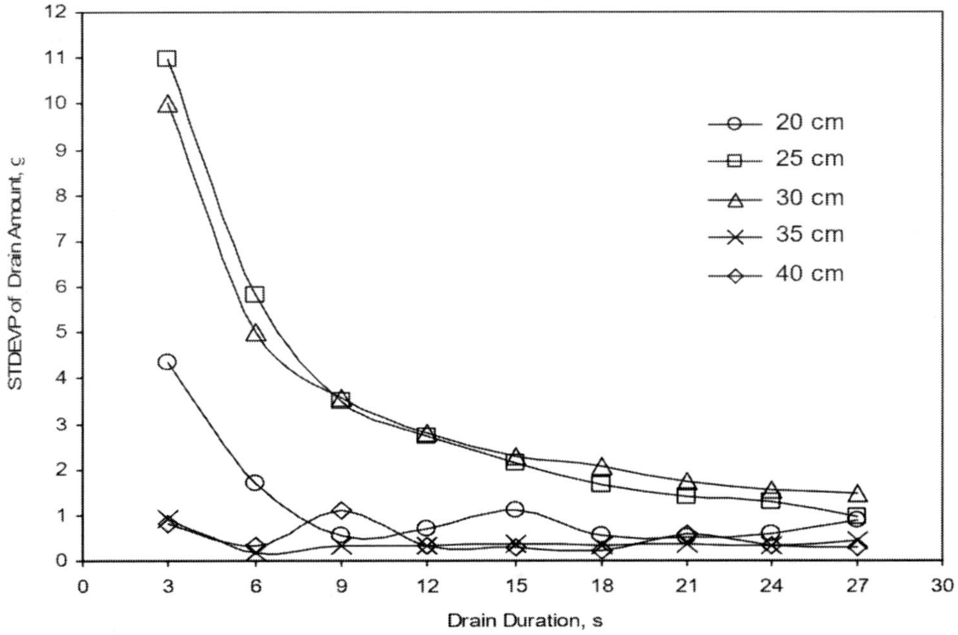

Hershey's Syrup

Quintuple experimental data of accumulated drain amount vs. drain duration for Hershey's syrup at 25.97°C are applied to the left-hand side (LHS) and the right-hand side (RHS) variable term of the Non-Newtonian viscosity equation (NNVE) (Eq. 3), as shown in Fig 8. The average LHS values, and their standard deviations and standard deviation % for quintuple experimental data obtained from the 20-cm drain tube, the 25-cm drain tube, the 30-cm drain tube, the 35-cm drain tube and the 40-cm drain tube at 25.97°C are shown in Tab. 3.

Table 3. Average quadruple left-hand side (LHS) values of non-Newtonian viscosity equation (NNVE), and their standard deviations (SDs) and percentages of standard deviations for average LHS values with Hershey's syrup.

Drain Tube Length, cm	Average LHS Value, ln(1/s)	SD of Average LHS Value, ln(1/s)	SD % of Average LHS Value
20	0.9961	0.0116	1.1679
40	0.5758	0.0144	2.5036
30	0.7803	0.0108	1.3781
35	0.6346	0.0190	2.9919
25	0.9225	0.0205	2.2237

Figure 8. Quadruple experimental data of drain amounts at various drain durations, applied to non-Newtonian viscosity equation (NNVE) to determine flow behavior index *n* and flow consistency index *k* of Hershey's syrup at 25.97°C.

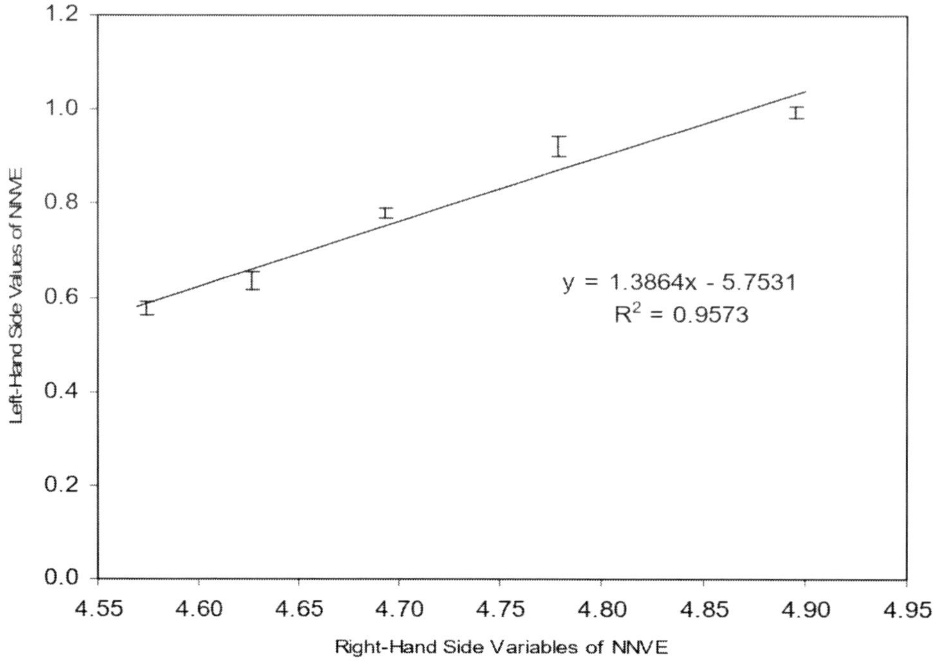

The slope value and the intercept value of this plot are obtained as shown in Fig. 8. The flow behavior index *n* and the flow consistency *k* of Hershey's syrup are obtained from the slope values and the intercept values of the plot. The average flow behavior index, and its standard deviation and standard deviation % at 25.97°C are 0.7218, 0.0212, and 2.9376, respectively, whereas the average consistency coefficient, and its standard deviation and standard deviation % at 25.97°C are 2.1825 Pa.s^n, 0.0874 Pa.s^n, and 4.0058, respectively, as shown in Tab. 4.

Table 6. Average flow behavior index and consistency coefficient of Hershey's syrup at 25.97°C, using quintuple experimental data.

Flow Behavior Index, *n*	Average Flow Behavior Index (*n*)	Standard Deviation of Average *n*	Standard Deviation % of Average *n*	Consistency Coefficient (k), Pa.s^n	Average Consistency Coefficient (k), Pa.s^n	Standard Deviation of Average *k*, Pa.s^n	Standard Deviation % of Average *k*
0.7582				2.0298			
0.7154				2.2354			
0.7198				2.1917			
0.7119				2.2165			
0.7036	0.7218	0.0212	2.9376	2.2394	2.1825	0.0874	4.0058

CONCLUSIONS

A tank-tube viscometer and its Newtonian and non-Newtonian viscosity equation were developed to determine flow characteristics of non-Newtonian fluids. Sucrose aqueous solution containing 60-w% sucrose as a reference/calibration fluid was chosen to calibrate the fabricated tank-tube viscometer with its non-Newtonian viscosity equation. The dynamic viscosity values of 60-w% sucrose aqueous solution were obtained from both this calibrated tank-tube viscometer and its Newtonian viscosity equation. The viscosity values of 60-w% sucrose aqueous solution obtained from this study are very agreeable with the literature viscosity value at 25.97°C with an average error of -0.48 %, which is way below the literature accuracy ±1 %.

Non-Newtonian behaviors of Hershey's syrup at 25.97°C were characterized with the various lengths of the drain tube of the tank-tube viscometer calibrated with 60-w% sucrose aqueous solution and its non-Newtonian equation to test capabilities of the tank-tube viscometer in determining flow characteristics of non-Newtonian fluids, and reproducibility of experimental data.

NOMENCLATURE

N	kg-m/s^2
Pa	N/m^2
mPa-s:	Pa-s x 10^{-3}
g:	acceleration of gravity, 9.81 m/s^2
H:	initial height of the level of a fluid in a reservoir tank, m or mm
L:	length of a vertical tube, mm
k:	consistency index of shear-thinning fluids, Pa-s^n
LHS:	left-hand side
m:	accumulated amount of a fluid drained at t, kg
n:	flow behavior index of shear-thinning fluids
R:	inside radius of a cylindrical reservoir tank, m or mm
R_o:	inside radius of a vertical tube, m or mm
t:	drain duration, s
μ:	dynamic viscosity of Newtonian fluids, Pa-s or mPa-s
ρ:	density of fluids, kg/m^3

ACKNOWLEDGMENTS

The authors thank for Chemical Engineering students to conduct experiments and obtain experimental data for this research project. The P&G curriculum grant supports this research project.

REFERENCES

[1] Shackleton ME and Green RG, On-line viscometer for measurement in the range 1 to 100 Pa, Measurement Science & Technology, 4 (1993) 395-404.

[2] Brodkey RS and Hershey HC, Transport phenomena: A Unified Approach, New York, McGraw-Hill, 1988.

[3] Barnes HA, Schimanski H, and Derek B: 30 years of progress in viscometers and rheometers, Applied Rheology 9 (1999) 69 – 76.

[4] Barnes HA: An examination of the use of rotational viscometers for the quality control of non-Newtonian liquid products in factories, Applied Rheology 11 (2001) 89 – 101.

[5] Coulson JM and Richardson JF, Chemical Engineering: Chemical Reactor Design, Biochemical Reaction Engineering including Computational Techniques and Control, Vol 3, Second Edition, Pergamon Press, New York, 1984.

[6] Skelland AHP., Non-Newtonian Flow and Heat Transfer, John Wiley & Sons, Inc. New York London Sydney, 1967.

[7] James C J, Mulcahy DE and Steel BJ, Viscometer calibration standards: viscosities of water between 0 and 60 °C and of selected aqueous sucrose solutions at 25 °C from measurements with a flared capillary viscometer, J. Phys. D: Appl. Phys., 17 (1984) 225-230.

[8] Sirivat A, Rajagopal KR and Szeri AZ, An experimental investigation of the flow of non-Newtonian fluids between rotating disks, *J. Fluid* Mech., 186 (1988) 243-256.

[9] Jimeneza JA and Kostic M., A tank-tube computerized viscometer/rheometer, Rev. Sci Instrum. 65 (1994) 229-241.

[10] Bird RB, Stewart WE, and Lightfoot EN, Transport Phenomena, John Wiley & Sons, 1960.

[11] Kwon KC, Pallerla S and Roy S, Experiments on Viscosity of Aqueous Glycerol Solutions using a Tank-Tube Viscometer, Chemical Engineering Education, Pg. 232-237, Summer 1999.

[12] Kwon KC and Pallerla S, Viscosity of Glycerol and its Aqueous Solutions Measured by a Tank-Tube Viscometer, Chemical Engineering Communications, 183 (2000) 71-97.

[13] U. S. Patent Application Number 11/439,699, 2006.

[14] Dean, J. O., Lange's Handbook, 14th Edition, McGraw Hill, 1992

436a Session Introduction - New Experiments/approaches

Marina Miletic, Department of Chemical and Biomolecular Engineering, University of Illinois Urbana-Champaign, 206 Roger Adams Lab, Urbana, IL 61801 and John Corn, Chemical & Biomolecular Engineering, Ohio State University, 416 Koffolt Lab, 140 W. 19th Ave., Columbus, OH 43210.

Session Introduction - 5 Minutes

436b Lactose Intolerance: a Simple Study of Kinetic Parameters Governing Lactose Conversion in Dairy Products

Jimmy L. Smart, Chemical & Materials Engineering, University of Kentucky, 4810 Alben Barkley Drive, Paducah, KY 42001

Lactose intolerance is a condition suffered by an estimated 50 million Americans. Certain ethnic and racial populations are more widely affected than others. As many as 75 percent of all African-American, Jewish, Native American and Mexican-American adults, and 90 percent of Asian-American adults are lactose intolerant (www. gastro.org). Lactose, or milk sugar, is a naturally occurring dissacharide found in dairy products and is normally converted to simpler sugars in the human digestion process. The lactase enzyme (β-Galactosidase) hydrolytically cleaves the lactose molecule to form the monosaccharides, galactose, and glucose. This enzyme is secreted by cells along the lining of the small intestine during normal digestion. In some people, this enzyme is absent or is present in reduced concentrations. Interestingly, most other mammals completely stop producing lactase at weaning, and are thereafter intolerant of milk. If, during the human digestion process, lactose is not converted to these simpler sugars, people may suffer symptoms of bloating, nausea, cramps, and diarrhea. Hence the term, lactose intolerance. Fortunately, it is generally not life-threatening and can be controlled through proper diet. Also, there are various commercial lactase products available on the market such as Lactaid™ and Dairy-Ease™. Lactase-deficient people can take these supplements orally with their meal to aid in digestion of dairy products. In the interest of appealing to high school outreach programs and funding-challenged college undergraduate departments, an inexpensive means was sought to monitor the kinetic reaction of conversion of lactose to glucose in milk. The obvious choice to monitor the reaction would be to adopt a standard glucose analyzer (Beckman, YSI et al.) commonly found in research and hospital laboratories. However, this complete set-up would cost over $10,000. Other methods considered were based upon refractive index, freeze point depression, use of a lactose probe, chromatography, and a colorimetric procedure. Finally, an inexpensive blood glucose monitoring meter (cost: $60), commonly used by diabetics to measure glucose levels in blood, was selected as a candidate for monitoring lactose conversion in milk. The commercial lactase enzyme, Lactaid™, was added to skim milk and lactose conversion to glucose was studied as a function of time with the blood glucose monitoring meter. Conditions of temperature, enzyme concentration, and pH were investigated. Michaelis-Menten parameters for these enzymatic reactions were also evaluated. These results will be presented at the 2007 AIChE Meeting.

436c An Inexpensive And Versatile Drying And Convective Heat Transfer Apparatus

David R. Mills, Chemical Engineering, Auburn University, 212 Ross Hall, Auburn, AL 36849-4827

Versatile and robust equipment for the study of convective heat transfer and drying of porous media can be easily and cheaply constructed using commonly available heating and cooling ducting. For the drying experiment, inexpensive USB temperature/humidity sensors and data acquisition boards make it possible to collect accurate data not only in the dryer, but also at various depths in an insulated tray containing porous solids so as to perform an in-depth analysis of the drying process. Electronic balances with RS-232 interfaces are also now available at reasonable prices and are easily interfaced to give mass data. Using this equipment, remarkably good data can be collected to calculate heat transfer coefficients to a tray, as well as effective diffusivity in the second falling rate regime. This is an excellent experiment for the application of mass and energy balances to a real system. Mass data is differentiated using TableCurve 2-D ®software and yields drying curves that illustrate constant, first falling rate and second falling rate periods. This experiment is easily adapted to an "on-line" format since sensors are interfaced to a PC and the equipment is reliable with few moving parts. Other simple transport experiments using the same apparatus and instrumentation include the study of transient center-line temperatures and heat transfer coefficients to materials of various geometries under various flow conditions. A compact design for this unit will be presented along with sample data and student performance.

436d Efficiently Expanding A Unit Operations Laboratory

David L. Silverstein, University of Kentucky, 4810 Alben Barkley Drive, P.O. Box 7380, Paducah, KY 42002-7380 and Jimmy L. Smart, Chemical & Materials Engineering, University of Kentucky, 4810 Alben Barkley Drive, Paducah, KY 42001.

Part of the continuous improvement process common to all chemical engineering programs is the need to update and improve laboratory experiences for undergraduate students. At the University of Kentucky Extended Campus in Paducah, much of the existing equipment is bench scale due to space constraints not uncommon to engineering programs. Most lab space is single-story with limited utility capacity. Existing equipment is fairly new (less than 10 years old), but consists of prefabricated experiments purchased from educational laboratory suppliers.

One particular area of concern to faculty in the chemical engineering program was the lack of a larger scale heat transfer experiment. The existing experiment included a shell and tube exchanger with a tube length of about four inches and diameter of three inches. Upgrading this experiment to facilitate the program's drive to incorporate laboratory experiences throughout the curriculum was a priority, but budget and space constraints were a concern. The first part of this paper describes the process leading to purchase and installation of an industrial scale heat exchanger at a "bargain" price using modern alternative purchase sources (i.e. internet-based brokers and auction houses). Reminders of key considerations in the selection/specification process and infrastructure requirements when selecting an exchanger for educational use will be discussed. The second part of this paper describes why we believe the shell and tube heat exchanger to be the single most valuable investment that can be made in an undergraduate laboratory. Experiments useful in classes from introductory course to process control will be presented, with details provided for selected experiments. This includes standard heat transfer experiments in addition to experiments and applied design opportunities in process control, fluid dynamics, mass balances, thermodynamics, and equilibrium staged separations. A selection of previously published laboratory experiences are cited to supplement the novel applications provided.

436e Complex Biological Concepts In A Hands On Laboratory Course For Engineers

Greg Sitton and Friedrich Srienc. Biotechnology Institute / Chemical Engineering, University of Minnesota, 1479 Gortner Ave, 140 Gortner Labs, St Paul, MN 55108

Meaningful biological experiments often require elaborate biological preparation techniques and significant time investment that preclude untrained engineering students to address relevant problems in a laboratory setting. While biologists typically focus on the structural components of a cell, the engineering interest is usually related to the dynamics of a cell system since the kinetics determines the process design. A good example for this is cell growth and the heterogeneity of a cell population in a growing culture. We have developed a cell preparation system that carries out the labor intensive cell preparation task in an automated way at frequent sampling intervals and injects the prepared cell samples into a flow cytometer. With this device samples can be withdrawn from a reactor and prepared for analysis by untrained persons in a highly reproducible way. We have implemented this device for a laboratory course for undergraduate chemical engineering students that are not specifically trained in biological techniques. The students were able to generate detailed data on growing cell cultures within the time frame of a typical Unit Operations laboratory experiment. They were able to focus their data analysis on the growth dynamics and how it is affected by culture heterogeneity due to the operation of the cell cycle and due to genetic instabilities. The approach is very valuable as laboratory experiments can be carried out that were impossible before. Students get exposed to complex biological concepts in a hands-on laboratory environment that complement the bioengineering lecture courses. It is planned to develop additional laboratory experiments on cellular dynamics for undergraduate students using this platform.

436f Process Control Laboratory Using Unit Ops Equipment And Honeywell Plantscape

Laura P. Ford and Christi L. Patton. Chemical Engineering, University of Tulsa, 600 S. College Ave., Tulsa, OK 74104

The University of Tulsa has recently revised its process controls class from one 3-hour course to one 2-hour course covering transient modeling and one 3-hour course in advanced control techniques with laboratory experiments. While many departments have experiments in process controls, ours is different in that we use equipment from our unit operations laboratory which is controlled with Honeywell Plantscape software. The students have the benefit of familiarity with the equipment, and they gain experience with software used in industry.

The first experiment is tuning a PID feedback controller for the flow manifold used with the double pipe heat exchanger experiment. The second experiment is tuning a feedforward controller (lead-lag unit) for the liquid level at the bottom of the packed tower absorber. The third experiment is cascade control of the temperature of an oil bath, and this is the only experiment not part of the unit operations lab. A fourth experiment, demonstration only for this first year, is multivariable control of a distillation column.

This presentation will discuss the challenges associated with using the unit operations equipment and Honeywell Plantscape for process controls experiments and how we overcame those challenges.

462a Workshop Laboratories And Experiments For K-12 Educators And Students

Steve R. Duke and Gopal Krishnagopalan. Chemical Engineering, Auburn University, 230 Ross Hall, Auburn University, AL 36849-5127

Hands-on experiments and laboratories have been developed and used for workshops for high school students and K-12 and community college educators as part of recruiting strategies to bring people into chemical and paper technology and engineering degrees and industries. Summer institute and workshop activities have included large pilot-scale experiments conducted in technology training or engineering laboratory facilities and small-scale experiments that can be integrated into the classroom. The large scale experiments have a strong impact on the workshop participants; however they cannot be readily transferred back into classroom and career counseling situations. Small scale experiments tend to not provide the wonder and excitement of new technologies. Workshop best practices and assessments will be discussed with particular emphasis on the effectiveness of large and small scale experiences, and the proper balance between them.

Five-day, three-day, and one-day workshops were conducted in 2005, 2006, and 2007 at Alabama Southern Community College (Thomasville AL), Lower Columbia College (Longview WA), and Auburn University (Auburn AL). Faculty and industry partners from these institutions developed and delivered the workshops as part of the National Network for Pulp and Paper Technology Training (npt) 2, a National Science Foundation Advanced Technology Education Center.

Web Page: www.npt2.org

462b Build Your Own Water Gun - A Project To Introduce Chemical Engineering To Students

Keith Hohn, Chemical Engineering, Kansas State University, Manhattan, KS 66502

One of the challenges in introducing chemical engineering to high school students or entering freshman is to give them a concrete picture of the types of problems chemical engineers solve. It is difficult to develop experiments or demonstrations that are relevant to students with limited knowledge, safe for students, fun, and effective. For the past two years, I have required students in the freshman chemical engineering course at Kansas State University to engage in a semester-long team project in which they design and build a water gun. I have also used this project in a summer program that exposes junior high and high school young women to science and engineering. Students are required to build a water gun that shoots water as far as possible, powered only by hand. They are allowed to use no more than 10% of an existing water gun to build this water gun. This project forces students to think about many concepts relevant to chemical engineering. Students have to consider flow, friction, momentum transfer, and PVT behavior of gases in their design. In addition, they need to incorporate valves, switches, and nozzles into their water gun, providing some hands-on experience with process equipment. Furthermore, this project engages students in a true team design experience. The response to this activity has been extremely positive. Students have been very successful in building water guns that rival or are superior to commercial water guns. For example, the best design in 2006 shot water seventy feet. A semester-end survey also indicated that students valued the project as a way of getting a better understanding of chemical engineering. They also rated it as their favorite activity in the course.

462c Creating the Scientist and Engineers of the Future Using Inquiry Based Learning

Nyrée V. McDonald, Civil Engineering, University of Alaska Anchorage, 3211 Providence Drive, Engineering Building Room 112, Anchorage, AK 99508

The goal of this project is to build the skills, interest and knowledge K-12 students need to be successful in the science, mathematics and technology fields. In an effort to address this issue, engineering and science departments are working with K-12 schools to develop new skills in field-based research, new technology skills and inquiry-based teaching strategies. Inquiry-based teaching is a method that seeks to actively involve students in the learning process and encourages them to seek knowledge through questions. Three K -12 school districts in the state of Alaska - Highland Technical School, Chugach School District and Denali Borough School District - have developed core curriculum that allows for the implementation of science and engineering exploration into their classrooms. This core curriculum includes field trips designed to expose students to practicing engineers and scientists who are available to discuss their work and its significance. The students are required to develop their own projects. Each group has selected topics of interest to their community - while some students are studying whale migration, other groups are examining changes in carbon dioxide concentration in the atmosphere and another is studying groundwater quality and migration. They are using cutting edge technology and learning the importance of data collection, quality and analysis. Each group of students is required to maintain a webpage with an ongoing discussion and presentation of their work. While this project is in its initial phases, students and their teachers have been exposed to more science and engineering related topics and the importance of these fields to their every-day life. This project is also examining the changes in the students' and teachers' perception of scientists and engineers and our impact on society through a series of surveys and video interviews at pre-planned intervals throughout the study. To date, it is clear that inquiry-based learning/teaching is an excellent mechanism for encouraging science and engineering exploration in K-12 students.

Oral Drug Delivery: Introductions to Methods and Practices in Chemical Engineering

Oral drug delivery – it's easily explained – all the members of our audience will taken a drug in a pill form, however very few have truly considered the implications and complications of this simple act. The basis of our common experience provides the starting point from which we, as chemical engineers, can discuss and examine our field with numerous audiences. It's easy and possibly entertaining for introducing chemical engineering to high school students (K-12). We can discuss material and energy balances in our introductory class (freshmen, sophomores) effectively. The oral drug system can provide a framework in kinetics and reaction engineering by providing a system of interest to students (sophomores, juniors) with long-range complications for the class to consider. In our transport course(s) (juniors, seniors) the oral drug in pill form provide a basis for complex models that are easily linked to real systems and practical goals. The oral drug allows other important possibilities including basic modeling methods, complex system analysis and design, the possibilities and limitations of computational and experimental methods, as well as providing a connection to economics (profits) and ethics (risks and safety). In this presentation we'll discuss how the topic can be arranged and some of the methods that have proven invaluable in teaching to achieve the aims outlined. Relevant lectures have been presented to high school students (K-12), freshmen (general engineering), and sophomores and juniors majoring in chemical engineering, and also provided a case study and lecture material in a graduate student mathematical modeling course.

462e Summer Research For High School Students

Muthanna H. Al-Dahhan, Energy, Environmental and Chemical Engineering, Washington University in Saint Louis, One Brookings Drive, Saint Louis, MO 63130

Since 1995, every summer I have supervised a high school student during the summer on various research projects as a part of STAR program (high school students and teachers Research program). This program is currently directed by Dr. Kenneth Mares at University of Missouri – St. Louis and is sponsored by Pfizer and Solutia with the goal to encourage these students to study science and engineering. A large number of high school students and teachers are placed during the summer in many research laboratories in the local universities and research centers to do research in summer in addition to other activities such as seminars and field trips. Many of my students won awards and scholarships as a result of their achievement on the projects they worked on with me. Also some of the work resulted into publications jointly with the high school student. The students worked and interacted with the doctoral students and work with them while they do their own research. Most of these students went to very good schools and joined various science and engineering departments achieving the goal set for this summer research program.

In this presentation I will discuss the overall STAR program with the focus on my experience with the high school students. I will summarize the projects they worked on and their achievement.

547a How The Ch E Curriculum Is Failing Students, Industry And Society (And Some Things That Can Be Done About It)

Gavin P. Towler, UOP LLC, 25 E Algonquin Rd, Des Plaines, IL 60017-5017

Demographic changes in the profession are creating a growing demand for chemical engineering graduates. Starting salaries are increasing and there are currently abundant opportunities for new chemical engineers.

Unfortunately, many new graduates are very poorly prepared to enter the workplace. Educators clearly struggle to find the right balance between teaching fundamental theories, practical examples, and soft skills such as teamwork and communication. From the perspective of industry, the result is often that students graduate with a weak to non-existant grasp of some of the most important basic skills and practices of the profession. Few professors have significant industrial experience, and many appear to be unaware of the inadequacy of their teaching.

The presentation will describe some of the obvious gaps in the traditional ChE curriculum from an industrial standpoint. Some suggestions will be given for ways in which Universities could enhance their undergraduate programs and give their students a better preparation for a chemical engineering career.

547b Chemical Engineering Curriculum - Spiraling Out Of Control Or Spiraling Into Control?

Vinay K. Gupta, Chemical Engineering, University of South Florida, 4202 E Fowler Avenue, ENB 118, Tampa, FL 33620

Chemical engineering education today faces an assortment of forces from globalization of the chemical industry, rapid growth in science and technology, and the needs of a diverse range of careers. As chemical engineering departments attempt to balance the impact of these forces, there can be a sense that the unified and efficient chemical engineering curriculum is spiraling out of control. In this presentation, we will discuss whether alternative models such as Bruner's "spiral curriculum" can be used to effectively adapt chemical engineering education. We will outline the central ideas behind the curriculum reform at University of South Florida and describe an approach using three interlocking spiral paths to deliver a student-centered curriculum that also allows integration of core chemical engineering courses, incorporation of traditional and new technological applications, and the threading of process and product design concepts over the complete curriculum.

UNDERGRADUATE CHEMICAL ENGINEERING: BIOLOGY TRACK

Stanley M. Barnett
University of Rhode Island

AIChE Nov 4-9, 2007

NEED

- Enrollment
- Support local industry
- Active faculty and research efforts
- Support of advisory board
- Student issues: P.Chem I redundant

What are we doing at URI?

- 6 out of 8 faculty work in nanotechnology or biotechnology
- Curriculum is being modified to reflect this opportunity
- Undergraduates are involved in research
- Cutting edge

TRADITIONAL 1st YEAR CURRICULUM

Freshman, First Semester		Freshman, Second Semester	
CHM 101 Gen. Chem Lec. I	3	CHM 112 Gen. Chem	3
CHM 102 Chem. Lab	1	CHM 114 Chem. Lab	1
EGR 105 Fundam. of Engr. I	1	EGR 106 Fundam. of Engr. II	2
MTH 141 Intro. Calc.	4	MTH 142 Interm. Calc.	4
PHY 203 Elem. Physics I	3	PHY 204 Elem. Physics	3
PHY 273 Elem. Phys. Lab. I	1	PHY 274 Elem. Phys. Lab. II	1
Gen. Ed. Requirement (1,2)	3	ECN 201 Econ. Prin.	3
total:	**16**	*total*:	**17**

Sophomore, First Semester

CHE 212 Chem. Proc. Calc.	3
CHM 291 or 227 Organic Chem.	4 or 3
MTH 243 Multivariable Calc.	3
Gen. Ed. Requirements (2)	6
	total: **15** or **16**

Sophomore, Second Semester

CHE 272 Intro. ChE Calcs.	3
CHE 313 ChE Thermo I	3
CHE 332 Physical Metallurgy	3
CHM 292 Organic Chem. or approved Adv. Chem. course	4 or 3
MTH 244 Diff. Eq. or MTH 362 Adv. Eng. Math I	3
	total: **15** or **16**

Junior, First Semester

CHE 314 ChE Thermo. II	3
CHE 347 Transfer Operations I	3
CHM 431 Physical Chem.	3
CHM 335 Phys. Chem. Lab	2
Approved math elective	3
Gen. Ed. Requirement (2)	3
total:	**17**

Junior, Second Semester

CHE 322 ChE MicroLab.	2
CHE 348 Trans. Oper. II	3
CHE 464 Indus. React. Kinetics	3
CHM 432 Phys. Chem. or	3
approved dept. elective (3)	
Gen. Ed. Requirements (2)	6
total:	**17**

Senior, First Semester

CHE 328 Industrial Plants	1
CHE 345 Chem. Eng. Lab.	2
CHE 349 Trans. Oper. III	2
CHE 351 Plant Design & Econ.	3
CHE 425 Process Dynamics & Control	3
ELE 220 Passive/Active Circuits	
or approved Prof. Elective	3
Approved Prof. Elective	3
total:	**17**

Senior, Second Semester

CHE 346 Chem. Eng. Lab.	2
CHE 352 Plant Design & Econ.	3
CVE 220 Mech. Mat'ls or approved Prof. Elective (3)	3
Approved Prof. Elective (3)	3
Gen. Ed. Requirements (2)	6
	total: **17**

Freshman, Second Semester

CHM 112 Gen. Chem	3
CHM 114 Chem. Lab	1
EGR 106 Fundam. of Engr. II	2
MTH 142 Interm. Calc.	4
BIO 101 Principles Bio I	4
ECN 201 Econ. Prin. or Gen. Ed. Requirement(2)	3
total:	**17**

Sophomore, Second Semester

CHE 272 Intro. ChE Calcs.	3
CHE 313 ChE Thermo I	3
CHE 332 Physical Metallurgy	3
BIO 341 Cell Biology	3
MTH 244 Diff. Eq. or	3
MTH 362 Adv. Eng. Math I	
total:	**15**

Junior, First Semester

CHE 314 ChE Thermo. II	3
CHE 347 Transfer Operations I	3
BCH 311 Biochemistry	3
PHY 204 Elem. Physics	3
PHY 274 Elem. Phys. Lab. II	1
Gen. Ed. Requirement (2)	3
total:	**16**

Junior, Second Semester

CHE 348 Trans. Oper. II	3
CHE 464 Indus. React. Kinetics	3
BIO 352 Genetics	3
MIC 211 Intro. Microbiology	4
Gen. Ed. Requirements (2)	6
total:	**19**

Senior, First Semester

CHE 328 Industrial Plants	1
CHE 345 Chem. Eng. Lab.	2
CHE 349 Trans. Oper. III	2
CHE 351 Plant Design & Econ.	3
CHE 425 Process Dynamics & Control	3
Approved Prof. Elective	3
Gen. Ed. Requirement (2)	3
total:	**17**

Senior, Second Semester

CHE 346 Chem. Eng. Lab.	2
CHE 352 Plant Design & Econ.	3
BIO 437 Fundam. Molec. Bio.	3
Approved Prof. Elective (3)	3
Approved math elective	3
Gen. Ed. Requirement (2)	3
total:	**17**

CHANGES

- TRADITIONAL
  - 0 biology courses
  - 7 chemistry courses
  - 2 bioprocess electives
    - biochemical engineering
    - bioseparations
- BIO TRACK
  - 5 biology courses
  - 3 chemistry courses
  - 2 bioprocess electives
    - biochemical engineering
    - bioseparation

CHANGES for PREMED's

- Add 2nd organic chemistry and laboratory
- Additional advisor: Health Professions

ENROLLMENT IMPACT

- Prior Year: 10 ('08)
- Announcement Year 26 ('09)
  - transfers in 13
    - 5 biomed eng, 3 chem, 1civil, 3 undec, 1 soc
- First Full Year 36 ('10)
  - transfers in 9
    - 2 biomed eng, 2 chem, 2 undec, 3 double majors

CONCLUSIONS

- Adding biology courses forced us to make curriculum changes. Dropped redundant first physical chemistry, plus two other chemistry courses.
- Student response has been excellent. First graduates this year. Enrollment jump expected.
- Gaining pre-meds and students willing to double major: CHE and microbiology and CHE and BIO.
- Enrollment: one-half traditional and one-half biology track

547d Should We Teach Ethics In Chemical Engineering?

R. Ocone, Chemical Engineering, Heriot-Watt University, School of Engineering and Physical Sciences,, Edinburgh, EH14 4AS, United Kingdom

The Royal Academy of Engineering (RAEng) has been working with the Engineering Council UK (ECUK) to identify the ethical issues at the core of the Engineering profession; a working party of the RAEng and the Engineering Professor's Council is looking specifically at the issues of teaching Ethics in Engineering. Two documents have been produced: the "Statement of Ethical Principles" and "An Engineering Ethics Curriculum Map". By referring to those two documents, I will try to address the teaching of Ethics within the context of the Chemical Engineering curriculum.

One can think of a number of ways to teach Ethics in Engineering (Davis, 1999, reports eight different ways); personally, I see all the possible ways reduced to just two main broad routes: one consists of devising and teaching a specific module on Ethics, the other involves integrating Ethics into the curriculum. The former poses the obvious questions about "when" (at which year in the programme) to present the Ethics module and "who" should teach it. The latter approach, suggested by the Map, is perhaps more challenging. In addition to the "when" and "who", it poses a question about "where" in the curriculum it should be integrated, and about the degree of training in applied ethics Engineering teaching staff will need. However, this second option, which I refer to as "integrated" teaching of Ethics, presents a number of advantages: it gives the students the opportunity to see "Ethics in action"; it shows that Ethics is intrinsic to the discipline; it demonstrates that Engineering is essentially an ethical profession.

The Map proposes a framework (location, content, learning outcome, process) for the teaching of Ethics in Engineering. It represents a good starting point to introduce, in a systematic way, the teaching of Ethics, but of course it needs to be "adapted" to the wide variety of courses and programmes. As with all novel working tools, it would need to be continuously under "revision". In this paper, I will discuss the work to be done at Departmental level to adapt the Map for Chemical Engineering programmes, and illustrate this for a specific university curriculum.

Reference M Davis, "Teaching Ethics Across the Engineering Curriculum" OEC International Conference on Ethics in Engineering and Computer Science, 1999

547e Developing Design Projects That Scan The Length Scale

Richard Turton and Joseph Shaeiwitz. Chemical Engineering, West Virginia University, 453 Engineering Sc. Building, P.O. Box 6102, Morgantown, WV 26506

Although the design of traditional petrochemical-based chemical plants still provides an important role in the education of chemical engineers, the construction of such plants is scarce in the US. More and more, chemical engineers, especially in Europe, the US, and Japan, are involved in the design and manufacture of new, chemical-based products. The design of these new products, often involving biologically derived chemicals, requires the chemical engineer to consider phenomena occurring at multiple length-scales. For example, in the manufacture of a drug patch, a chemical engineer might be involved in determining the action of enhancers and excipients on the surface of the skin (nano-scale), what adhesives are suitable to adhere to human skin (colloidal-scale), the transport phenomena involved in the movement of APIs (active pharmaceutical ingredients) across the skin and the bioavailability in the body (micro-scale), and finally the manufacture of the patch itself (macro-scale). These phenomena occur over a wide range of length scales and an understanding of each is crucial to the development of a successful product. Examples developed at West Virginia University including this and other product designs involving multiple length-scales will be discussed.

547f A Micro-Macro Transport Sequence for the Che Curriculum: Role of Scaling

Pedro E. Arce[1], Joseph J. Biernacki[1], Ileana C. Carpen[2], and Venkat Subramanian[1]. (1) Chemical Engineering, Tennessee Technological University, Department of Chemical Engineering, P.O. Box 5013, Cookeville, TN 38505, (2) Chemical Engineering, Tennessee Tech University, P.O. Box 5013, Cookeville, TN 38505

The transport sequence, i.e. momentum, mass and energy, in many departments of chemical engineering is taught following a unit operation-based approach (MacCabe et al., 2004)or, alternatively, by using concepts already scaled-up (Felder and Rousseau, 2004) from microscopic variables giving the impression to students that every case is "divorced" from a general type of conservation equation. Thus, students perform analysis by using macro-balances for every process they face without a proper up-scaling (Arce et. al., 2005; Oyanader et al., 2007) from the microscopic concepts so important to achieve a masterful command of the concepts and to understand limitations of process descriptions. Moreover, the introduction course on mass and energy balances follows the same type of approach without, for example, selecting proper control volumes or surfaces in connection with the fundamentals conservation equations (Higgins et al., 2007). One of the results is that when students need to apply these concepts to the micro (and even nano) scales, they often get confuse since it looks like that an entirely new discipline must be learned.

In the approach we are experimenting with at Tech, students follow a different sequence. This is rooted on fundamentals conservation principles (based on microscopic variables) for the energy, momentum, and then mass processes. Students are introduced to heat transfer by radiation (no "media") first; then they are exposed to conduction and diffusion (media without a bulk motion); afterwards, they learn momentum conservation in order to describe convective-based transport and, finally they are exposed to diffusive-convective transport. All the concepts are taught in an integrated sequence that allow students to sequentially learned concepts and build knowledge. In addition, concepts are introduced at a microcospic level and, by spatial scaling to various levels, i.e. averaging in lines, surfaces, or volumes, students are automatically introduced to the key up-scaling concepts; these are needed to describe any process ranging from micro-level to a macro-level. The first introductory course is now centered on process analysis from a fundamental point of view and with the identification of the proper averaging surfaces or volumes.

In this presentation, details about the sequence and illustrative examples will be offered. In addition, future adjustment will be also discussed. The role of multi-scale systems (including molecular-level descriptions) now of paramount importance in chemical engineering system will be integrated in the discussion.

References:

1. McCabe, W.; J. Smith and P. Harriot, "Unit Operations of Chemical Engineering", McGraw Hill Chemical Engineering Series, 2004, N. Y., New York. 2. Felder, R. and R. Rosusseau, "Elementary Principal of Chemical Processes", J. Wiley & Son, (2004), N.Y., New York. 3. Arce, P.; M. Quintard and S. Whitaker, 2005, "The Art and Science of Up-scaling," Chapter 1 in Chemical Engineering: Trends and Developments, edited by M.A. Galán and Eva Marin de Valle, John Wiley & Sons, Ltd., England, (2005). 4. Oyanader, M.; P. Arce and S. Whitaker, "The Catalytic Pellet: A Rich Learning Environment for Up-Scaling", Chemical Engineering Education, (2007, under review). 5. Higgins, B.; R. Cerro and S. Whitaker, "Material Balances for Chemical Engineers"; web site http://www.higgins.ucdavis.edu/

582a Curriculum And Course Assessment At Ohio University

Michael E. Prudich, Darin Ridgway, Kendree J. Sampson, and Valerie L. Young. Chemical and Biomolecular Engineering, Ohio University, 172 Stocker Center, Athens, OH 45701

Developing a comprehensive, integrated system to assess chemical engineering program outcomes can be a difficult and time consuming process. Maintaining and continuing the assessment system, once it has been developed, can also be problematic. This paper describes the continuing curriculum and course assessment system currently used by the Department of Chemical and Biomolecular Engineering at Ohio University. This system consists of a set of individual course-level assessments which integrate together to constitute the backbone of a program assessment. Senior exit interviews and periodic surveys of recent graduates and alumni round out the assessment portfolio. Annual evaluations of the report set by a faculty Committee-of-the-Whole identifies both program successes and challenges and serves to focus the program on continuous improvement. The annual summary report, as well as the individual course assessment reports, are available to the faculty and selected constituencies (departmental Board of Advisors) on a secure web page. The accumulated annual summary reports and individual course reports provide the basis for the program self-study required for accreditation.

582b Progress In Developing A Web-Based Database-Driven Curriculum Assessment Tool

David A. Kofke, Jeffrey R. Errington, and Andrew Schultz. Chemical and Biological Engineering, University at Buffalo, 303 Furnas Hall, Buffalo, NY 14260

We describe progress in developing a tool to manage the assessment of student performance in our courses and to aid in collecting information needed to design and update our curriculum. The tool employs a web-based architecture, fronting a database that contains information about knowledge areas in chemical engineering and skills associated with each. The database maps these onto each other, and connects them to the courses in our curriculum. We developed a dynamic web page that permits instructors to input assessment data, which gets folded into the database in a way that permits analysis across different courses. The instructor selects a exam problem to analyze, and uploads the analysis information to the database via this page. In future developments we will add web portals through which our constituencies (employers and alumni) can provide feedback regarding the skills expected of our graduates and how well they are conveyed by the curriculum.

582c Continuous Improvement Across The Curriculum

Joseph J. Biernacki, Tennessee Technological University, Department of Chemical Engineering, P.O. Box 5013, Cookeville, TN 38505

How can we integrate elements of continuous improvement across the curriculum, capturing elements of course- and program-level assessment? Effective reporting of assessment data must involve a strategy that permits information from the course-level to "communicate" with the program-level assessment, i.e. the program outcomes and program educational objectives as defined by ABET. One possible strategy being used by the Department of Chemical Engineering at Tennessee Technological University (TTU) will be presented. Experiences in excess of six years with this methodology will be shared to illustrate that an effective process must be adaptive and evolving.

582d "Ways To Succeed At Assessment Of Educational Objectives From An Evaluator's Viewpoint"

Gary K. Patterson, Chemical and Biological Engineering Dept., University of Missouri-Rolla, Schrenk Hall, Rolla, MO 65401

The assessment of educational objectives in engineering programs for the purpose of satisfying ABET Criterion 2 has become progressively more difficult. ABET has over the last six years required that the objectives be more and more measurable. This means that highly generalized objectives for the program and its graduates are no longer acceptable because there is no way to measure whether they are being met. More specific educational objectives must be established that reflect what kinds of successes graduates should attain as a result of their education. Education objectives should not be aimed at operational aspects of the program such as "excellent relations with recruiting companies" or "service to the state and the nation". This paper is aimed at stimulating a discussion of what measurable educational objectives can be, how to establish them and how to measure the success of the graduates in meeting them.

Paper #611a: Engaging the Undergraduate Through Research and Mentoring

Nyrée McDonald, Ph.D., Assistant Professor Civil Engineering, University of Alaska Anchorage, 3211 Providence Drive, Anchorage, AK 99508
Ron Spatz, Dean Honors College, University of Alaska Anchorage, 3211 Providence Drive, Anchorage, AK 99508

National and State Needs

The National Academy is chartered by Congress to advise the government on matters of science and technology. The National Academy's Committee on Prospering in the Global Economy of the 21st Century was given "the task of assessing the top ten actions that federal policy makers could take to enhance the science and technology enterprise so that the United States can successfully compete, prosper, and be secure in the global community of the 21st Century". The Academy responded with a 500 page document entitled, "Rising Above the Gathering Storm: Energizing and Employing America for a Brighter Economic Future" which states: "...the committee is deeply concerned that the scientific and technological building blocks critical to our economic leadership are eroding at a time when other nations are gathering strength..." Their most striking conclusion is, "The nation must prepare with great urgency to preserve its strategic and economic security". The document includes many alarming statistics along with four excellent recommendations, each with a list of implementation actions. One of these recommendations is to encourage students to obtain post baccalaureate degrees in the science, engineering and technology fields.

Unfortunately the state of Alaska is also in need of well educated scientist and engineers. The 2000 census data for Alaskans between the age of 20 and 34 showed a decline in the number of young adults (approximately 36,000) as compared to the same demographic in 1990. While the number is relatively small compared to the nation's population, Alaska has about 660,000 residents. This decrease is expected to be multiplied as an excessively large number of older Alaskans retire and create openings that can not be filled by the existing workforce because many of the workers in these fields are over the age of 50 and the current work force will lack the education and or experience to fill these positions (Hadland and Williams, 2000). Moreover these authors have reported that the oil and gas extraction, water transportation sector, Pipeline, electric, gas and sanitary services and engineering accounting research management & services firms, have some of the largest percentage of workforce over the age of 50, at 22.1%, 18.2%, 35.7%, 25.3%, and 18.3%, respectively. With these projections the state has an urgent need for well trained scientist and engineers. As a state institution the University of Alaska Anchorage is charged with addressing this situation.

The University of Alaska Anchorage

The University of Alaska Anchorage (UAA) is accredited by the Northwest Association of Schools and Colleges. It is the largest higher education institution in the state of Alaska, with community campuses in the; Kenai Peninsula, Mat-Su, Kodiak, and Prince William Sound. The institution is about 50 years old and is experiencing staggering growth. UAA offers 1 AA, 31 AAS, 52 BS, 34 certificates, 29 Masters, in addition to 1 collaborative Ph.D. program with University of Alaska Fairbanks. Since 2005, the institution has experienced tremendous expansion in its degree programs offering an A.A.S. degree in Construction Management, a B.S. in Geological Sciences, a B.S. in Health, Physical Education and Adventure Leadership, a B.S. in Engineering where students may specialize in Electrical, Mechanical or Computer Systems Engineering and a joint Ph.D. in Psychology. (UAA website)

UAA is primarily an undergraduate institution with a relatively diverse population. For the Fall of 2006, UAA served 19,921 students, of this total about one fourth (24.8%) are minorities such as African American, Alaska Native, American Indian, Hispanic, Asian, and Pacific Islanders. Women represent the largest undergraduate group with 12, 046 students. Undergraduates make up 95.6% of the student body with a total of 19,057 undergraduate students and 858 graduate students. (UAA Office of Planning Research and Assessment website)

The disparity between the graduate student and undergraduate is exacerbated by the fact that most of the graduate students are part time students with full time jobs. For example, the School of Engineering offers Masters degrees in Arctic Engineering, Civil Engineering and Applied Environmental Science and Technology. There are over 30 part time graduate students and 4 fulltime students. The part time students are working professionals whose companies are usually paying some portion of their tuition, therefore many of them are not able to make the kind of commitment required for full scale research. The undergraduate researcher has become more important to the school of engineering faculty. Recently, the Civil Engineering faculty has decided to incorporate undergraduate research through an honors thesis of its own design.

Even those departments without graduate degree programs are heavily involved in research and employ undergraduate students to do a significant amount of their work. This is positive for both the faculty and the student because undergraduate research helps maintain students in STEM pipeline and faculty derive personal satisfaction from working with undergraduate researchers, though they feel it is not rewarded in the tenure process (Russel et al. 2006). The B.S. in Geological Sciences is only two years old but the majority of the professors in this department are doing research, and they are using undergraduates within their department to accomplish this task. According to Dr. LeeAnn Munk (2007), Chair of the Geological Sciences Department, "The OURS grants have made it possible for students to become involved in original research at the undergraduate level. The students have the opportunity to participate in a complete process of applying for funding, conducting the research and presenting it to UAA as well as at the national level. These opportunities would be almost impossible if it were not for the OURS grants. The OURS program is pioneering the effort to bring UAA closer to being able to provide the margin of excellence that many of our students are striving for. The grants have been the catalyst to inspire students to go on to graduate school and where they can continue to make their own discoveries and contribute to the global scientific community."

The Office of Undergraduate Research and Scholarship (OURS) in the University Honors College is a major advocate of undergraduate research and is eager to assist faculty who wish to include undergraduates in their research. Many of the grants provide a stipend and monies to purchase the necessary equipment and for other expenses to perform the research. Grants also fund student travel to present their research.

Initiated by Chancellor Edward Lee Gorsuch, the University Honors Program was founded in 1998. Undergraduate research was a central component to the program's design. We founders of the program strongly believe that the integration of research, scholarship, engagement and learning at the undergraduate level, and across all disciplines, is a powerful mechanism to enhance disciplinary and multi-disciplinary excellence. This view was validated in the 2000 ten-year accreditation review of University of Alaska Anchorage. The Northwest Association of Schools and Colleges described the University Honors Program as a "small but energetic program" and singled it out for special commendation (the only such commendation for a university academic program) for providing "a locus of activity to support student inquiry and research." That commendation led directly to an expansion of the University Honors Program's undergraduate research focus. During the 2001-2002 academic year the Office of Undergraduate Research & Scholarship was established within the honors framework with the objective of providing access to undergraduate research opportunities to all students at the University. The Office of Undergraduate Research & Scholarship (including the University Honors Program) has been and continues to be the tip of the spear to change the culture of learning at UAA to one of research, discovery, and public service as inherent and inseparable parts *of the undergraduate experience, and student success*. The ability to include undergraduates in significant and meaningful research and to foster engaged and active learning through creative opportunities and public service is among UAA's greatest potential strengths within the University of Alaska System. Developing these strengths will attract more top Alaskan students to UAA, reducing the "brain drain" of students choosing to pursue their education outside of the State. In addition, it will also attract more undergraduate students to attend UAA from the rest of the U.S. and abroad.

In March 2007 the University Honors Program and the Office of Undergraduate Research were elevated to college status. The University Honors College was established with a mission to be a catalyst for scholarly excellence in undergraduate education. The College advances, coordinates, and administers active learning and undergraduate research opportunities for students across the campus. The College houses the Office of Undergraduate Research and Scholarship, and three University Honors academic programs: the University Honors Core Program, the Natural and Complex Systems Program, and the Forty-Ninth State Fellows Program.

Background and Data

Prior to 2001 there was no university-wide infrastructure to foster and support undergraduate research at UAA. Consequently, despite some small pockets of activity within programs, there was very little undergraduate research activity university-wide.

- Since fall of 2001 there have been 52 University Honors undergraduate theses (39 in the past three years).

- The Office Undergraduate Research & Scholarship Division (OURS) was established in 2001. OURS created ten university-wide competitive awards programs (four of them in the past year) funded exclusively through private donations.

- OURS awards program has involved faculty mentoring data for the past three years: student submissions/awards (2004-05) were 63/27; student submissions/awards (2005-06) were 63/34; student submissions/awards (2006-07) were 100/52. Total for three years: 226 undergraduate research proposals; 113 awards.

- In the 2006 -2007 academic year, OURS sponsored an estimated 100 undergraduate research students. The corresponding increase in faculty mentors across the Anchorage campus schools and colleges and Kenai Peninsula College include approximately 100 faculty members supporting undergraduate research either as mentors or as facilitators.

- In 2007, the annual Faculty Exemplar Awards were established by the Office of Undergraduate Research. These awards recognize faculty members who have demonstrated long-term, exemplary mentoring of undergraduate research and creative projects. Their contributions to student success are hallmarks of inspired teaching and active learning.

The Council for Undergraduate Research (CUR) recognized UAA's Office of Undergraduate Research and Scholarship's web presence in its national journal, *CUR Quarterly*, for the fact that there is a direct link from UAA's home page to the Undergraduate Research homepage, and for the OURS "unique approach" and "firm accountability" in not paying students a stipend until they have publicly presented their research results. We are also increasing our footprint in major scholarships for undergraduates as UAA has won its third Truman in five years and its second Fulbright in three years.

Student Success as measured by acceptance to graduate programs (incl. doctoral) and professional schools (incl. M.D. programs) are hallmarks of the University Honors Program. Additionally this year, University Honors students have been recipients of the Undergraduate Research Grant, Discovery Grant, Parasca Science Research Award, Alaska Heart Institute Fellowship, Policy Research Internship Award, a Truman Scholarship, a full scholarship for a semester at The Washington Center in Washington DC. They have been selected as delegates to the Student Conference on United States Affairs held at West Point, selected to participate in the Conference of Young Alaskans in Fairbanks, selected as the student representative to the Oxford University Ethical Leadership Enclave, and honored at the Joint Meeting of the American Mathematical Society and the Mathematical Association of America in San Antonio with the Mathematical Association of America Award for Outstanding Undergraduate Research. A University Honors student was invited to speak at the United States Business Leadership Network's National Conference, another's research was selected for presentation at the Council for Undergraduate Research "Posters on the Hill" in Washington DC, two were selected to participate in research projects in Greenland and Hungary over the summer, and a University Honors student's essay was selected as a finalist in the Elie Weisel Prize for Ethics. Finally, a University Honors student was named "Alaskan of the Year" by the Governor's Committee on Employment and Rehabilitation of People with Disabilities.

Student Success Stories

The following narratives are provided by students who have received a research award from the Office of Undergraduate Research and Scholarship. Each story is prefaced by a few comments which are intended to give the reader more insight into the student's undergraduate research experience.

Student #1

In 2004, this student was diagnosed with myasthenia gravis (an autoimmune neuromuscular disease) and systemic lupus erythematosus. She has learned how to cope with constant muscle weakness and sometimes brings her walker with her to campus. Consequently, she understands the obstacles students with disabilities must face everyday. Her current medical challenges have only strengthened her desire to become a physician. Her undergraduate research activities with faculty mentoring became a strong motivator and she credits undergraduate research as helping her gain admittance into medical school.

"During my junior year of college at UAA, I took a molecular biology course and fell in love with the subject. My professor was amazing and I asked her if I could work in her lab and complete an Honors Thesis under her supervision. She agreed and I started my research project. I wrote a research proposal for my project entitled "Maintenance of copper homeostasis: Effects of Cup1p overexpression, Mac1p-dependent gene expression and survival of S. cerevisiae." I applied for grants and was awarded the UAA Parasca Science Research Award for Undergraduates and an Experiemntal Program to Stimulate Competitive Research (EPSCoR) Undergraduate Summer Research Experience Award. I was recently awarded a UAA Office of Undergraduate Research and Scholarship Discovery Grant Award and an EPSCoR Student Travel Award for a trip to a research conference in California. At the conference, I began to collaborate with a world-renowned expert in my field. He agreed to help me interpret some data I obtained from recent experiments."

“Through research, I have had the opportunity to explore the world of scientific inquiry. I have learned many different facets of research and have dispelled some of my preconceived notions about research. For example, I assumed my research would run smoothly as planned, without interruption or delays of the timeline I had set for myself. I learned that many unexpected events occur during research, such as flasks breaking or equipment becoming suddenly inoperable. Research has given me the chance to apply knowledge I have acquired from various undergraduate courses such as general biology, cell biology, genetics, physics, molecular biology, organic chemistry and biochemistry. I am learning practical applications for materials and procedures learned in lectures and labs. I have realized how different disciplines within the sciences are integrated. In the future, I look forward to conducting research as a medical student. As a doctor, I hope to use the practical application of knowledge to best help my patients. In addition to learning the “hard science” of conducting lab research, I learned the “soft science” of being a lab member. I learned about different roles each member in a lab plays. I imagine that this “soft science” will be applicable in the field of medicine: each healthcare professional will have distinct roles to play and individual responsibilities to fulfill. My research experience as an undergraduate has left a lasting impression on my goals for the future.”(UAA University Honors College and Biological Sciences Student, Currently Medical School Student)

Student #2

The following student used his undergraduate honors thesis to develop his doctoral dissertation topic. His dissertation earned an award and he recently completed his Ph.D. In the Fall of 2007 he returned to University of Alaska Anchorage as an assistant professor. One of the first things he did was to contact the OURS and volunteer to help with the continued growth and development of the program.

“Getting the opportunity to participate in the conceptualization, planning, implementation, and reporting of a research project as an undergraduate at UAA benefited me immensely. It exposed me to the rigors of research early in my educational development, which prepared me to tackle the strong research emphasis of graduate school. This early involvement in every phase of the research process was also enhanced by the quality of mentoring and support I received from UAA faculty. They were able to strike a good balance between providing guidance and encouraging scholarly independence. As important as learning about the research process, however, is that undergraduate research gave me a taste of how rewarding research can be, especially when you realize that your research has the potential to improve the lives of the people and community you care about. This piece is key, as research is supposed to be intended to serve the general society and not just the scientific community. Although publications and other academic and scholastic endeavors are nice, I learned that as a researcher I should never forget the real purposes of my projects. This research perspective, I believe, has contributed to the accomplishment of my personal and career goals: come back to UAA and make a difference in the Alaska community by training doctoral-level mental health practitioners and by conducting research on issues relevant to Alaskans.” (Former UAA University Honors College and Psychology Student, Currently an Assistant Professor in Psychology)

Student #3

As a junior this student received an undergraduate research grant that utilized social science methods. Once that project was successfully completed, she "globalized" and adapted it for a research Fulbright application. She was awarded a research Fulbright and this was the first time in the history of the University of Alaska Anchorage that an undergraduate student had been the recipient of a research Fulbright. Undergraduate research has played a key role in her success. She is continuing undergraduate research, this time in biological sciences as she completes a second baccalaureate degree in preparation for medical school.

"As a member of the honors program I was required to submit an undergraduate thesis. I started working on my undergraduate research project during the summer between my junior and senior year. My goal was to use my Spanish degree to enter medic al school so I decided to look at access to healthcare for Spanish speaking Anchorage residents. My senior research grant was funded and I spent a about two months trying to figure out what I proposed to do. Luckily the support network of faculty members who encourage and support undergraduate researchers helped me develop my survey for the twelve Spanish speakers and eight doctors. While I was working on my research grant, I was asked to consider applying for a Research Fulbright Scholarship. I spent a year in Ecuador, working with researchers, professionals, and students. Initially, I was allowed struggle, and I did with many aspects of my research including networking and research development. Fortunately, I was not allowed to linger indefinitely, eventually I was given additional guidance so that I may complete y research. I completed my Fulbright in August 2006, and returned to UAA as a pre-med student. Even though my program is very different form my undergraduate degree, it seems simpler and comes more naturally, because of my undergraduate research experience Research Fulbright Scholarship." (Former UAA Honors College and Language Arts, Currently UAA Premedicine Major)

Student #4

This student has been very enthusiastic throughout her career at UAA. Professors from her freshman year still remember and have a high regard for this young lady. She has an interest in everything and while she has thought about graduate school she did not know how to get there or what to do. Her research experience afforded more contact with faculty and integration into the lab group. This has helped her clarify her objectives; she will be attending graduate school in the Fall of 2008.

"My properties of materials professor offered extra credit for submitting a proposal to the Office of Undergraduate Research and Scholarship. I decided to write a grant proposal using alternative energy as my main idea. My Fluid Mechanics instructor agreed to advise me on this project. I wrote the grant and was awarded an Alex Hills Engineering Research Award."

"My undergraduate degree will be in Civil Engineering; this is a very broad field. The experience helped me identify my passion and has clarified some of my long term goals. The cool thing about undergraduate research is that it provides students with a way to learn how to research and general guide lines for how to write a proposal. Bu the best part of undergraduate research is that it made me want to do more research. I have decided to attend graduate school in the Fall of 2008, with a major in Mechanical Engineering, with a focus in Fluid Mechanics." (UAA Senior Civil Engineering)

Student #5

This student is an excellent theoretician but she is uncertain about graduate school. The primary goal with this student was to improve her writing skills and build her confidence level so that she would be successful in graduate school.

"I received an email from a former professor suggesting that I submit an undergraduate research proposal. I thought this would be a great idea because it would 1) be preparation for graduate school, 2) apply what I learned in classes and 3) provide real world hands on experience. When I was awarded my Parasca Science Award I felt privileged, and motivated to perform the research because a group of UAA faculty believed that I could do this work. I knew that the Fall 2007 would be a challenging semester since I am carrying 19 credit hours, preparing for the

Fundamentals of Engineering Exam, completing graduate school applications, and trying to start my research. But I am not overwhelmed because I have a strong support network, of graduate and undergraduate students, and faculty mentors. Even though I have weaknesses, I am aware of my strengths and will use my skills to be successful in graduate school or in industry." (UAA Senior BS Engineering)

Conclusion

These stories share many common threads but the underlying theme in all of these experiences is that the students are benefiting from exposure to research. The value of mentoring is prevalent in all of these stories, they are using graduate students and faculty to help with the development of their career paths. The students have used their research experience to prepare them for graduate school and have built relationships that extend outside of UAA. They are learning that faculty have very diverse careers, and in fact have used faculty to help them prepare to enter academia as colleagues. The undergraduate research program at UAA provides research opportunities for students in all disciplines. Students have gained immense opportunities and the state of Alaska has also benefited.

The development of the University Honors College and its OURS program are crucially important to the future of the University of Alaska Anchorage. This investment has been valuable to the university as students are acquiring national fellowships like the Fulbright. The support of the OURS program has given faculty more opportunities to incorporate undergraduates in their research groups. This exposure is invaluable to the undergraduate thinking about a post baccalaureate degree and the faculty who feel they may be able to help students evaluate and develop their professional goals. In her September 7, 2007 evaluation report on the University Honors College, Dr. Diane Lee, Dean of Undergraduate Education and Vice Provost at the University of Maryland Baltimore County wrote: "The Honors College has certainly incorporated known practices that foster student success. Its programs are oriented to the development and integration of knowledge, intellectual and practical skills, and individual civic and social responsibility. Importantly, these programs cross all disciplines, colleges, and schools at UAA. It has also become a model of distinction for other institutions."

References

Holland, J. and G. Williams, "The Aging of Alaska's Workforce", Alaska Economic Trends, 20, no.9, pp. 3-9, September 2000.

Munk, L., (personal communication), University of Alaska Anchorage, Department of Geological Sciences, September 16, 2007.

National Academy of Sciences, National Academy of Engineering, and Institute of Medicine, Rising Above the Gathering Storm: Energizing and Employing America for a Brighter Economic Future, (National Academies Press, Washington D.C., 2007).

Russell, S., Hancock, M. and J. McCullough, "Evaluation of NSF Support for Undergraduate Research Opportunities: Draft Synthesis Report", SRI international, July 2006.

The University of Alaska Anchorage, (2007), About UAA,
http://www.uaa.alaska.edu/aboutuaa/index.cfm, Sept.12, 2007.

The University of Alaska Anchorage Office of Planning Research and Assessment, (2007), Trendbook, http://www.uaa.alaska.edu/opra/index.cfm. Sept. 12, 2007.

611b Interdisclipinary Research: An Reu Site Perspective

Mark E. Byrne, Biomimetic & Biohybrid Materials, Biomedical Devices, and Drug Delivery Laboratories, Department of Chemical Engineering, Auburn University, Auburn, AL 36849-5127 and Steve R. Duke, Chemical Engineering, Auburn University, 212 Ross Hall, Auburn University, AL 36849-5127.

This talk will focus on successful strategies and components of our REU site based on student reviews and exit interviews. It will also compare and contrast other research experiences with the REU experience. Providing undergraduate students with research experiences has been shown to lead to an increase in student perseverance in higher education, higher graduation rates, and a greater number of students pursuing bachelor and graduate studies. The mission of the Auburn University REU site is to provide community college and university undergraduates a research experience that will prepare and motivate them for further education and careers in science and engineering. The specific objectives are to (i) promote REU opportunities to the targeted students, (ii) present well-focused current projects that appeal to the students and faculty mentors, (iii) provide a ten-week program for 10 students per year with meaningful research and comprehensive professional development experiences, (iv) provide a positive social experience with diverse grouping of student participants and scientists, and (v) create a cohort experience between participants and among faculty mentors. The research projects are multidisciplinary in scope and have a strong intellectual focus on micro and nano-material science and technology.

611c Strategies For Increasing The Participation Of Women And Underrepresented Minority Students In Undergraduate Research

Lealon L. Martin, Department of Chemical and Biological Engineering, Rensselaer Polytechnic Institute, 110 8th st, Troy, NY 12180

Academic institutions have demonstrated a strong commitment towards increasing the number of women and underrepresented minorities with advanced degrees. However, efforts stemming from this commitment oftentimes have little affect on the successful recruitment and retention of women and underreprented minorities in graduate (PhD) programs. In this talk we will explore strategies, rooted in the promotion of research participation among women and underrepresented minorities at the undergraduate level, that may prove viable to this end. We will analyze and address issues surrounding conventional approaches to inclusion and diversity in undergraduate research programs and suggest non-conventional approaches that may be effective.

Use of Comsol Multiphysics in Undergraduate Research Projects to Solve Real-life Problems

Bruce A. Finlayson, University of Washington, Seattle, WA, USA

Abstract

This paper shows how undergraduates are introduced to computational fluid dynamics (CFD) in research projects. The students first solve problems in their textbook, then solve extensions to them (i.e. they remove assumptions). Then they solve a simple version of their ultimate problem to learn how to verify their solution. Finally they solve the research problem.

Introduction

While transport phenomena textbooks are good at presenting the fundamentals, most of the problems are one-dimensional, since that is the limit of the mathematical ability of most undergraduates. Today's students are motivated by real-life examples, but they have limited time. With the advent of sophisticated software, however, it is possible for undergraduates to solve meaningful transport and flow problems in two and three dimensions. This talk presents the methods used to introduce undergraduates to Comsol Multiphysics and the problems they solve in a research project format.

The learning occurs in three stages. First, the students learn to solve problems in their textbooks and learn to validate the solution. This gives confidence that the computer program is solving the right equations. Next, they solve more complicated 2D problems, which go beyond their textbooks by removing assumptions. In this stage, students are confronted with the necessity of translating their problem into the notation of the computer program and proving they have solved the problem correctly even though there is no analytic solution. They learn how to check their data input to the program, look for artifacts in the solution, and use mesh refinement to estimate the numerical error. They explore the many ways to analyze and view the results, from streamlines, contour plots, integrals, etc. of the dependent variables as well as derived quantities (defined in terms of the dependent variables). For example, is the total flow in equal to the total flow out? – not a straightforward question when there are multiple inputs and outputs and the solution is numerical. The second step is illustrated in detail to show the breadth of analysis techniques.

The third stage in the learning process is to solve their research problem. Examples done during the past year (many suggested by research groups or companies) are shown in Table 1.

Valuable features of Comsol Multiphysics include the graphical user interface, the tools for creating the geometry and internal boundaries and domains, automatic mesh generation and refinement, the ability to solve different equations on different meshes (all in the same problem), the multi-physics capability which permits addition of equations to represent additional phenomena, the ability easily to make parameters depend upon the solution, the parametric solver, and the post-processing graphical features.

Table 1. Research Problems Tackled in 2006-7 by Undergraduate Chemical Engineering Students

Thermal field flow fractionation – the effect of entry effects;
Use of ferrofluid to remove arsenic from water supplies;
Flow of viscoelastic fluids in contractions;
Movement of nanoparticles in gels;
Mixing of pharmaceuticals in orbital mixers and injection/remove devices;
Mixing and flow in Swagelok and Circor Tech devices;
Flow in expansions – pressure drop and entry/recovery lengths;
Flow of water in porous media;
Perturbation method applied to the flow of a ferrofluid in an oscillating magnetic field;
Hole pressure problem – used to measure normal stresses in polymers;
Calculation of vortex viscosity of ferrofluids.

Organization of the Research Projects

The procedure for teaching students how to use CFD programs is described. An essential element of the process is to make the steps small and manageable, with the eventual goal clearly specified. The quarter is 10 weeks long. The first two weeks are spent introducing students to the program and having them solve the textbook problems, and extensions. Most of the students have already taken a computer course in which they have used the program some, but only in problems that were well defined. Weeks 3 and 4 are devoted to solving two-dimensional problems that are simplifications of their final project. The problems contain the essential features of the final project so that they have to address those features. For example, many three-dimensional problems have features that can be studied in two dimensional cases, too. Since the two-dimensional cases run much faster, they are used first when the students are learning. Then, when the three-dimensional real case is studied, fewer mistakes will be made and the project will run more smoothly. Weeks 5 to 9 are spent on the project. The project is defined in a broad context, as illustrated in Figure 1. The tenth week is devoted to finishing the work, preparing and giving the oral presentation, providing information for a project web page, and writing the final report. Sometimes the final report is finished after finals. During the quarter, the student meets with the instructor once per week to show their current work. In addition, the entire group of students meets together to discuss mutual problems and solutions. Progress reports are given, which is a motivating factor. One arrangement that works very well is to have several students working on similar and related projects. Then they can learn from each other and interact with each other. Progress made by one student is then a motivating factor for their partners. For example, in Autumn, 2007, correlations are being developed for pressure drop in flow devices: tees, ells, and devices with multiple input/output ports. Thus, each student is working on a different device, but they are all doing basically the same thing.

Hole Pressure for Newtonian Fluids
Spring, 2007, Ch.E. 499 Project for Stephanie Yuen

Project description: The project simulates the flow of a Newtonian fluid between flat plates. A hole is drilled in one of the plates and the pressure is measured at the bottom and across from the hole. That pressure can be related to polymer properties if the fluid is viscoelastic. This is a problem one of Prof. Finlayson's graduate students did in 2D when finite elements were just beginning. Can you believe we used only 28 elements? Now it is time to do it with more elements and in 3D and compare with published experimental results.

First steps:

Do the problem in Table 10.1
Read the paper by my students
Meet with group on first two Tuesdays.
First week – I'll demonstrate Comsol Multiphysics
Second week – Have done the problem in Table 10.1; read the paper by my students; have read about your problem and have questions.

Milestones

Run the existing code, from the polymer paper by BAF
Change the dimensions and parameters to do the exact specs in the experiment
Prepare a 3D geometry
Solve the problem in 3D

At the end of the quarter you need to submit: written report, PowerPoint presentation, Comsol Multiphysics dataset, and suggested web page.

Here is the **checklist** for a proper submission of results.

1. Say what problem you are solving;
2. Give the shape and dimensions, number of elements, degrees of freedom;
3. Give the parameters in the equation and identify the boundary conditions;
4. Tell how you solved it;
5. Give checks to your answer (previous similar results, etc.);
6. Give your results, including pertinent plots and integrals;
7. Your report should have an Appendix with sample calculations.

Figure 1. Sample Statement of Goals

Step One – Solve Problems with Known Solution

Shown in Figures 2 and 3 are two problems that are assigned. The student is to solve them, make plots of them, and report back to the instructor. This exercise gives the student confidence, since they are able to solve the problems, although maybe the comparison to their textbook is difficult the first time. It also gets them 'over the hump' on how to access the program, use it, save their results, print figures, etc. A printed copy of the mesh is always required, to emphasize the importance of mesh refinement. Figure 2 involves cylindrical geometry, and the student learns about inlet/outlet flow conditions. The pressure is set at the outlet and the velocity is set at the inlet. The solution gives the pressure drop in the device, which can be compared with the Hagen-Poiseuille law. The student also learns how to setup problems in cylindrical geometry. Figure 3 involves geometries that are not quite square, and boundary conditions that are not uniform. Thus, the student learns how to insert such boundary conditions into the computer code, and how to draw the boundaries.

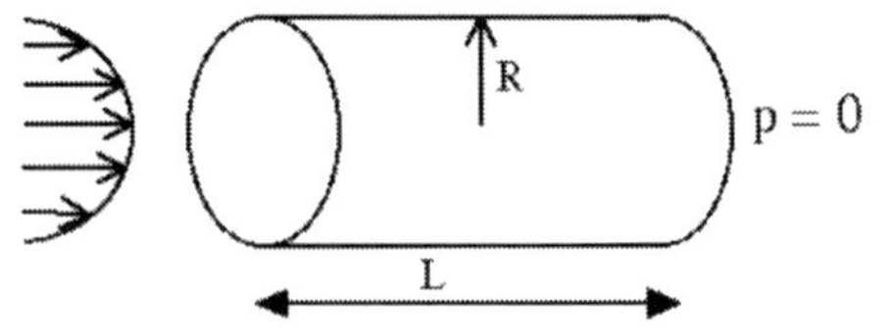

Put $u = u_0\left[1 - \left(\frac{r}{R}\right)^2\right]$ as boundary condition.

Compare with Hagen-Poiseuille law in your textbook.

Figure 2. Flow in Pipe

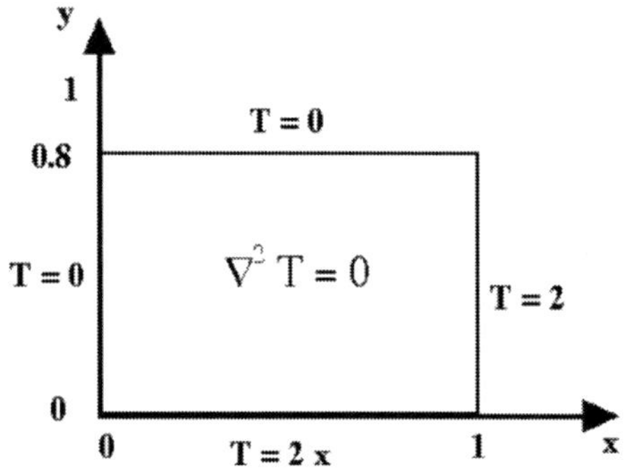

Put 2 * x into the BC window rather than a number.

Figure 3. Heat Transfer Problem

The use of the computer program, Comsol Multiphysics®, is aided by reference to the book, *Introduction to Chemical Engineering Computing*, which has step by step instructions (1). Table 10.1 from that reference is shown in Figure 4. Reference (1) also has screen shots of the different steps: draw the domain, pick the equation and set the parameters, set the boundary conditions, create the mesh, solve the problem, and examine the solution. When drawing the domain, one learns how to resize objects, reproduce repeating patterns easily, create multiple domains with different properties. When setting the equations, one has complete flexibility including the Navier-Stokes equation, the convective diffusion equation, the energy equation, in two or three dimensions, as well as other modes, such as Nernst-Planck equation. The parameters can be functions of the solution: one merely needs to be able to type the function in a box. There are a variety of choices for boundary conditions; this is a good lesson because some combinations are physically realizable and some are not; thought is required! The mesh is created automatically to a scale which you set, and it can be refined automatically or adaptively or in regions the student specifies. If the mesh is not sufficiently fine, either the solution method does not converge (infrequently) or oscillations are observable. Thus, one must be alert. Postprocessing options are illustrated below.

Open FEMLAB (Note: FEMLAB with MATLAB® makes and reads .mat and .fl files. FEMLAB stand-alone only makes and reads .fl files.)

- Choose Axial Symmetry 2D
- Choose Chemical Engineering/Momentum/Incompressible Navier-Stokes and click OK

Draw
- Click on the square icon and draw a rectangle
- Double click on the object to set exact dimensions
- Use Option/Axis/Grid settings to set the plotting range on the screen

Physics/Subdomain Settings
- Select domain 1; set $\rho = 0$, $\eta = 1180$ (Newtonian), $F_x = F_y = 0$

Physics/Boundary Settings
- Click on a boundary number (1 through 4) (Note: the corresponding boundary is highlighted in red
- Set the boundary condition for each boundary segment
  1 – slip/symmetry; 2 – inflow/outflow, v = 0.02; 3 - Outflow/pressure; 4 - no slip

Mesh
- Click once on triangle icon or select mesh/initialize mesh
- Note how many elements are used (you should report this)
- Click on divided triangle icon to refine the mesh if desired
- To refine locally, click on the 'refine selection' icon, select some elements

Solve
- Click on '=' to solve the problem (click on 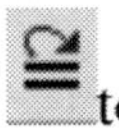 to re-start from the last solution)

Postprocessing
- Choose Domain Plot Parameters
- Select the desired quantities for contour plots and surface plots
- Or, click on the arrow plot icon, or streamline plot icon (flow plot)
- Choose Cross Section Plot Parameters to make line plots.
- Plot the v velocity along a line by selecting v velocity, put in the (r,z) coordinates of the beginning and ending points.

You can refine the mesh (click the more refined mesh symbol) and re-solve the problem. This gives you an indication of the accuracy of the solution. When you have a figure showing, you can export it in a figure format you choose such as jpeg. You always want a figure showing the mesh, as well as parts of the solution.

To document your work, show the domain and mesh, give the dimensions, identify the boundary conditions, give values of parameters you used in the subdomain options, list how many elements and degrees of freedom were used, and indicate how solutions with different number of elements compare. Then give the results and indicate why they are reasonable.

Table 10-1. Using FEMLAB to solve transport problems
(This example is for a 2D flow problem.)

Figure 4. Table 10.1 from *Introduction to Chemical Engineering Computing*

Step Two – Solve More Complicated Problems

The next lesson that must be learned is how to make the equations nondimensional, and how to report results that others can use. An important checklist is:

- What velocity is 1.0 in the computer?
- What distance is 1.0 in the computer?
- What boundary conditions did you use?
- What is the Reynolds number, and where did you use it (see below)?

The Navier-Stokes equation is

$$\rho \mathbf{u} \bullet \nabla \mathbf{u} = -\nabla p + \nabla \bullet [\eta(\nabla \mathbf{u} + \nabla \mathbf{u}^T)]$$

If one uses u_s, p_s, and x_s for the velocity, pressure, and distance standards, the dimensionless equations can be rewritten as

$$\mathbf{u}' \bullet \nabla' \mathbf{u}' = -\nabla p' + \frac{1}{Re} \nabla' \bullet [(\nabla' \mathbf{u}' + \nabla' \mathbf{u}'^T)] \text{ if } p_s = \rho u_s^2.$$

or as

$$Re\,\mathbf{u}' \bullet \nabla' \mathbf{u}' = -\nabla p'' + \nabla' \bullet [(\nabla' \mathbf{u}' + \nabla' \mathbf{u}'^T)] \text{ if } p_s = \eta u_s / x_s.$$

One must be careful to make sure the dimensions drawn for the problem are consistent with this non-dimensionalization. In both cases, the Reynolds number is given by $Re = \rho u_s x_s / \eta$. The first equation is used for high flow rates while the second one is used for low flow rates, since convergence is easier in those cases, respectively. In microfluidics, too, often the Reynolds number is low, and non-dimensional velocity solutions for all cases with Reynolds number less than one practically coincide. Thus, only one solution need be done for a range of Reynolds numbers.

In order to bring the lesson home, the students are asked to take a given non-dimensional solution along with specified standard velocity, density, size, and viscosity values (with dimensions) and provide the dimensional values derived from the non-dimensional solution. It is instructive to do this with both forms of Reynolds numbers to show that the velocities are the same (since the non-dimensional velocities are the same), and the dimensional pressure drops are the same even though the non-dimensional pressure drops differ in the two formats.

The solutions are found using the finite element method. While undergraduates are not expected to understand the details of the finite element method, it is possible to show them the essential features: dividing a region into smaller regions, approximating a function on each small region, having some means to determine the numerical values for the approximation, and seeing the effect of using more smaller finite elements. All of these steps are illustrated in a PowerPoint presentation; the one that is hard for undergraduates is the third one: determining the numerical values. This is done in the computer program by using the Galerkin method, which involves integrating by parts and some simple functional analysis. The undergraduates are not expected to understand these details, although the instructor is happy to discuss them, but it is essential that the students realize that an approximation is being made and the only way to test the accuracy of the approximation when the solution is not known is to examine solutions obtained on finer and finer meshes.

Once the solutions are found, they should be examined in great detail. Available in Comsol Multiphysics are a variety of tools. One can plot a variable as a contour plot, or a color plot, or in a line drawing across some part of a two- or three-dimensional domain. One can also integrate variables over boundary segments (or all of them) or over the domain (or parts of it if it is constructed that way). One can plot and integrate not only the variable itself, but variables defined as equations. For the problem of flow in a pipe, Figure 2, students are asked to calculate the flow rate in and the flow rate out; they should be the same (and they are to six or more significant figures). The pressure drop for a given length must be compared with the analytical solution, and this can be done in both dimensionless terms and dimensional terms. Two or more meshes should be tried, and the pressure drop compared. Streamlines can be plotted, and in this case they should be straight lines. Other complications involve different boundary conditions (slip or symmetry conditions on the axis in axi-symmetric geometry), no slip or velocity specified, specifying the pressure or velocity (but not both). When doing convective diffusion problems, if the Peclet number is large the mesh must be small, or a Petrov-Galerkin option should be used. These matters are discussed if the problems involve convection and diffusion.

Step Three – Solve Research Problem

The research problem obviously involves a problem whose solution that is not known. It is helpful, though, if the student can solve similar problems whose solution is known. This provides a check and is a confidence builder. As an example, suppose a student is going to determine a correlation for pressure drop in a flow element such as a tee (in laminar flow). At low Reynolds number (below 1.0) the form of the correlation is

$$\frac{\Delta p\, x_s}{\eta v_s} = K,$$

and one only needs to solve the flow problem to determine K. This has been done for other cases, which are illustrated in Table 2 (2). This table is part of a chapter in a book; it includes work done by ten prior undergraduate researchers.

Table III. Coefficient K_L for contractions and expansions for *Re* negligbily small	Picture	K_L
2:1 pipe/planar	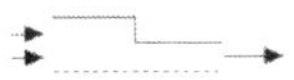	7.3/3.1
3:1 pipe/planar	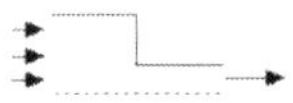	8.6/4.1
4:1 pipe/planar	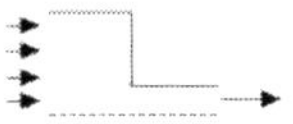	9.0/4.5
45 degrees tapered, planar, 3:1	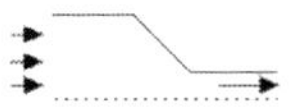	4.9
28.07 degrees tapered, planar, 3:1	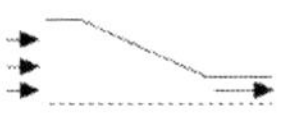	10.8
3:1 square (quarter of the geometry)	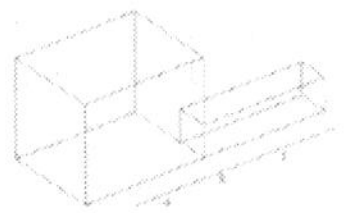	8.1

v_s = average velocity, x_s = thickness or diameter, both in the small section

Table 2. Coefficient K for contractions and expansions at small Reynolds number (2)

Example Results

Case 1 - Flow in small orifices

When an orifice is used in a small device, the thickness of the orifice affects the pressure drop, particularly in laminar flow. Febe Kusmanto did simulations of this as shown in Figure 5. The dotted lines are an analytical solution due to Dagan, *et al.* (3) for Stokes flow. This work was published (4) and also corrects a misleading impression in the literature (5). Once a solution is obtained, it is possible to gain insight into the solution, by looking at the streamlines, as in Figure 6. Additional complications can also be considered, as shown in Table 3.

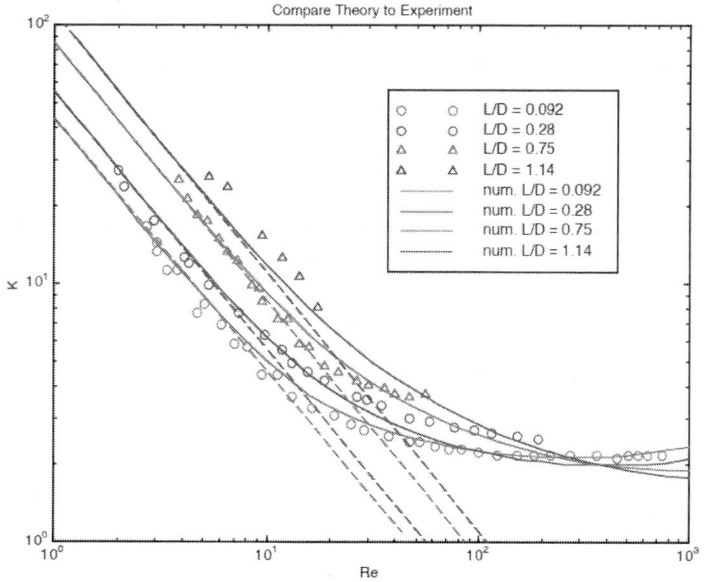

Figure 5. Excess pressure for orifices with a non-negligible thickness

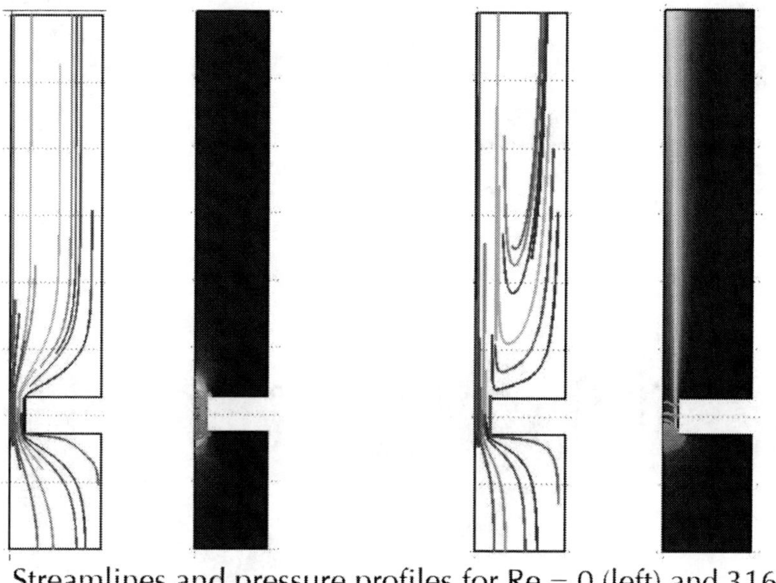

Figure 6. Streamlines and pressure profiles for Re = 0 (left) and 316 (right)

Table 3. Extensions to Orifice Problem

Does the temperature rise enough to affect the solution?
Just add in the temperature equation.
An adiabatic situation shows a temperature rise of less than one degree at the highest Reynolds number.
Work done by Yuli Tan.

Case2 – Mixing in a Serpentine Microfluidic Mixer

In microfluidic applications, mixing is slow because it is done mostly by diffusion unless the flow geometry causes the fluid paths to mix. The serpentine mixer shown in Figure 7(a) has been designed to mix better in a smaller total length. This problem was solved by Zachery Tyree; first the Navier-Stokes equation was solved, and the streamlines are shown in Figure 7(b). Then the convective diffusion equation was solved, as shown in Figure 7(c). Since the Peclet number is large (2200) it was necessary to refine the mesh for the convective diffusion equation. Thus, the finite element problem is being solved on two meshes, one sufficient for the flow and the other sufficient for the convective diffusion equation. Furthermore, the convection-diffusion problem is solved after the flow is solved. This flexibility in Comsol Multiphysics makes the solution much faster than it would be if all equations had to be solved simultaneously on the same mesh. Figure 8 shows the comparison with experiment (6) and Figure 9 shows the concentration distribution internal to the device, which shows how the mixing actually takes place.

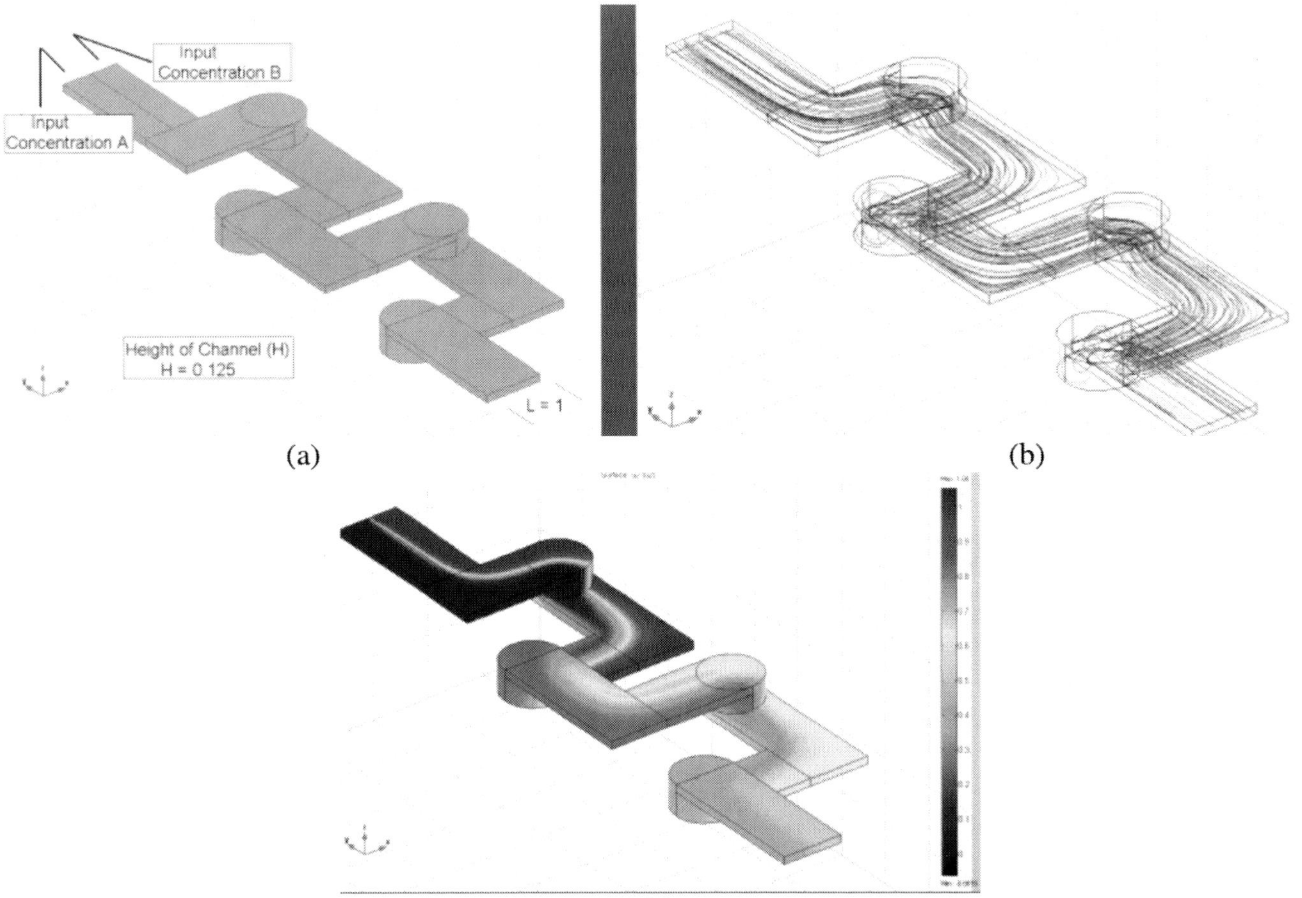

(a) (b)

(c)

Figure 7. Serpentine mixer

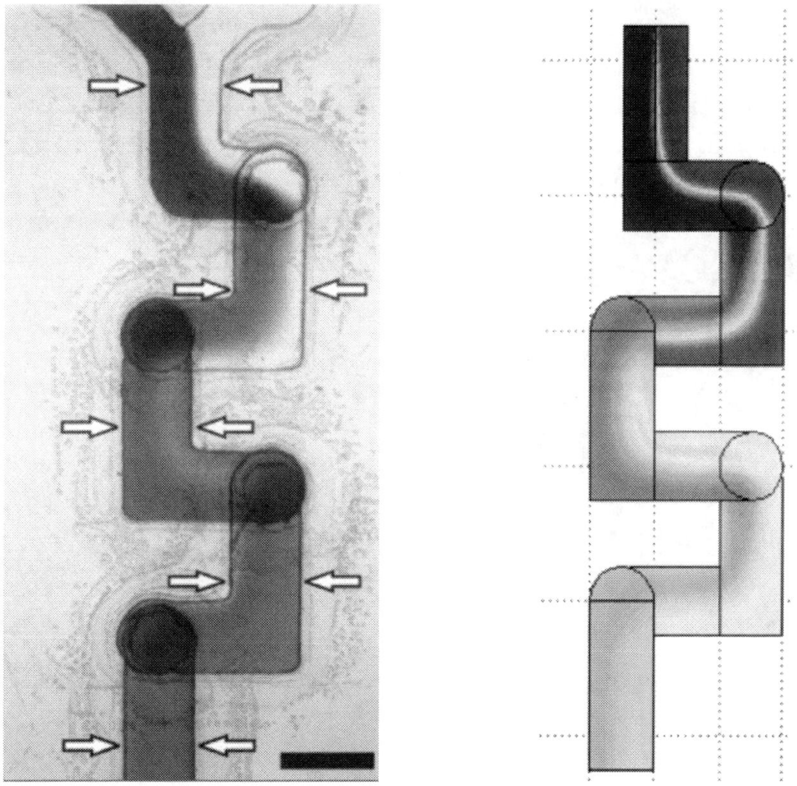

Figure 8. Comparison with experiment (6)

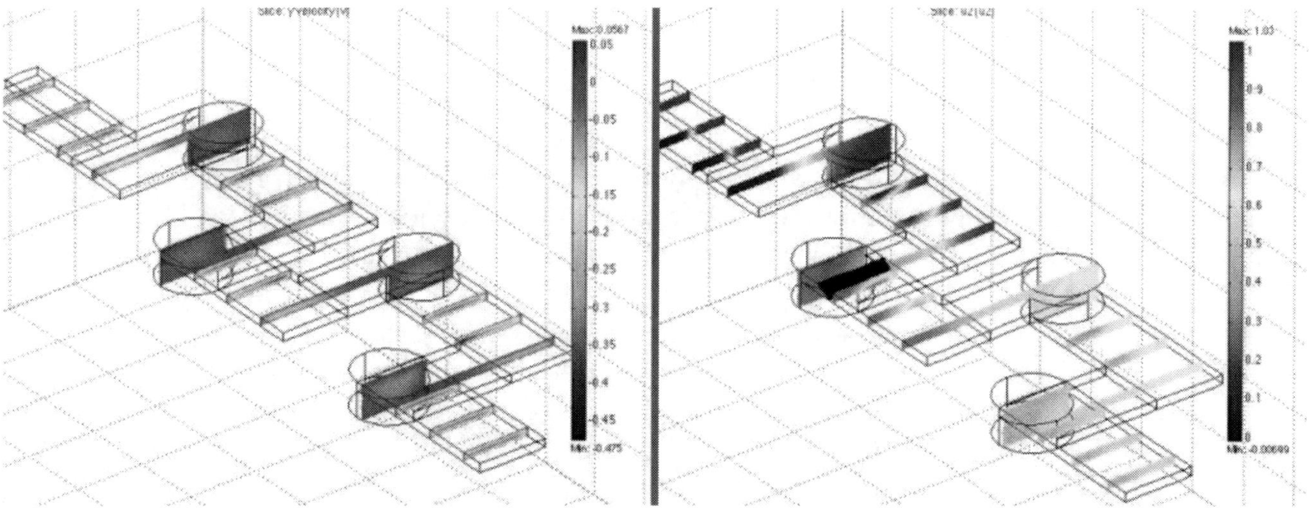

Figure 9. Velocity profiles and concentration profiles inside serpentine mixer

Case3 – Mixing in a Pharmaceutical Device

One method of mixing chemicals for analysis is to use a process which injects one chemical into a solution and then removes part of the solution (aspiration). Nick Cox studied such a device based on suggestions from Dr. Mark Petrich, Rosetta Inpharmatics, Inc. The talk will show a movie of the mixing that takes place.

Case4 – Examination of magnetic field for removal of arsenic from water

Jonathan Lundt, an undergraduate physics student, studied the problem of removing arsenic from water. The method is to add a magnetic material on which the arsenic absorbs and then to remove the magnetic material (7,8). The removal is achieved by passing the magnetic solution through a bed (or steel wool) that is in an applied magnetic field. The magnetic particles then adhere to the steel wool. His role was to find a geometry that would give the maximum magnetic force. One geometry, and its force magnitude, is illustrated in Figure 10.

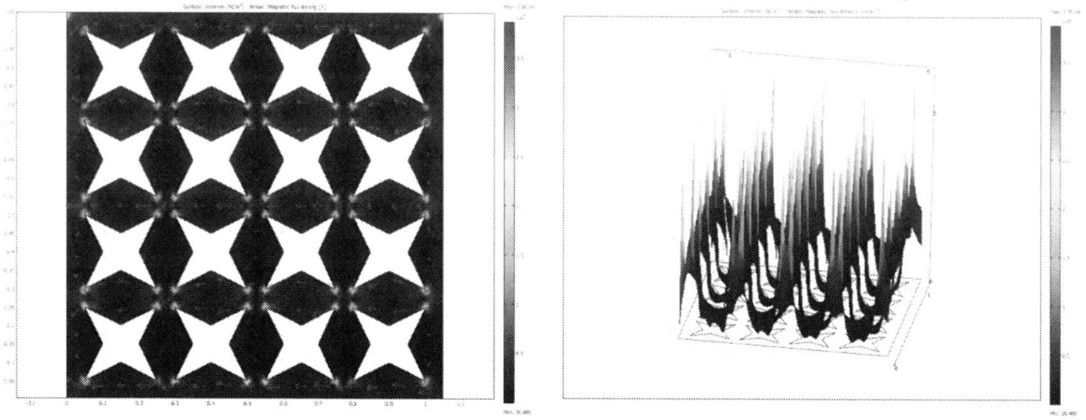

Figure 10. Geometry of packing and magnetic force field for removal of arsenic-coated magnetic particles

Case 5 – Polymer mixing

In the Renton Wastewater Treatment plant near Seattle, a polymer is added to a solution of sludge to encourage it to flocculate. Then the stream goes to a centrifuge where the sludge is removed and sent over the Cascade mountains for fertilizer in Eastern Washington. To enhance the mixing, a mixer was installed, but the mixing was still insufficient. Sharpe Mixers helped develop a project that eleven students worked on. Shown in Figure 11 are the streamlines for a simulation when a polymer is injected into the main stream. The interesting feature discovered by the students is that the power-law index of the polymer solution is very low (by measurements) and turbulent flow only occurs at a higher Reynolds number for low power-law indices. Thus, the flow was in a regime which could have been laminar. Figure 11 shows that the polymer solution is not well mixed, and other simulations over eight feet show it still isn't well mixed, when the flow is laminar. In this case, the instructor was willing to settle for a Newtonian flow, since the problem is so complicated, but the students proceeded to use a non-Newtonian model, not knowing it might be hard. But they succeeded!

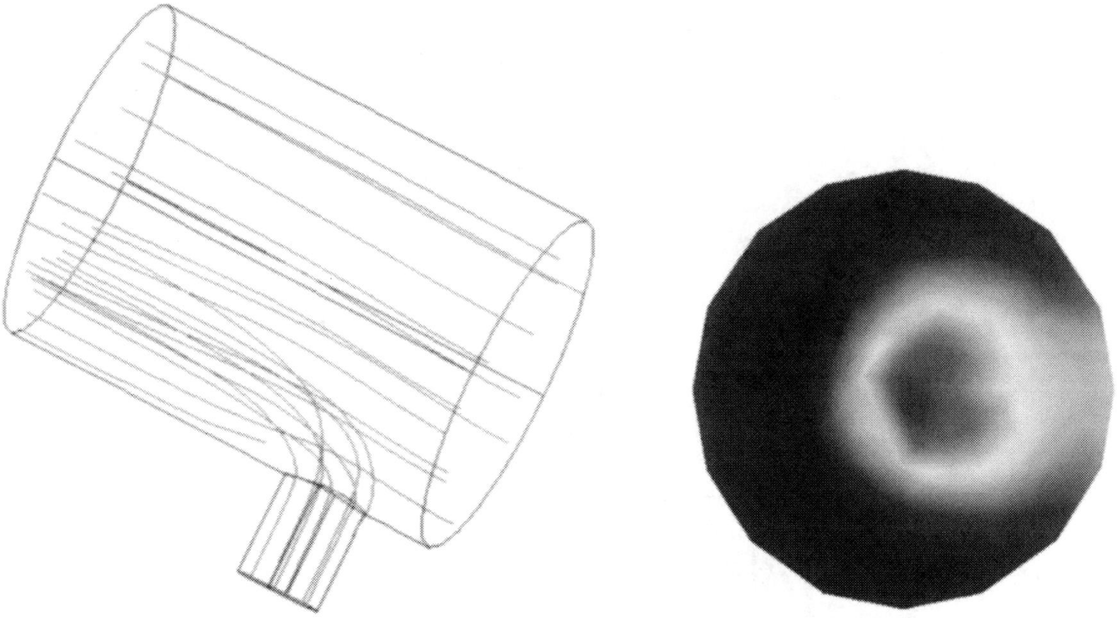

Figure 11. Streamlines and polymer concentration at the exit

Case 6 – Transport Effects in Thermal Flow Field Fractionation

This case is treated in another presentation, paper 304c, with co-authors Nick Cox and Pawel Drapala.

Case 7 – Three-dimensional hole pressure problem

Stephanie Yuen solved the two- and three-dimensional hole pressure problem for a Newtonian fluid. This geometry can be used with polymers to deduce the first normal stress difference. Shown in Figure 12 are the velocity magnitudes for the two-dimensional problem and for the center of the three-dimensional problem. As can be seen, the two results are very similar.

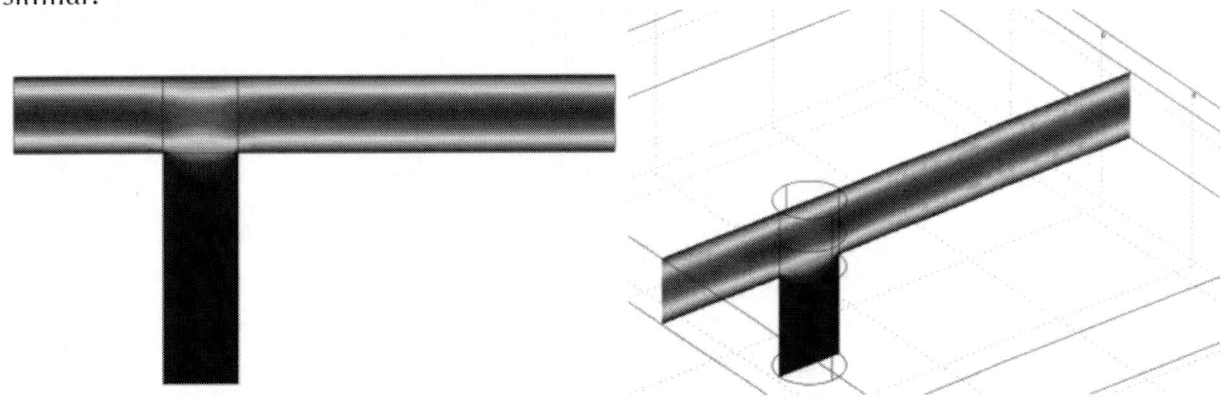

Figure 12. Flow profiles in the hole pressure problem

Case 8 – Mixing in a Three-dimensional T

It has long been recognized that in a two-dimensional T-sensor the mixing of two fluids is entirely by diffusion. Previous students have solved the problem and measured the mixing using a mixing cup concentration and a variance defined below.

$$c_{\text{mixing cup}} = \frac{\int_A c\,\mathbf{u} \bullet d\mathbf{A}}{\int_A \mathbf{u} \bullet d\mathbf{A}}, \quad variance = \frac{\int_A (c - c_{\text{mixing cup}})^2\,\mathbf{u} \bullet d\mathbf{A}}{\int_A \mathbf{u} \bullet d\mathbf{A}}$$

Daniel Kress studied the same situation but in a three dimensional case with two circular channels coming together and forming a T with a third channel as the outlet. There was no problem creating the geometry because Comsol Multiphysics handles the complications of two circular pipes joining in sideways to another circular pipe. The inherent three-dimensional nature of the problem is illustrated in Figure 13, where it is seen that regions near the walls have more time to diffuse sideways because their velocity is smaller. Figure 14 shows variance for a variety of conditions. Plotted there are two-dimensional variances for Peclet numbers from 14 to 288 as well as three-dimensional variances. Figure 14 shows that there is a universal mixing curve that can be determined even from a two-dimensional simulation.

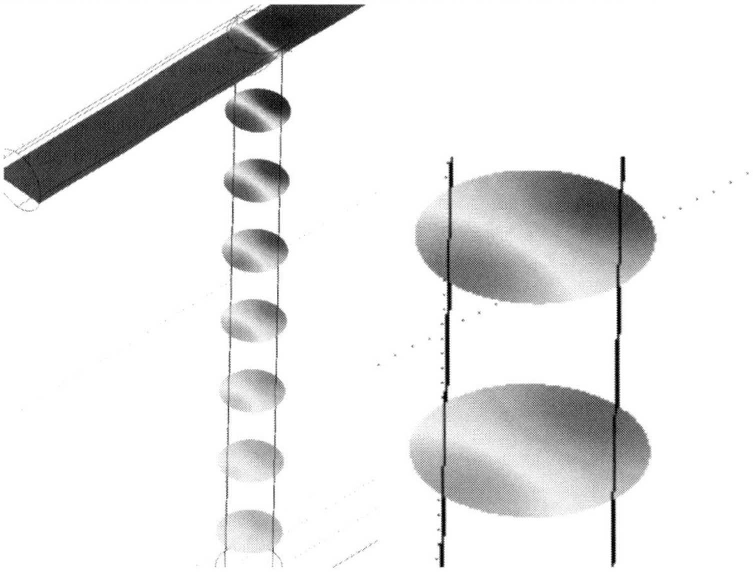

Figure 13. Concentration profiles in mixing device

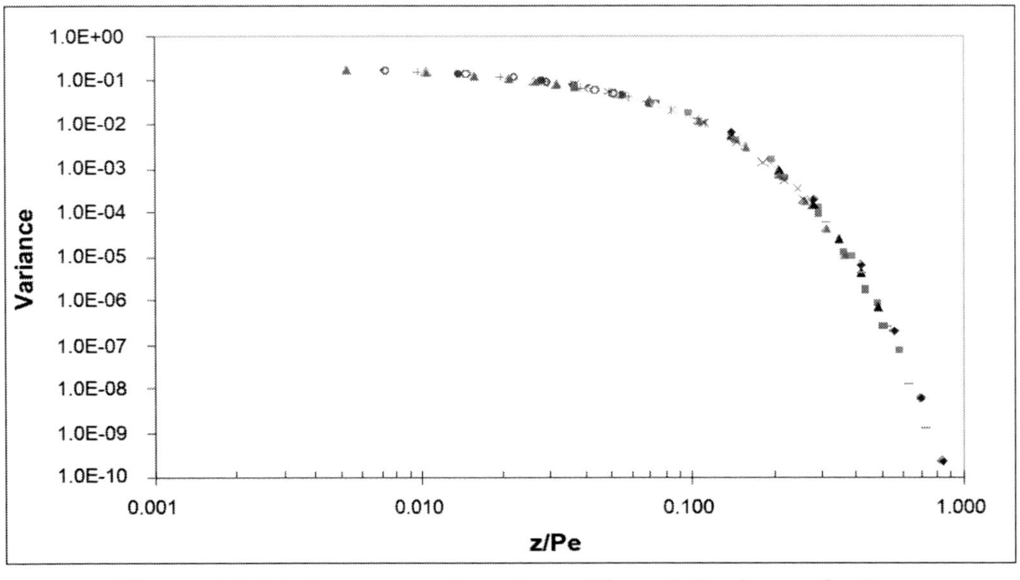

Figure 14. Variance as a function of length in the outlet leg

Case 9 – Flow of Water in Partially Saturated Soils

Comsol Multiphysics also has equations for flow through porous media. Anna Moon solved a one-dimensional case in which water flowed under pressure through a series of four different soils. The pressure profiles are shown in Figure 15. This problem was one the author originally tackled when studying the flow of water in dry soils, as in the Hanford Atomic Energy Complex. Once the one-dimensional problem was solved, it was easy to extend it to two dimensions, as shown in Figure 16.

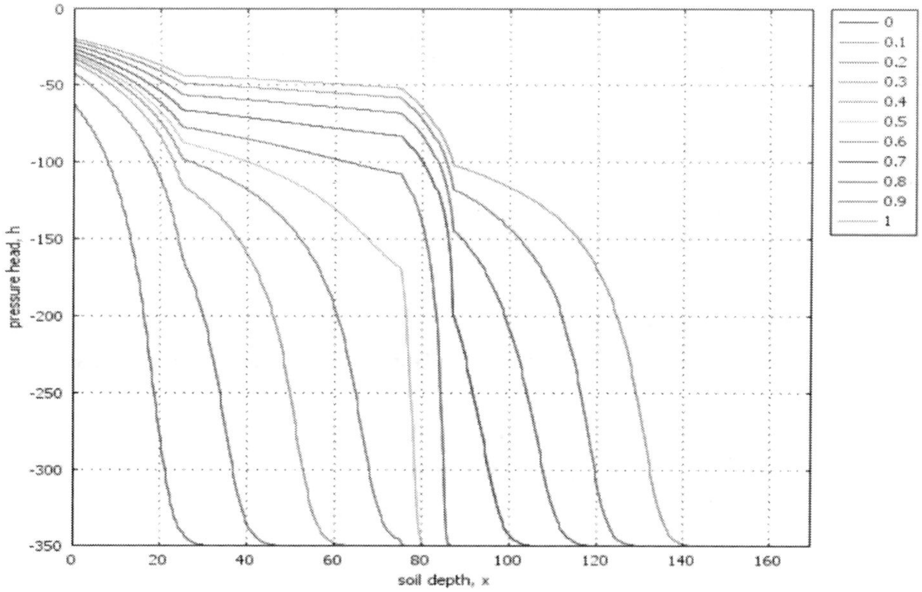

Figure 15. Pressure change when water flows through different soils

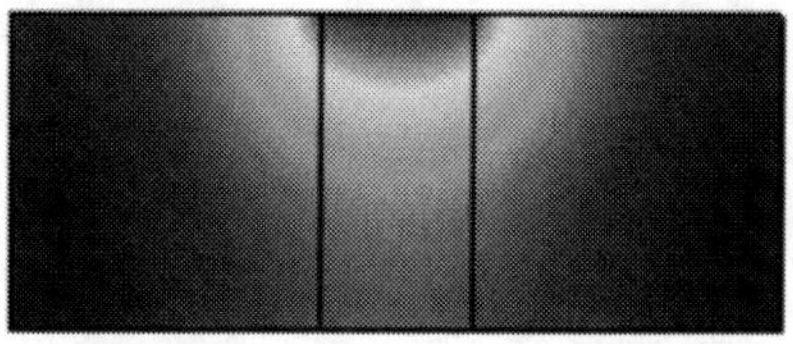

Figure 16. Two-dimensional injection into a porous media

Conclusions

The program Comsol Multiphysics is relatively easy to use for quite complicated problems. This allows undergraduates to perform meaningful simulations that go beyond their textbook examples, challenge their creativity, and improve their problem-solving skills. Results of prior quarters are described on the web site:

http://courses.washington.edu/microflo/

Acknowlegement

Recent undergraduate projects have been supported by the Dreyfus Senior Mentor Award, which provides partial tuition payments to students doing undergraduate research.

References

1. Finlayson, B. A., *Introduction to Chemical Engineering Computing*, Wiley (2006).
2. Bruce A. Finlayson, Pawel W. Drapala, Matt Gebhardt, Michael D. Harrison, Bryan Johnson, Marlina Lukman, Suwimol Kunaridtipol, Trevor Plaisted, Zachary Tyree, Jeremy VanBuren, Albert Witarsa, "Micro-component flow characterization," Ch. 8 in Micro-Instrumentation, (M. Koch, K.Vanden Bussche, R. Chrisman (ed.), Wiley, 2007).
3. Dagan, Z., Weinbaum, S. and R. Pfeffer, "An infinite-series solution for the creeping motion through an orifice of finite length," J. Fluid Mech. **115** 505 (1982).
4. Febe Kusmanto, Elissa L. Jacobsen, Bruce A. Finlayson "Applicability of Continuum Mechanics to Pressure Drop in Small Orifices", Phys. Fluids **16** 4129-4134 (2004).
5. Hasegawa, T., Suganuma, M., and H. Watanabe, "Anomaly of Excess pressure drops of the flow through very small orifices," Phys. Fluids **9** 1 (1997).
6. Christopher Neils, Zachary Tyree, Bruce Finlayson, Albert Folch, "Combinatorial mixing of microfluidic streams", Lab-on-a-Chip **4** 342-350 (2004).
7. Yavuz, C. T., *et al.* "Low-Field Magnetic Separation of Monodisperse Fe3O4 Nanocrystals," *Science* **314** 964-967 (2006).
8. Halford, Bethany. "Cleaning Water with 'Nanorust.'" *Chemical & Engineering News,* **84(46),** p.12, November 13, 2006

Ferreting Factors That Lead to Positive Outcomes for Undergraduate Researchers

Tamara Floyd Smith
Chemical Engineering Department, Tuskegee University

Abstract

This work describes the results of a survey administered to individuals involved in administrating and supervising undergraduate research. The purpose of the survey is to understand what factors lead to positive outcomes for undergraduate researchers. Positive outcomes include undergraduates pursuing graduate degrees, pursuing research careers and reporting their work at conferences and in journal publications. The factors investigated include various topics such as how candidates are identified, factors considered most important when selecting an undergraduate, defining a project for an undergraduate and assessing the undergraduate.

Introduction

Conducting undergraduate research (UR) has been shown to have a number of positive outcomes for the undergraduate researcher including increasing confidence and understanding, clarifying interests in science, technology, engineering and math (STEM) careers and increasing the anticipation of a Ph.D [1]. Russell et al. recently performed an extensive survey for the National Science Foundation which included 3600 faculty, post doctoral and graduate student mentors of undergraduate research [1]. These mentors were identified because they were funded by one or more of the eight NSF programs with an undergraduate research focus. Although this study was comprehensive from the standpoint of NSF, mentors without funding from NSF may have been overlooked. This number could be significant given that many UR mentors (corporate) don't write proposals for funding and the funding levels of other organizations such as the National Institutes of Health, the Department of Energy, the Department of Defense, the Environmental Protection Agency and the United States Department of Agriculture are significant.

The inclusion of undergraduate researchers is a typical component of a successful research proposal. However, little guidance is given on how to ensure the best outcomes for undergraduate researchers. Given the importance of UR to the pipeline of graduate researchers, it is imperative that undergraduate research mentors are aware of the factors that lead to positive outcomes for undergraduate researchers.

Thus, the purpose of this work is two-fold. First, this study includes mentors that were contacted based on their affiliation with chemical engineering without regard to the agencies that fund their research. Consequently, this study should provide a different perspective than the Russell et al. study. Second, this study seeks to understand what factors from identification to assessment of undergraduate researchers effect positive outcomes.

Methodology

A web-based survey was formulated and administered to undergraduate research mentors affiliated with chemical engineering. The survey, consisting of ten questions listed in Table 1, was designed by the author with varying degrees of input from other chemical engineering UR mentors. Electronic messages were sent directly to over 150 chemical engineering undergraduate research mentors with a request to forward to applicable parties. The data were analyzed using primarily single variable statistics [2]. Forty eight mentors completed the survey.

Table 1: Survey Questions and Responses

Question	Example Responses
Please choose the option that best describes your institution.	Research, PUI, HBCU, Hispanic serving, Corporation, Government Lab, Other
Identify the undergraduate population at your institution.	Numbers ranging from 0-10,000
Identify the graduate population at your institution.	Numbers ranging from 0-10,000
What is your position?	Administrator, professor, post doc. etc.
With which of the following disciplines are you most closely affiliated?	Chemical engineering, environmental engineering, materials engineering, etc.
What is the primary mode by which you identify undergraduate researchers?	Word of mouth, formal application, informal application, from class, etc.
What factor do you consider most important when selecting an undergraduate researcher?	Enthusiasm for research, GPA, previous experience, academic level, etc.
Which of the following best describes the most important factor when defining a project for an undergraduate?	Ability of the student to perform tasks independently, opportunity to reinforce a class taken, etc.
On what primary basis do you assess undergraduate research?	Quality of results, critical thinking, enthusiasm, etc.
What has been the impact of the undergraduate research that you have supervised or been involved with? Please assess by identifying the approximate percentage of undergraduates that you supervised that are/were involved in the following activities:	Attending graduate school, pursuing careers in research, presenting at conferences for undergraduates, etc.

Results and Discussion

The focus of this study was chemical engineering UR mentors. Consequently, it was expected that the majority of the forty eight respondents would identify chemical engineering (73%) as the discipline with which they are most closely affiliated. Other disciplines with representation were chemistry (10%), materials engineering (4%), bioengineering (4%), other (4%), mechanical engineering (2%) and environmental engineering (2%). Table 2 shows the results for the institution type and position of the UR mentors. Over half of the respondents were from research institutions and greater than 80% are professors. Although corporate researchers were contacted, none participated in the survey. Similarly, although post doctoral associates and graduate students make up a significant fraction of UR mentors, none participated in the survey.

Table 3 shows the results for the undergraduate and graduate populations of the institutions. Half of the respondents are from institutions with more than 10,000 undergraduates. In contrast, most of the respondents have fewer than 3,000 graduate students at their institution.

Table 2: Responses for Institution Type and UR mentor position (% of 48)

Institution Type	%	Position	%
Research Institution	56	Administrator	4
PUI	17	Professor	88
HBCU	13	Post doctoral associate	0
Hispanic Serving	4	Graduate student	0
Corporation	0	Corporate researcher	0
Government Lab	6	Government research	6
Other	2	Other	2

Table 3: Responses for Undergraduate and Graduate Population (% of 48)

Undergraduate	Undergraduate (%)	Graduate (%)
Fewer than 1,000	6	29
1,000-1,999	4	17
2,000-2,999	10	6
3,000-4,999	8	25
5,000-10,000	21	13
More than 10,000	50	4
N/A	0	2

Table 4 lists the results for the methods that UR mentors use to identify undergraduate researchers. The most popular method is to choose students from undergraduate courses. Informal application and word of mouth were the next most popular methods. Interestingly, less than 20% of UR mentors use a formal application process. This process would include undergraduates from Research Education for Undergraduates (REU) programs and other structured programs. Table 4 also lists the selection criteria that UR mentors consider most important. Overwhelmingly, UR mentors consider enthusiasm for research the most important factor for selecting an undergraduate.

Table 4: Methods for Identifying URs and Criteria for Selection (% of 48)

ID Mode	%	Selection Criterion	%
Word of Mouth	23	Enthusiasm for research	73
Formal application	17	GPA	10
Informal application	27	Previous experience	2
From class	29	Academic level	4
Mining student transcripts	0	Conceptual understanding	6
Other	4	Other	4

Table 5 lists what UR mentors identified as the most important factor when defining a project for an undergraduate. Almost half of the respondents indicated that it is most important that the undergraduate can perform the tasks independently. One explanation for this result is that there is no value added from the productivity standpoint if the undergraduate needs a significant amount of assistance beyond initial training. Another significant fraction indicated that the opportunity to learn something completely new is most important. Table 5 also lists the results for the primary basis that UR mentors use to assess undergraduate research. The most popular method was critical thinking and quality of results was second. Given the uncertainty of research, it seems reasonable that the most popular method of assessment is critical thinking. One avenue for further analysis would be to determine if there is a link between defining a project that the student can conduct independently and assessing the undergraduate based on the quality of the results.

Table 5: Defining Projects and Assessing Undergraduates (% of 48)

Defining a Project	%	Assessment	%
Ability of student to perform tasks independently	48	Quality of results	33
Opportunity to reinforce a class taken	4	Critical thinking	42
Opportunity to learn something completely new	25	Enthusiasm	13
Ability to be completed in a short time frame	13	Attendance	4
Likelihood of yielding good results	10	Other	8

The following outcomes were listed for UR mentors to evaluate: attending graduate school (Outcome 1), pursuing careers in research (Outcome 2), presenting at conferences for undergraduates (Outcome 3), presenting at general conferences (Outcome 4), authoring journal articles (Outcome 5) and authoring conference proceedings (Outcome 6). UR mentors were asked to indicate the percentage of undergraduates supervised who had participated in the activities listed as outcomes. The results are not weighted based on the number of students supervised. Figure 1 shows the averages for each response. There was no statistical difference for each of the outcomes. This is believed to be caused in part by the misinterpretation of the question. In particular, several responses were very low integers (1, 2, 3, etc.) suggesting that the question was answered in absolute terms rather than on a % basis. Based on the average alone, attending graduate school and presenting at conferences for undergraduates were the most popular outcomes at 44% and 46% respectively.

Going forward, the contribution of each factor to the percentage of undergraduates attending graduate school will be investigated. This outcome was chosen as the focus because it is assumed to be of more interest than undergraduates presenting at undergraduate conferences.

Figure 2 shows the relationship between the method of identifying undergraduates and the likelihood that they will attend graduate school. The number in parentheses refers to the number of respondents without the inclusion of the "other" category. The results suggest that undergraduates who formally apply are more likely to attend graduate school. Figure 3 shows the relationship between the method of choosing undergraduates and the likelihood that they will attend graduate school. Most respondents indicated enthusiasm for research as most important, but there may be a more positive relationship for indicating conceptual understanding as the most important factor. The number of respondents (3) is low which complicates that inference. Figure 4 shows the relationship between the method of defining a project and the likelihood that the undergraduate will attend graduate school. Defining a project that can be completed in a short time frame appears to be the least effective method to ensure that an undergraduate researcher will attend graduate school. Finally, Figure 5 shows the relationship between the method of assessment and the likelihood that an undergraduate will attend graduate school. Assessing an undergraduate based on critical thinking appears to have the most favorable relationship to an undergraduate attending graduate school.

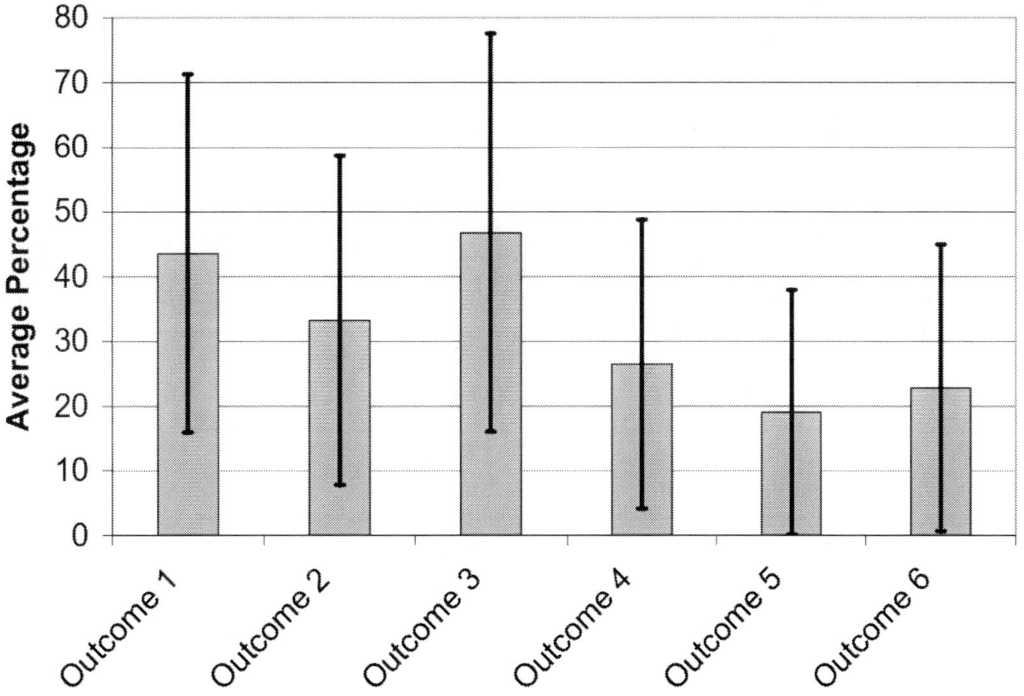

Figure 1: Outcomes Reported for Undergraduate Research

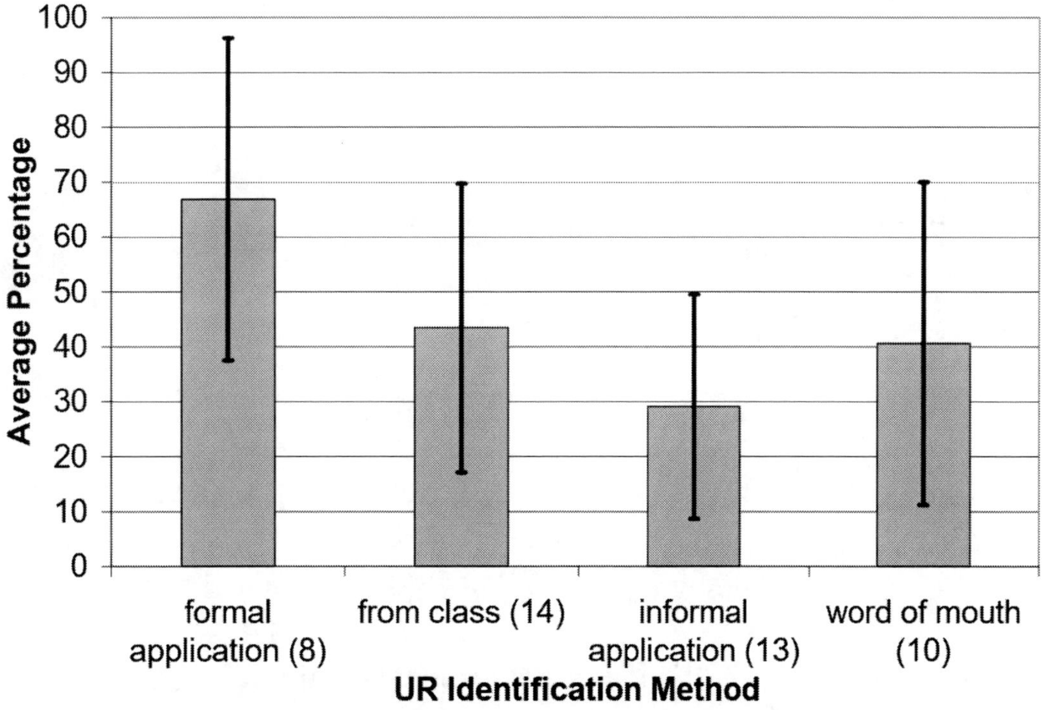

Figure 2: Relationship Between Method of Identifying Undergraduates and Attending Graduate School

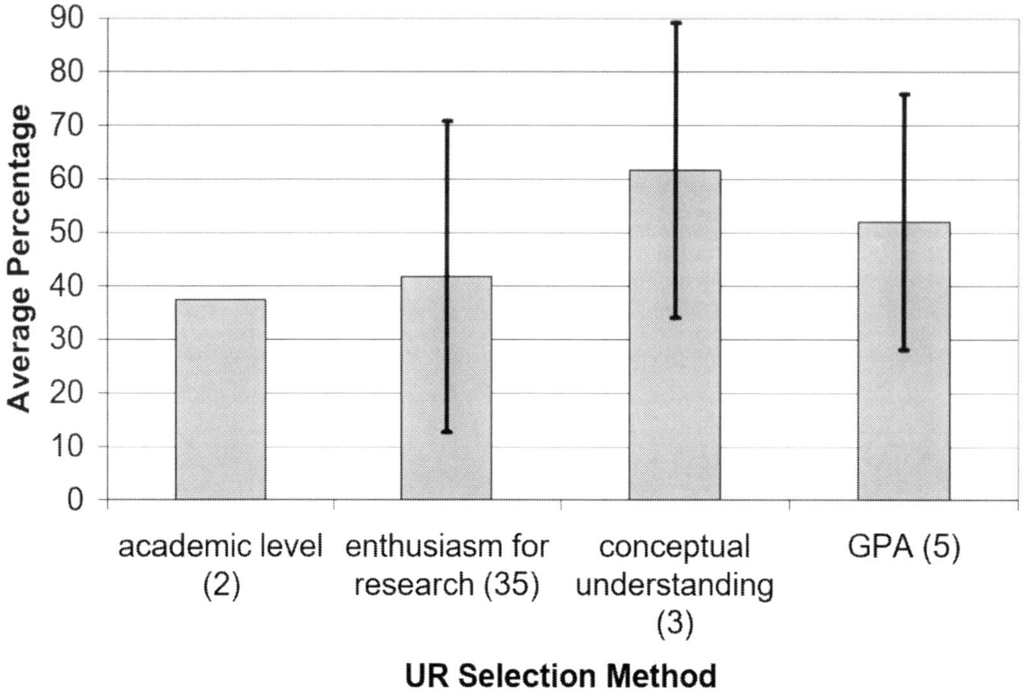

Figure 3: Relationship Between Method of Choosing Undergraduates and Attending Graduate School

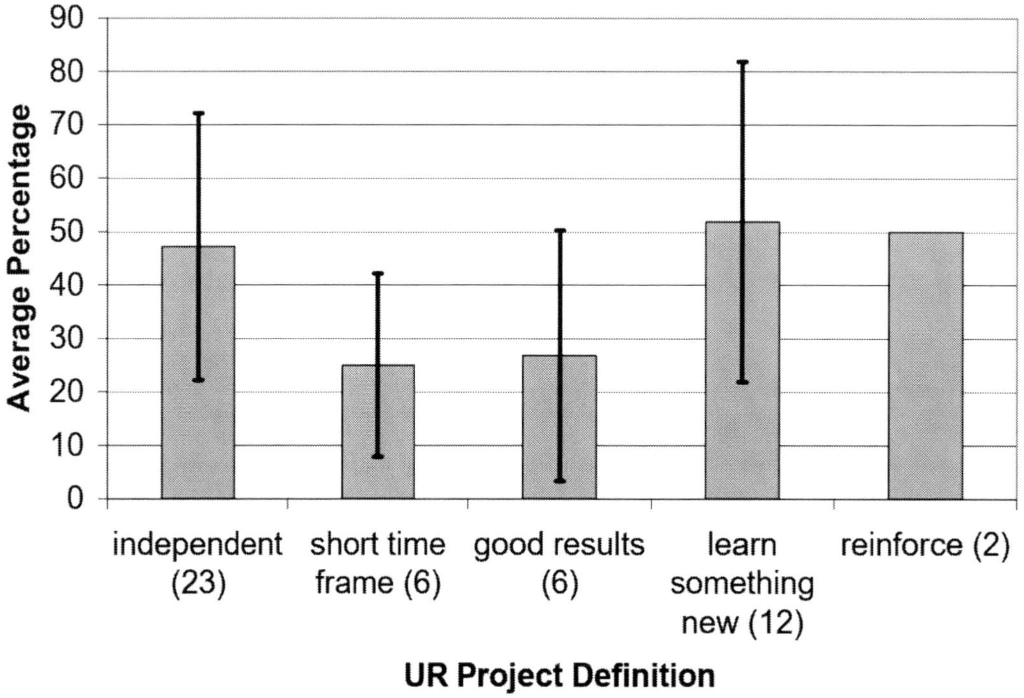

Figure 4: Relationship Between Method of Defining a Project and Attending Graduate School

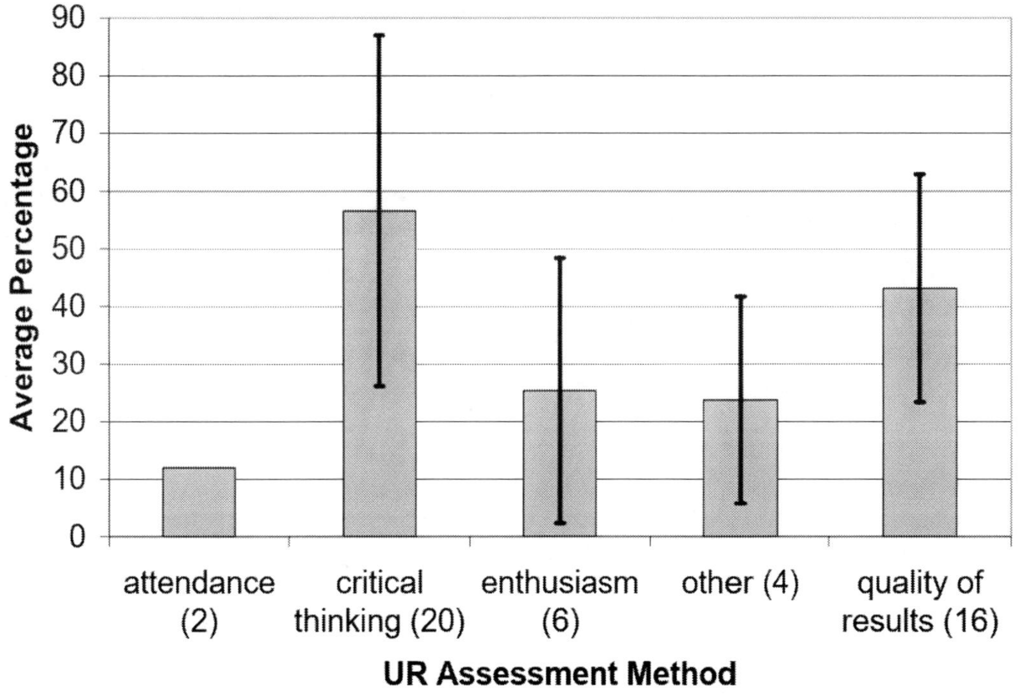

Figure 5: Relationship Between Method of Assessment and Attending Graduate School

Conclusions

A survey was conducted to ferret factors that lead to positive outcomes for undergraduate researchers. Forty eight respondents, the majority of whom were professors from research institutions, completed the survey. Although there were not any statistically significant conclusions, the results suggest that (1) having a formal application procedure, (2) choosing undergraduates based on conceptual understanding and (3) assessing undergraduates based largely on critical thinking lead to more undergraduates attending graduate school.

Acknowledgements

The author would like to thank Professors William Krantz (University of Cincinnati), Christine Grant (North Carolina State University) and Willie Harper Jr. (Auburn University) for feedback on the survey.

References

1. Russell, S. H., Hancock, M. P. and McCullough, J., "Benefits of Undergraduate Research Experiences," *Science*, Vol. 316, 2007, pp. 548-549.
2. Weisberg, H. F., Krosnick, J. A., Bowen, B.D., *An Introduction to Survey Research, Polling and Data Analysis*, 3rd *ed.*, London: Sage Publications, 1996.

AUTHOR INDEX

AUTHOR INDEX

AUTHOR INDEX

AUTHOR INDEX

AUTHOR INDEX

AUTHOR INDEX